Introduction to 3G Mobile Communications

Second Edition

For a listing of recent titles in the *Artech House Mobile Communications Series,*
please turn to the back of this book.

Introduction to 3G Mobile Communications

Second Edition

Juha Korhonen

Artech House
Boston • London
www.artechhouse.com

Library of Congress Cataloging-in-Publication Data
Korhonen, Juha.
    Introduction to 3G mobile communications / Juha Korhonen.—2nd ed.
        p. cm. — (Artech House mobile communications series)
    Includes bibliographical references and index.
    ISBN 1-58053-507-0 (alk. paper)
    1. Wireless communication systems.   2. Mobile communication systems.
    3. Universal Mobile Telecommunications System.   I. Title.   II. Series.

    TK5103.2.K67   2003
    384.5'3—dc21

                                                            2002043665

British Library Cataloguing in Publication Data
Korhonen, Juha
    Introduction to 3G mobile communications.—2nd ed.—(Artech House
mobile communications series)
    1. Mobile communication systems
    I. Title
    621.3'8456
    ISBN 1-58053-507-0

Cover design by Yekaterina Ratner. Text design by Darrell Judd.

© 2003 ARTECH HOUSE, INC.
685 Canton Street
Norwood, MA 02062

International Standard Book Number: 1-58053-507-0
Library of Congress Catalog Card Number: 2002043665

10 9 8 7 6 5 4 3

Contents

9   Network Planning   251

10   Network Management   279

Preface

The *third generation* (3G) mobile communication system is the next big thing in the world of mobile telecommunications. The first generation included analog mobile phones [e.g., Total Access Communications Systems (TACS), Nordic Mobile Telephone (NMT), and Advanced Mobile Phone Service (AMPS)], and the *second generation* (2G) included digital mobile phones [e.g., global system for mobile communications (GSM), personal digital cellular (PDC), and digital AMPS (D-AMPS)]. The 3G will bring digital multimedia handsets with high data transmission rates, capable of providing much more than basic voice calls.

This book was written to provide the reader with an information source that explains the principles and the basic concepts of the most important of the 3G telecommunications systems—universal mobile telecommunication system (UMTS) or Third Generation Partnership Project (3GPP)—in an easily understandable form. Some comparative information on the other 3G systems (the most important of which is CDMA2000) appears in the early sections of the text, but the UMTS/3GPP version of 3G is the largest and most important of the 3G initiatives, and it is the primary subject of the book. All the significant 3G versions serve to protect their corresponding 2G system investments. Since UMTS/3GPP is a GSM extension, and 2G is mostly about GSM [not code-division multiple access (CDMA) or time-division multiple access (TDMA)], UMTS plays a key role in 3G.

Numerous research papers and technical specifications about 3G are available, but these are generally quite difficult to understand, especially if the reader does not have substantial experience in telecommunications engineering. A typical specification contains exact rules on how a certain technical feature should be implemented. It does not explain why it is implemented in a certain way, nor does it tell us how this feature fits into the big picture, that is, into the entire 3G system. In this book I have deciphered that information, added my own analysis about the subject, and provided it to the reader in plain English. The result is an entry-level introduction to 3G, with an emphasis on the 3GPP-specified frequency division duplex (FDD) mode system, which will most probably be the most widely used 3G system.

It is not the intention of this book to go into great detail. 3G is a broad subject, and it would be impossible to provide a detailed analysis of every aspect in one volume. Instead, the basics are discussed and references to other information sources are provided so that interested readers can study specific subjects in more depth if they so wish. The Internet is also a very good source of information where telecommunications is concerned, and the references include appropriate Web site addresses.

I have also tried to avoid mathematics as much as possible in this book. I have found that mathematics most often prevents rather than furthers an understanding of a new subject. A theoretical approach is generally useful only when a topic is analyzed in depth, but not necessary when basic concepts are discussed.

The book starts with an overview of mobile communication systems. The history is briefly discussed, because an understanding of the past aids in the development of an understanding of the present. The 2G systems are briefly introduced here, and then the various proposals for 3G technology are explained. There are several different standards below the 3G banner, and these are also discussed in Chapter 1.

Most 3G networks will be based on the *wideband CDMA* (WCDMA) air interface, and thus a crash course on CDMA principles is given in Chapter 2. TDMA was the most popular technology in 2G systems, and this chapter concentrates especially on the differences between the CDMA and TDMA systems. Thus, a reader already familiar with 2G TDMA (especially GSM) systems will get intensive instruction on this new generation.

The WCDMA (as specified by 3GPP) air interface is an important component of the 3G system and it is discussed in several chapters. We start with a general physical layer presentation in Chapter 3, followed by a more detailed discussion about some special physical layer issues, such as modulation techniques (Chapter 4), spreading codes (Chapter 5), and channel coding (Chapter 6).

The WCDMA air interface protocol stack (layer 2 and 3 tasks) is discussed in Chapter 7. The most important functions of these protocols are explained briefly. What is new here are the *access stratum* (AS) protocols, or protocols specific to the WCDMA air interface. They include the layer 2 protocols, and the lower end of layer 3. The upper end of layer 3 forms the *nonaccess stratum* (NAS), which is more or less a replica of GSM/general packet radio system (GPRS) systems.

The network (both the radio access and the core network) is discussed in three chapters. Chapter 8 covers the architecture of the network. Network planning and network management are both difficult arts, and they are discussed in Chapters 9 and 10, respectively.

Chapter 11 presents the most common signaling procedures of the 3G system. Signaling flow diagrams are given for each procedure, as this is the most efficient way to describe the functionality. Again, it is impossible to include all signaling procedures in a work of this scope, but the cases discussed comprise the most common and interesting scenarios.

Chapter 12 contains a selection of new and interesting concepts in the 3G system. The list of issues handled here is by no means exhaustive, but I have tried to choose a few interesting concepts that cannot be found in the current 2G systems and that are likely to raise questions in the mind of the reader. Note that the core network to be used in most 3G networks is an evolved GSM/GPRS core network, and thus many of these concepts can also be used in the future GSM networks.

3G services and applications are discussed in Chapters 13 and 14, respectively, although these are closely related subjects. Applications are very important for every communication system, especially for 3G. They are the reason why consumers buy handsets and consume services. Without good applications, even the most advanced and technically superior telecommunications system is useless. In 3G systems many of the applications will be totally new; they will not have been used or tested in any other system. Finding the right application and service palette will be important as well as challenging for operators and service providers.

In Chapter 15 we take a look into the future and try to see what comes after the 3G as we know it today. This item includes 3G enhancements and *fourth generation* (4G). (There is no official definition for 4G yet, and as a result, system developers are keen on naming their new inventions 4G.) This chapter tries to predict what kind of telecommunication systems and services we will be using in 2010. The development cycle of a new mobile telecommunications system is around 10 years. The development work of UMTS (3G) began in the beginning of the 1990s, and the first systems were launched in 2001 and 2002. Work towards the 4G has already started, but it will be around 2010 before the 4G is actually in use.

Chapter 16 explains how 3G standards are actually made. It seems that even within the telecommunications industry there is some uncertainty about this process. This chapter first presents the structure of 3GPP organization, and then discusses the standardization process, and finally introduces the specification-numbering scheme.

The book also includes a set of interesting appendixes. Among these, standardization organizations and the most important industry groups are presented briefly here. We also have interesting cellular subscriber statistics and a list of useful Web addresses classified by subject.

....
Acknowledgments

The person who has suffered most from this book project, and deserves the most acknowledgements, is my wife Anna-Leena. She has had to live with a grumpy old man for some time now. During this time I have spent all my free time, including many long nights, with the manuscript. She has taken it all remarkably well.

I am very grateful to my colleagues at TTPCom for the support I received while I was writing this book. I have had many long discussions with Dr. John Haine, Mr. Stephen Laws, and Mr. Neil Baker. They have spent a great number of hours of their own time while reviewing my drafts. Many embarrassing errors were found and removed by them.

I would especially like to thank my teddy bear, Dr. Fredriksson, for his steadfast support during the preparation of this manuscript. He kept me company during the late-night writing sessions without making a single complaint, although I think his nose is a bit grayer now.

At Artech House, I would especially like to thank Dr. Julie Lancashire and Ms. Tiina Ruonamaa. They have been remarkably patient with my slipping deadlines, although they must have heard all the excuses many times before.

CHAPTER 1

Overview

1.1 History of Mobile Cellular Systems

1.1.1 First Generation

The first generation of mobile cellular telecommunications systems appeared in the 1980s. The first generation was not the beginning of mobile communications, as there were several mobile radio networks in existence before then, but they were not cellular systems either. The capacity of these early networks was much lower than that of cellular networks, and the support for mobility was weaker.

In mobile cellular networks the coverage area is divided into small cells, and thus the same frequencies can be used several times in the network without disruptive interference. This increases the system capacity. The first generation used analog transmission techniques for traffic, which was almost entirely voice. There was no dominant standard but several competing ones. The most successful standards were *Nordic Mobile Telephone* (NMT), *Total Access Communications System* (TACS), and *Advanced Mobile Phone Service* (AMPS). Other standards were often developed and used only in one country, such as C-Netz in West Germany and Radiocomm 2000 in France (see Table 1.1).

NMT was initially used in Scandinavia and adopted in some countries in central and southern Europe. It comes in two variations: NMT-450 and NMT-900. NMT-450 was the older system, using the 450-MHz frequency band. NMT-900 was launched later and it used the 900-MHz band. NMT offered the possibility of international roaming. Even as late as the latter half of the 1990s, NMT-450 networks were launched in several Eastern European countries. TACS is a U.K. standard and was adopted by some Middle Eastern countries and southern Europe. It is actually based on the AMPS protocol, but it uses the 900-MHz band. AMPS is a U.S. standard that uses the 800-MHz radio band. In addition to North America, it is used in some countries in South America and the Far East, including Australia and New Zealand. NTT's MCS was the first commercial cellular network in Japan.

Note that although the world is now busy moving into 3G networks, these first-generation networks are still in use. Some countries are even launching new first-generation networks, and many existing networks are

1

TABLE 1.1   FIRST-GENERATION NETWORKS

SYSTEM	COUNTRIES
NMT-450	Andorra, Austria, Belarus, Belgium, Bulgaria, Cambodia, Croatia, Czech Republic, Denmark, Estonia, Faroe Islands, Finland, France, Germany, Hungary, Iceland, Indonesia, Italy, Latvia, Lithuania, Malaysia, Moldova, Netherlands, Norway, Poland, Romania, Russia, Slovakia, Slovenia, Spain, Sweden, Thailand, Turkey, and Ukraine
NMT-900	Cambodia, Cyprus, Denmark, Faroe Islands, Finland, France, Greenland, Netherlands, Norway, Serbia, Sweden, Switzerland, and Thailand
TACS/ETACS	Austria, Azerbaijan, Bahrain, China, Hong Kong, Ireland, Italy, Japan, Kuwait, Macao, Malaysia, Malta, Philippines, Singapore, Spain, Sri Lanka, United Arab Emirates, and United Kingdom
AMPS	Argentina, Australia, Bangladesh, Brazil, Brunei, Burma, Cambodia, Canada, China, Georgia, Guam, Hong Kong, Indonesia, Kazakhstan, Kyrgyzstan, Malaysia, Mexico, Mongolia, Nauru, New Zealand, Pakistan, Papua New Guinea, Philippines, Russia, Singapore, South Korea, Sri Lanka, Tajikistan, Taiwan, Thailand, Turkmenistan, United States, Vietnam, and Western Samoa
C-NETZ	Germany, Portugal, and South Africa
Radiocom 2000	France

growing. However, in countries with more advanced telecommunications infrastructures, these first-generation systems will soon be, or already have been, closed, as they waste valuable frequency spectrum that could be used in a more effective way for newer digital networks (e.g., the NMT-900 networks were closed at the end of 2000 in Finland). The history of mobile cellular systems is discussed in [1–3].

1.1.2   Second Generation

The *second-generation* (2G) mobile cellular systems use digital radio transmission for traffic. Thus, the boundary line between first- and second-generation systems is obvious: It is the analog/digital split. The 2G networks have much higher capacity than the first-generation systems. One frequency channel is simultaneously divided among several users (either by code or time division). Hierarchical cell structures—in which the service area is covered by macrocells, microcells, and picocells—enhance the system capacity even further.

There are four main standards for 2G systems: *Global System for Mobile* (GSM) communications and its derivatives; *digital AMPS* (D-AMPS); *code-division multiple access* (CDMA) IS-95; and *personal digital cellular* (PDC). GSM is by far the most successful and widely used 2G system. Originally designed as a pan-European standard, it was quickly adopted all over the world. Only in the Americas has GSM not reached a dominant position yet. In North America, Personal Communication System-1900 (PCS-1900; a

GSM derivative, also called GSM-1900) has gained some ground, and in South America, Chile has a wide-coverage GSM system. However, in 2001 the North American *time-division multiple access* (TDMA) community decided to adopt the *Third Generation Partnership Project* (3GPP)-defined *wideband CDMA* (WCDMA) system as its 3G technology, and as an intermediate solution in preparation for WCDMA many IS-136 systems did convert to GSM/GPRS.

The basic GSM uses the 900-MHz band, but there are also several derivatives, of which the two most important are Digital Cellular System 1800 (DCS-1800; also known as GSM-1800) and PCS-1900 (or GSM-1900). The latter is used only in North America and Chile, and DCS-1800 is seen in other areas of the world. The prime reason for the new frequency band was the lack of capacity in the 900-MHz band. The 1,800-MHz band can accommodate a far greater user population, and thus it has become quite popular, especially in densely populated areas. The coverage area is, however, often smaller than in 900-MHz networks, and thus dual-band mobiles are used, where the phone uses a 1,800-MHz network when such is available and otherwise roams onto a 900-MHz network. Lately the *European Telecommunications Standards Institute* (ETSI) has also developed GSM-400 and GSM-800 specifications. The 400-MHz band is especially well suited for large-area coverage, where it can be used to complement the higher-frequency-band GSM networks in sparsely populated areas and coastal regions. However, the enthusiasm towards GSM-400 seems to have cooled down, and there were no operational GSM-400 networks by the end of 2002. GSM-800 is to be used in North America.

Note that GSM-400 uses the same frequency bands as NMT-450:

GSM-400: 450.4–457.6 [uplink (UL)] 0/460.4–467.6 [downlink (DL)] MHz and 478.8–486.0 (UL)/488.8–496.0 (DL) MHz;

NMT-450: 453–457.5 (UL)/463–467.5 (DL) MHz.

Therefore, countries using NMT-450 have to shut down their systems before GSM-400 can be brought into use.

D-AMPS (also known as US-TDMA, IS-136, or just TDMA) is used in the Americas, Israel, and in some countries in Asia. It is backward compatible with AMPS. AMPS, as explained earlier, is an all-analog system. D-AMPS, as defined in standard IS-54, still uses an analog control channel, but the voice channel is digital. Both of these control channels are relatively simple *frequency shift keying* (FSK) resources, while the D-AMPS version has some additional signaling to support the *digital traffic channel* (DTC). D-AMPS was first introduced in 1990. The next step in the evolution was an all-digital system in 1994. That was defined in standard IS-136. AMPS and D-AMPS are operating in the 850-MHz band, but the all-digital IS-136 protocol can also operate in the 1,900-MHz band. US–TDMA and GSM do not have common roots, although both are based on the TDMA technology. Note that the term TDMA may cause some misunderstanding, as sometimes it may be used to refer to all time division multiple access systems, including GSM, and sometimes it is used to refer to a particular TDMA system in the United States, either IS-54 or IS-136.

CDMA, and here we mean the IS-95 standard developed by Qualcomm, uses a different approach to air interface design. Instead of dividing a frequency carrier into short time slots as in TDMA, CDMA uses different codes to separate transmissions on the same frequency. The principles of CDMA are well explained later on, as the 3G *Universal Terrestrial Radio Access Network* (UTRAN) uses wideband CDMA technology. IS-95 is the only 2G CDMA standard so far to be operated commercially. It is used in the United States, South Korea, Hong Kong, Japan, Singapore, and many other east Asian countries. In South Korea especially this standard is widely used. IS-95 networks are also known by the brand name cdmaOne.

PDC is the Japanese 2G standard. Originally it was known as *Japanese Digital Cellular* (JDC), but the name was changed to *Personal Digital Cellular* (PDC) to make the system more attractive outside Japan. However, this renaming did not bring about the desired result, and this standard is commercially used only in Japan. The specification is known as RCR STD-27, and the system operates in two frequency bands: 800 MHz and 1,500 MHz. It has both analog and digital modes. Its physical layer parameters are quite similar to D–AMPS, but its protocol stack resembles GSM. The lack of success of PDC abroad has certainly added to the determination of the big Japanese telecommunications equipment manufacturers to succeed globally with 3G. Indeed, they have been pioneers in many areas of the 3G development work. PDC has been a very popular system in Japan. This success has also been one of the reasons that the Japanese have been so eager to develop 3G systems as soon as possible, as the PDC system capacity is quickly running out.

Note that quite often when 2G is discussed, digital cordless systems are also mentioned. There are three well-known examples of these: CT2, *Digital Enhanced Cordless Telecommunications* (DECT), and *Personal Handyphone System* (PHS). These systems do not have a network component; a typical system configuration includes a base station and a group of handsets. The base station is attached to some other network, which can be either a fixed or mobile network. The coverage area is often quite limited, consisting of town centers or office buildings. Simpler systems do not support any handover (HO) techniques, but PHS is an advanced system and can do many things usually associated with mobile cellular systems. However, these systems are not further discussed here, as they are not mobile cellular systems as such. Excellent reviews of DECT can be found in [4] and of PHS can be found in [5].

Recently there has been an attempt in the GSM community to enhance GSM to meet the requirements of cordless markets. *Cordless Telephone System* (CTS) is a scheme in which GSM mobiles can be used at home via a special home base station, in a manner similar to the present-day cordless phones. This scheme can be seen as an attempt of the GSM phone vendors to get into the cordless market.

1.1.3   Generation 2.5

"Generation 2.5" is a designation that broadly includes all advanced upgrades for the 2G networks. These upgrades may in fact sometimes provide almost the same capabilities as the planned 3G systems. The boundary line between 2G and 2.5G is a hazy one. It is difficult to say when a 2G becomes a 2.5G system in a technical sense.

Generally, a 2.5G GSM system includes at least one of the following technologies: *high-speed circuit-switched data* (HSCSD), *General Packet Radio Services* (GPRS), and *Enhanced Data Rates for Global Evolution* (EDGE). An IS-136 system becomes 2.5G with the introduction of GPRS and EDGE, and an IS-95 system is called 2.5G when it implements IS-95B, or CDMA2000 1xRTT upgrades.

The biggest problem with plain GSM is its low air interface data rates. The basic GSM could originally provide only a 9.6-Kbps user data rate. Later, 14.4-Kbps data rate was specified, although it is not commonly used. Anyone who has tried to Web surf with these rates knows that it can be a rather desperate task. HSCSD is the easiest way to speed things up. This means that instead of one time slot, a mobile station can use several time slots for a data connection. In current commercial implementations, the maximum is usually four time slots. One time slot can use either 9.6-Kbps or 14.4-Kbps speeds. The total rate is simply the number of time slots times the data rate of one slot. This is a relatively inexpensive way to upgrade the data capabilities, as it requires only software upgrades to the network (plus, of course, new HSCSD-capable phones), but it has drawbacks. The biggest problem is the usage of scarce radio resources. Because it is circuit switched, HSCSD allocates the used time slots constantly, even when nothing is being transmitted. In contrast, this same feature makes HSCSD a good choice for real-time applications, which allow for only short delays. The high-end users, which would be the most probable HSCSD users, typically employ these services in areas where mobile networks are already congested. Adding HSCSD capability to these networks certainly will not make the situation any better. An additional problem with HSCSD is that handset manufacturers do not seem very interested in implementing HSCSD. Most of them are going to move directly to GPRS handsets, even though HSCSD and GPRS are actually quite different services. A GPRS system cannot do all the things HSCSD can do. For example, GPRS is weak with respect to real-time services. It can be seen that HSCSD will be only a temporary solution for mobile data transmission needs. It will only be used in those networks where there is already a high demand for quick data transfer and something is needed to ease the situation and keep the customers happy while waiting for 3G to arrive.

The next solution is GPRS. With this technology, the data rates can be pushed up to 115 Kbps, or even higher if one can forget error correction. However, with adequate data protection, the widely quoted 115 Kbps is the theoretical maximum in optimal radio conditions with eight downlink time slots. A good approximation for throughput in "average" conditions is 10 Kbps per time slot. What is even more important than the increased throughput is that GPRS is packet switched, and thus it does not allocate the radio resources continuously but only when there is something to be sent. The maximum theoretical data rate is achieved when eight time slots are used continuously. The first commercial launches for GPRS took place in 2001. GPRS is especially suitable for non-real-time applications, such as e-mail and Web surfing. Also, bursty data is well handled with GPRS, as it can adjust the assigned resources according to current needs. It is not well suited for real-time applications, as the resource allocation in GPRS is contention based; thus, it cannot guarantee an absolute maximum delay.

The implementation of a GPRS system is much more expensive than that of an HSCSD system. The network needs new components as well as modifications to the existing ones. However, it is seen as a necessary step toward better data capabilities. A GSM network without GPRS will not survive long into the future, as traffic increasingly becomes data instead of voice. For those operators that will also operate 3G networks in the future, a GPRS system is an important step toward a 3G system, as 3GPP core networks are based on combined GSM and GPRS core networks.

The third 2.5G improvement to GSM is EDGE. Originally this acronym stood for Enhanced Data rates for GSM Evolution, but now it translates into Enhanced Data rates for Global Evolution, as the EDGE idea can also be used in systems other than GSM [6]. The idea behind EDGE is a new modulation scheme called *eight-phase shift keying* (8PSK). It increases the data rates of standard GSM by up to threefold. EDGE is an attractive upgrade for GSM networks, as it only requires a software upgrade to base stations if the RF amplifiers can handle the nonconstant envelope modulation with EDGE's relatively high peak-to-average power ratio. It does not replace but rather coexists with the old *Gaussian minimum shift keying* (GMSK) modulation, so mobile users can continue using their old phones if they do not immediately need the better service quality provided by the higher data rates of EDGE. It is also necessary to keep the old GMSK because 8PSK can only be used effectively over a short distance. For wide area coverage, GMSK is still needed. If EDGE is used with GPRS, then the combination is known as *enhanced* GPRS (EGPRS). The maximum data rate of EGPRS using eight time slots (and adequate error protection) is 384 Kbps. Note that the much-advertised 384 Kbps is thus only achieved by using all radio resources of a frequency carrier, and even then only when the mobile station is close to the base station. ECSD is the combination of EDGE and HSCSD and it also provides data rates three times the standard HSCSD. A combination of these three methods provides a powerful system, and it can well match the competition by early 3G networks.

This chapter has so far discussed the upgrade of GSM to 2.5G. There are, however, other types of 2G networks in need of upgrading. IS-136 (TDMA) can be upgraded using EDGE, and the schedules for doing that are even quicker than in the GSM world. In addition, GPRS can be implemented in IS-136 networks.

The IS-95 (CDMA) standard currently provides 14.4-Kbps data rates. It can be upgraded to IS-95B, which is able to transfer 64 Kbps with the use of multiple code channels. However, many IS-95 operators have decided to move straight into a CDMA2000 1xRTT system. 1xRTT is one of several types of radio access techniques included in the CDMA2000 initiative. The North American version of 3G, CDMA2000, is in a way just an upgrade of the IS-95 system, although a large one. The IS-95 and CDMA2000 air interfaces can coexist, so in that sense the transition to 3G will be quite smooth for the IS-95 community. There are several evolution phases in CDMA2000 networks, and the first phase, CDMA2000 1xRTT, is widely regarded to be still a 2.5G system.

Qualcomm has its own proprietary high-speed standard, called *High Data Rate* (HDR), to be used in IS-95 networks. It will provide a 2.4-Mbps data rate. A standard for HDR has been formulated in IS-856. The *1x Evolved Data Optimized* (1xEV-DO) term is used when referring to the nonproprietary form of this advanced CDMA radio interface. The 1xEV-DO adds a TDMA component beneath the code components to support highly asymmetric, high-speed data applications. A more detailed discussion on how the IS-95 system is evolved into a full CDMA2000 system, with all the intermediate phases, can be found in Section 1.5.

PDC in Japan has also evolved to provide faster data connections. NTT DoCoMo has developed a proprietary service called *i-mode*. It uses a packet data network (PDC-P) behind the PDC radio interface. Customers are charged based on the amount of data retrieved and not on the amount of time spent retrieving the data, as in typical circuit-switched networks. The i-mode service can be used to access wireless Internet services. In addition to Web surfing, i-mode provides a good platform for wireless e-mail service. In a packet-switched network the delivery of e-mails over the radio interface is both economical and quick. Each i-mode user can be sent e-mail simply by using the address format <*mobile_number*>@*docomo.ne.jp*.

The i-mode Internet Web pages are implemented using a language based on standard HTML. So in that sense, the idea behind i-mode is similar to the *Wireless Application Protocol* (WAP). This similarity becomes even more evident once GPRS networks are used and WAP can be used over packet connections. Indeed, NTT DoCoMo's competitor in Japan, KDDI, is offering a WAP-based Internet service.

The i-mode has been a true success story. The system was launched in February 1999, and in June 2002, it already had more than 33 million subscribers. In fact, the demand for i-mode has been so overwhelming that DoCoMo has had to curb new subscriptions at times. This proves that there is a market for WAP-like services, but they will require a packet-based network, like GPRS, to be feasible and affordable for users.

It seems that NTT DoCoMo has made a conscious decision to introduce new services as early as possible, even if that may require proprietary solutions. The i-mode is one example, and WCDMA is another. NTT DoCoMo was first to start 3G services before other operators, using a proprietary version of 3GPP WCDMA specifications. This gave them a few months' head start, even though the launch was a bit rocky, as a new complex system always includes new problems.

1.2   Overview of 3G

The rapid development of mobile telecommunications was one of the most notable success stories of the 1990s. The 2G networks began their operation at the beginning of the decade (the first GSM network was opened in 1991 in Finland), and since then they have been expanding and evolving continuously. In September 2002 there were 460 GSM networks on air worldwide, together serving 747.5 million subscribers.

In the same year that GSM was commercially launched, ETSI had already started the standardization work for the next-generation mobile telecommunications network. This new system was called the *Universal Mobile Telecommunications System* (UMTS). The work was done in ETSI's technical committee *Special Mobile Group* (SMG). SMG was further divided into subgroups SMG1–SMG12 (SMG5 was discontinued in 1997), with each subgroup specializing in certain aspects of the system.

The 3G development work was not done only within ETSI. There were other organizations and research programs that had the same purpose. The European Commission funded research programs such as *Research on Advanced Communication Technologies in Europe* (RACE I and II) and *Advanced Communication Technologies and Services* (ACTS). The UMTS Forum was created in 1996 to accelerate the process of defining the necessary standards. In addition to Europe, there were also numerous 3G programs in the United States, Japan, and Korea. Several telecommunications companies also had their own research activities.

An important leap forward was made in 1996 and 1997, when both the *Association of Radio Industries and Businesses* (ARIB) and ETSI selected WCDMA as their 3G radio interface candidate. Moreover, the largest Japanese mobile telecommunications operator, NTT DoCoMo, issued a tender for a WCDMA prototype trial system to the biggest mobile telecommunications manufacturers. This forced many manufacturers to make a strategic decision, which meant increasing their WCDMA research activities or at least staying out of the Japanese 3G market.

Later the most important companies in telecommunications joined forces in the 3GPP program, the goal of which is to produce the specifications for a 3G system based on the ETSI *Universal Terrestrial Radio Access* (UTRA) radio interface and the enhanced GSM/GPRS *Mobile Application Part* (MAP) core network. At the moment it is the 3GPP organization that bears the greatest responsibility for the 3G development work.

The radio spectrum originally allocated for UMTS is given in Figure 1.1. As can be seen, the allocation is similar in Europe and Japan, but in the United States most of the IMT-2000 spectrum has been allocated to 2G PCS networks, many of which are deployed on small 5-MHz sub-bands. Therefore, proposals like CDMA2000 are attractive to North American operators. This 3G proposal is backward compatible with the IS-95B system, and they can both exist in the same spectrum at the same time. The exact IMT-2000 frequency bands are 1,885–2,025 MHz and 2,110–2,200 MHz. From these the satellite component of IMT-2000 takes 1,980–2,010 MHz and 2,170–2,200 MHz. Note that these allocations were the original ones; later on the allocations were extended and the current situation is presented in Section 15.1.

In all, the 3G development work has shown that development of the new systems is nowadays done more and more within the telecommunications industry itself. The companies join to form consortia, which then produce specification proposals for the official standardization organizations for a formal approval. This results in a faster specification development process, as these companies often have more available resources than intergovernmental organizations. Also, the standards may be of higher quality (or at least

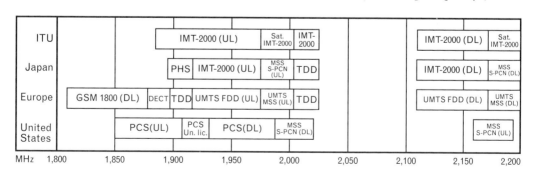

FIGURE 1.1   *IMT-2000 spectrum allocations.*

more suitable for the actual implementation) when they have been written by their actual end users. In contrast, this also means that the standardization process is easily dominated by a few big telecommunications companies and their interests.

1.3   Proposals for 3G Standard

There have been (and still are) several competing proposals for a global 3G standard. Below, these are grouped based on their basic technology, WCDMA, advanced TDMA, hybrid CDMA/TDMA, and *orthogonal frequency division multiplexing* (OFDM).

1.3.1   WCDMA

By definition, the bandwidth of a WCDMA system is 5 MHz or more, and this 5 MHz is also the nominal bandwidth of all 3G WCDMA proposals. This bandwidth was chosen because:

• It is enough to provide data rates of 144 and 384 Kbps (these were 3G targets), and even 2 Mbps in good conditions.

• Bandwidth is always scarce, and the smallest possible allocation should be used, especially if the system must use frequency bands already occupied by existing 2G systems.

• This bandwidth can resolve more multipaths than narrower bandwidths, thus improving performance.

The 3G WCDMA radio interface proposals can be divided into two groups: network synchronous and network asynchronous. In a synchronous network all base stations are time synchronized to each other. This results in a more efficient radio interface but requires more expensive hardware in base stations. For example, it could be possible to achieve synchronization with the use of *Global Positioning System* (GPS) receivers in all base stations, although this is not as simple as it sounds. GPS receivers are not very useful in high-block city centers (many blind spots) or indoors.

Other WCDMA characteristics include fast power control in both the uplink and downlink and the ability to vary the bit rate and service parameters on a frame-by-frame basis using variable spreading.

The ETSI/ARIB WCDMA proposal was asynchronous, as was Korea's TTA II proposal. Korea TTA I and CDMA2000 proposals included synchronous networks.

The ETSI/ARIB proposal was the most popular proposal for 3G systems. Originally it had the backing of Ericsson, Nokia, and the big Japanese telecommunications companies, including NTT DoCoMo. Later it was also adopted by the other European manufacturers, and was renamed as UTRAN, more precisely as the UTRAN FDD mode. It is an attractive choice for existing GSM operators because the core network is based on the GSM MAP network, and the new investments are lower than with other 3G system proposals. This also means that all the GSM services are available from day one via the new UMTS network. It would have been difficult to attract customers from existing 2.5G networks to 3G networks if the services in the new network were inferior to those in 2.5G. The specifications for this proposal are further developed by the industry-led 3GPP consortium.

The CDMA2000 proposal is compatible with IS-95 systems from North America. Its most important backers include the existing IS-95 operators, Qualcomm, Lucent, and Motorola. The specifications for this proposal are further developed by the *3G Partnership Project number 2* (3GPP2) consortium (see Section 1.5 for an introduction to 3GPP2). Although CDMA2000 clearly has less support than the 3GPP scheme, it will be an important technology, especially in areas where IS-95 networks are used. In the United States the 3G networks must use the existing 2G spectrum in many cases; thus, CDMA2000 offers an attractive technology choice, as it can coexist with IS-95 systems. Also, the core network is different from GSM MAP, as CDMA2000 uses the ANSI-41 core network. Since CDMA2000 employs a synchronous network, the increased efficiency is attractive to new operators, or existing GSM operators more concerned with deploying an efficient network than attending to the needs of their legacy subscribers. These operators may jump off the GSM track and deploy CDMA2000 instead of upgrading to the UTRAN-FDD mode.

1.3.2   Advanced TDMA

Serious research was conducted around advanced TDMA systems in the 1990s. For some time, the European 3G research was concentrated around TDMA systems, and CDMA was seen only as a secondary alternative. However, in the IMT-2000 process the UWC-136 was the only surviving TDMA 3G proposal, and even that one had backing only in North America. As of 2002, UWC-136 was no longer supported even by UWCC, but North American TDMA and GSM operators have decided to adopt the WCDMA system, that is, IMT-DS, as their 3G technology.

UWC-136 is a system compatible with the IS-136 standard. It uses three different carrier types: 30 kHz, 200 kHz, and 1.6 MHz. The narrowest bandwidth (30 kHz) is the same as in IS-136, but it uses a different modulation. The 200-kHz carrier uses the same parameters as GSM EDGE and provides data rates up to 384 Kbps. This carrier is designed to be used for outdoor or vehicular traffic. The 1.6-MHz carrier is for indoor usage only, and can provide data rates up to 2 Mbps. UWC-136 supporters included North American IS-136 operators. This system is called IMT-SC in IMT-2000 jargon.

However, when advanced TDMA is discussed, it must be noted that a GSM 2.5G system with all the planned enhancements (GPRS, HSCSD, EDGE) is also a capable TDMA system. It might not be called a 3G system, but the boundary between it and a 3G system will be narrow, at least during the first years after the 3G launch. There are still many possibilities to enhance the GSM infrastructure further. Also, the further specification work for GSM has been transferred into 3GPP work groups. Thus, it is likely that those new UTRAN features, which are also feasible in GSM networks, will be specified for GSM systems as well.

1.3.3 Hybrid CDMA/TDMA

This solution was examined in the European FRAMES project. It was also the original ETSI UMTS radio interface scheme. Each TDMA frame is divided into eight time slots and within each time slot the different channels are multiplexed using CDMA. This frame structure would have been backward compatible with GSM.

This particular ETSI proposal is no longer supported. However, the UTRAN TDD mode is actually also a hybrid CDMA/TDMA system. A radio frame is divided into 15 time slots, and within each slot different channels are CDMA multiplexed.

1.3.4 OFDM

OFDM is based on a principle of multicarrier modulation, which means dividing a data stream into several bit streams (subchannels), each of which has a much lower bit rate than the parent data stream. These substreams are then modulated using codes that are orthogonal to each other. Because of their orthogonality, the subcarriers can be very close to each other (or even partly overlapping) in the frequency spectrum without interfering each other. And since the symbol times on these low bit rate channels are long, there is no *intersymbol interference* (ISI). The result is a very spectrum-efficient system.

Digital audio broadcasting (DAB) and *digital video broadcasting* (DVB) are based on OFDM. It is also employed by 802.11a, 802.11g, and HiperLAN2 WLAN systems, and by *Asymmetric Digital Subscriber Line* (ADSL) systems. OFDM itself can be based on either TDMA or CDMA. The main advantages of this scheme are:

- Efficient use of bandwidth: Orthogonal subcarriers can partly overlap each other.

- Resistance to narrowband interference;

- Resistance to multipath interference.

The main drawback is the high peak to average power.

None of the chosen IMT-2000 technologies employ OFDM. However, as some WLAN technologies use OFDM, and WLAN—cellular interworking is the way of the future—it is quite possible that OFDM will enter the cellular world via a backdoor as part of an interworking WLAN system. Also, it is possible that some later HSDPA enhancement (see Section 12.2 on HSDPA) will include OFDM carriers.

1.3.5 IMT-2000

IMT-2000 is the "umbrella specification" of all 3G systems. Originally it was the purpose of the *International Telecommunication Union* (ITU) to have only one truly global 3G specification, but for both technical and political reasons this did not happen.

In its November 1999 meeting in Helsinki, the ITU accepted the following proposals as IMT-2000 compatible [7]:

- IMT Direct Spread (IMT-DS; also known as UTRA FDD);

- IMT Multicarrier (IMT-MC; also known as CDMA2000);

- IMT Time Code (IMT-TC; also known as UTRA-TDD/TD-SCDMA "narrowband TDD");

- IMT Single Carrier (IMT-SC; also known as UWC-136);

- IMT Frequency Time (IMT-FT; also known as DECT).

The number of accepted systems indicates that the ITU adopted a policy that no serious candidate should be excluded from the new IMT-2000 specification. Thus, the IMT-2000 is not actually a single radio interface specification but a family of specifications that technically do not have much in common.

Since then there has been lots of progress on the 3G system front. IMT-DS and IMT-TC proposals are both being developed by 3GPP consortium. IMT-MC is adopted by another industry consortium, 3GPP2. Doubtlessly the most important IMT-2000 system will be IMT-DS, followed by IMT-MC. The IMT-SC proposal was supported by UWCC, but this organization has made a decision to adopt IMT-DS (i.e., WCDMA) as its 3G technology. In December 2001 the UWCC organization was disbanded, and in January 2002 a new organization, 3G Americas, was founded. The mission of 3G Americas is to support the migration of GSM and TDMA networks into WCDMA systems in the Americas. IMT-TC is further divided into two standards: TDD and TD-SCDMA. Both standards are specified, but so far there has not been much commercial interest toward them.

1.4   3GPP

The 3GPP is an organization that develops specifications for a 3G system based on the UTRA radio interface and on the enhanced GSM core network. 3GPP is also responsible for future GSM specification work. This work used to belong to ETSI, but because both 3GPP and GSM use the same core network (GSM-MAP) and the highly international character of GSM, it makes sense to develop the specifications for both systems in one place. 3GPP's organizational partners include ETSI, ARIB, T1, *Telecommunication Technology Association* (TTA), *Telecommunication Technology Committee* (TTC), and *China Wireless Telecommunications Standard* (CWTS) group.

The UTRA system encompasses two modes: *frequency division duplex* (FDD) and *time division duplex* (TDD). In the FDD mode the uplink and downlink use separate frequency bands. These carriers have a bandwidth of 5 MHz. Each carrier is divided into 10-ms radio frames, and each frame further into 15 time slots. The UTRAN chip rate is 3.84 Mcps. A chip is a bit in a code word, which is used to modulate the information signal. Since they represent no information by themselves, we call them chips rather than bits. Every second, 3.84 million chips are sent over the radio interface. However, the number of data bits transmitted during the same time period is much smaller. The ratio between the chip rate and the data bit rate is called the *spreading factor*. In theory we could have a spreading factor of one, that is, no spreading at all. Each chip would be used to transfer one data bit. However, this would mean that no other user could utilize this frequency carrier, and moreover we would lose many desirable properties of wideband spreading schemes. In principle, the spreading factor indicates how large a chunk of the common bandwidth resource the user has been allocated. For example, one carrier could accommodate at most 16 users, each having a channel with a spreading factor of 16 (in practice the issue is not so straightforward, as will be shown in later chapters). The spreading factors used in UTRAN can vary between 4 and 512. A sequence of chips used to modulate the data bits is called the spreading code. Each user is allocated a unique spreading code.

The TDD mode differs from the FDD mode in that both the uplink and the downlink use the same frequency carrier. The 15 time slots in a radio frame can be dynamically allocated between uplink and downlink directions, thus the channel capacity of these links can be different. The chip rate of the normal TDD mode is also 3.84 Mcps, but there exists also a "narrowband" version of TDD known as TD-SCDMA. The carrier bandwidth of TD-SCDMA is 1.6 MHz and the chip rate 1.28 Mcps. TD-SCDMA can potentially have a large market share in China, and this technology is briefly discussed in Section 1.4.2.

UTRAN includes three types of channel concepts. A physical channel exists in the air interface, and it is defined by a frequency and a spreading code (and also by a time slot in the TDD mode). The transport channel concept is used in the interface between layers 1 and 2. A transport channel defines how the data is sent over the air, on common or on dedicated channels. Logical channels exist within layer 2, and they define the type of data to be sent. This data can be either control or user data.

In the beginning UTRAN was considered to be a Euro-Japanese system, with close connection to the GSM world, and CDMA2000 was supposed to rule in the Americas. This division is no longer valid, as North American TDMA operators are adopting UTRAN as their 3G system. Also, an increasing number of other operators in America have adopted GSM technology, and thus their 3G future is also linked with UTRAN. On the other hand, CDMA2000 has gained some foothold in East Asia.

This book is mostly about the UTRA FDD mode, so it will not be discussed further in this chapter.

1.4.1   TDD

If not otherwise stated, the text in this book generally refers to the FDD system in the 3GPP specifications. Thus, FDD functionality is explained throughout the other chapters. The basic principle of the FDD mode is that separate frequency bands are allocated for both the uplink and downlink directions, but in the TDD mode the same carrier is used for both the uplink and the downlink. Each time slot in a TDD frame can be allocated between uplink and downlink directions. The original ETSI/ARIB proposal for WCDMA was based on the FDD mode alone. The TDD mode was included to the UTRAN scheme later in the standards formulation process.

There are several reasons for using TDD systems. The first one is spectrum allocation. The spectrum allocated for IMT-2000 is asymmetric, which means that an FDD system cannot use the whole spectrum, as it currently requires symmetric bands. Thus the most obvious solution was to give the symmetric part of the spectrum to FDD systems, and the asymmetric part to TDD systems. The proposed spectrum allocations for UTRAN TDD are 1,900–1,920 MHz and 2,010–2,025 MHz. The first granted 3G

TDD licences have been 5 MHz per operator, so each TDD operator could only have one TDD carrier.

Second, many services provided by the 3G networks will require asymmetric data transfer capacity for the uplink and downlink, where the downlink will demand more bandwidth than the uplink. A typical example of this is a Web-surfing session. Only control commands are sent in the uplink, whereas the downlink may have to transfer hundreds of kilobits of user data per second toward the subscriber. As the TDD capacity is not fixed in the uplink and downlink, it is a more attractive technology for highly asymmetric services. The base station can allocate the time slots dynamically for the uplink or downlink according to current needs.

The third reason for TDD is easier power control. In the TDD mode both the uplink and downlink transmissions use the same frequency; thus, the fast fading characteristics are similar in both directions. The TDD transmitter can predict the fast fading conditions of the assigned frequency channel based on received signals. This means that closed-loop power control is no longer needed, but only open loop will be sufficient. However, open-loop control is based on signal levels, and if the interference level must be known, then this must be reported using signaling.

This "same channel" feature can also be used to simplify antenna diversity. Based on uplink reception quality and level, the network can choose which base station can best handle the downlink transmissions for the MS in question. This means less overall interference. Note that there is no soft HO (SHO) (see Section 2.5.1) in the TDD mode and all HOs are conventional hard HOs (HHOs) (similar to the ones in GSM).

Because the TDD mode is a TDMA system, an UE only has to be active (receiving or transmitting) during some of the time slots. There are always some idle slots during a frame and those can be used for measuring other base stations, and systems.

There are also problems with TDD. The first problem is interference from TDD power pulsing. The higher the mobile speed, the shorter the TDD frame so that fast open-loop power control can be used. This short transmission time results in audible interference from pulsed transmissions, both internally in the terminal and with other electronic equipment. Also, the timing requirements for many components are tighter. Both problems can be solved, but the solutions probably require more costly components.

The carrier bandwidth used in UTRA TDD is 5 MHz, and the chip rate used is 3.84 Mcps. The frame structure is similar to the FDD mode in that the length of a frame is 10 ms, and it consists of 15 time slots (see Figure 1.2). In principle, the network can allocate these timeslots freely for the uplink and the downlink. However, at least one time slot must be allocated for the uplink and one for the downlink, as the communication between a UE and the network always needs a return channel.

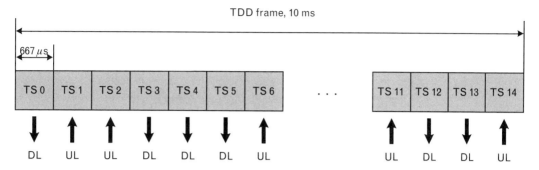

FIGURE 1.2   *An example of a TDD frame.*

Time slots are not exclusively allocated for one user, as in GSM. The TDD mode is a combination of TDMA and CDMA techniques, and each time slot can be accessed by up to 16 users. Different user signals sharing a time slot can be separated because they are modulated with user specific orthogonal chan0nelization codes. These codes can have *spreading factors* (SF) of 1, 2, 4, 8, or 16. The data rate of a user depends on the spreading factor allocated. A spreading factor of 1 gives a user all the resources of a time slot, a spreading factor of 2 gives half of them, and so forth. However, in the downlink only spreading factors 1 and 16 are allowed. A user can still be given "intermediate" data rates with the use of multicodes, that is, a user can be allocated several SF=16 spreading codes to be used in parallel. Also note that a user can be allocated different spreading factors in the downlink and in the uplink directions when there is a requirement for asymmetric data transmission.

A TDD system is prone to intracell and intercell interference between the uplink and downlink. The basic problem is that in adjacent cells, the same time slot can be allocated for different directions. It may happen that one UE tries to receive on a slot while another UE nearby transmits on the same slot. The transmission can easily block the reception attempt of the first UE. This problem can be prevented if all base stations are synchronized, and they all use the same asymmetry in their transmissions. However, this is costly (time-synchronous base stations), and also limits the usability of the system (fixed asymmetry).

Given these facts, it is most probable that FDD is used to provide wide-area coverage, and TDD usage will be limited to complement FDD in hot spots or inside buildings. TDD cells will typically be indoors, where they can provide high downlink data rates and the indoor nature of the system prevents the interference problems typical in TDD systems.

The UTRA TDD mode is especially well described in [8, 9]. The 3GPP specifications for the TDD mode radio interface are [10–14].

1.4.2 TD-SCDMA

In addition to standard UTRA TDD, there is also another TDD specification within the IMT–2000 umbrella. *Time-division synchronous CDMA* (TD-SCDMA) is a narrowband version of UTRA TDD developed by the *China Academy of Telecommunications Technology* (CATT) supported by Siemens. Within 3GPP this system is commonly known as *low chip rate* (LCR) TDD, or just as the 1.28 Mcps TDD option. Whereas the used carrier bandwidth in UTRA TDD is 5 MHz, in TD-SCDMA it is only 1.6 MHz. In some sources the 5-MHz TDD mode is called *high chip rate* (HCR) TDD mode to emphasize the difference between these two modes, but usually it is simply called the TDD mode. The used chip rates are 3.84 Mcps and 1.28 Mcps for the TDD and TDD-LCR systems, respectively. Both UTRA-TDD and TD-SCDMA (TDD-LCR) fit under the IMT–2000 IMT-time code (TC) banner.

The TD-SCDMA technology is promoted by TD-SCDMA Forum [15]. The TD-SCDMA standard drafts are submitted to 3GPP, where they are published as part of the TDD mode standards. In the 3GPP grand scheme the TD-SCDMA mode is thus seen as a submode of the TDD mode. Unofficially this system is also called the *narrowband TDD mode.* TD-SCDMA is quite similar to the mainstream TDD mode, especially in the higher layers of the protocol stack, but in the physical layer there are some fundamental differences.

First of all, the frame structure is different. The basic frame length is 5 ms, whereas in UTRAN-TDD it is 10 ms. To retain some similarity between the two TDD modes, this 5-ms frame is then called a *subframe*, and two subframes together make a 10-ms frame. One subframe consists of seven normal time slots and of three control slots. The duration of the normal time slot is 675 ms. Time slot 0 is always reserved for the downlink, and time slot 1 for the uplink. Other normal traffic time slots (2–6) can be freely allocated for the uplink or the downlink according to the traffic distribution by moving the location of the single additional switching point (the 5–MHz TDD mode can have multiple switching points). For example, in Figure 1.3 there are two uplink and five downlink slots, making this frame suitable for asymmetric downlink-heavy traffic. The only limitation for the time slot allocation is that there has to be one downlink (#0) and one uplink (#1) time slot.

The TD-SCDMA mode is similar to the TDD mode in that a time slot can be shared by up to 16 users. Spreading codes and spreading factors are similar to the TDD mode too, that is, spreading factors of 1, 2, 4, 8, or 16 can be used, but in the downlink only 1 and 16 are allowed. However, multicodes can be employed in the downlink to overcome this limitation.

One advantage of a TD-SCDMA system is that because of the narrower frequency carrier, an operator has more frequencies available for network

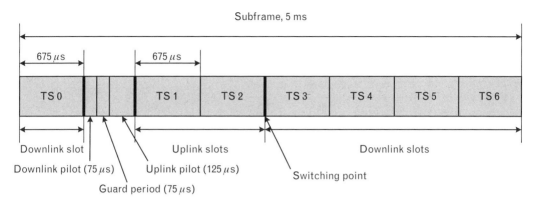

FIGURE 1.3 *TD-SCDMA subframe.*

planning purposes. This is an important factor, especially if the operator has been given only small spectrum allocations. For example a 2×10 MHz allocation can accommodate only two FDD mode carriers, four TDD mode carriers, but altogether 12 TD-SCDMA mode carriers. A typical TDD mode spectrum allocation in the first phase of 3G is only 5 MHz, and that could only accommodate either one TDD mode or three TD-SCDMA mode carriers.

Because there can only be a relatively limited number of users (and codes) in each time slot, and the chip rate is slower than in the TDD mode, it is possible to employ joint detection in TD-SCDMA receivers. The receiver can detect and receive all parallel codes and remove the unwanted signals that are declared to be interference from the result. This is not practical in the mainstream FDD mode because of the large number of parallel codes and the faster chipping codes. Joint detection is further discussed in Section 2.6 and in [9].

To make the migration from GSM into TD-SCDMA easier, an intermediate system called *TD-SCDMA System for Mobile Communication* (TSM) was developed. Whereas a genuine TD-SCDMA 3G system needs a new radio access network, TSM recycles the existing GSM/GPRS access network. In short, the TD-SCDMA physical layer is combined with the modified GSM/GPRS protocol stack. However, here we are combining CDMA technology (TD-SCDMA) with TDMA technology (GSM), which it is not a straightforward task. Figure 1.4 shows the GSM/GPRS air interface protocols that need modifications for the TSM system. In case of radio resources (RR) and radio link control/medium access control (RLC/MAC) these modifications are rather extensive. A TSM system can later be upgraded into a genuine TD-SCDMA system. TSM specifications are available from [16] (an all-Chinese site) and from [17] (an English mirror site).

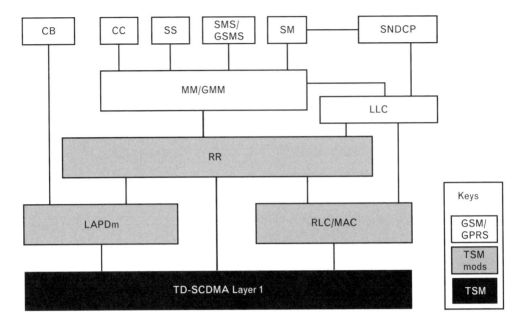

FIGURE 1.4   *TSM protocol stack.*

In 3GPP specifications [18, 19] the reader will find technical reports that explain the principles of TDD LCR. However, note they are not standards as such. The normative TD-SCDMA specifications are "embedded" into mainstream TDD mode specifications [10–14].

1.5   3GPP2

The 3GPP2 initiative is the other major 3G standardization organization. It promotes the CDMA2000 system, which is also based on a form of WCDMA technology. In the world of IMT–2000, this proposal is known as IMT-MC. The major difference between the 3GPP and the 3GPP2 approaches into the air interface specification development is that 3GPP has specified a completely new air interface without any constraints from the past, whereas 3GPP2 has specified a system that is backward compatible with IS-95 systems. This approach has been necessary because in North America, IS-95 systems already use the frequency bands allocated for 3G by the *World Administrative Radio Conference* (WARC). It makes the transition into 3G much easier if the new system can coexist with the old system in the same frequency band. The CDMA2000 system also uses the same core network as IS-95, namely, IS-41 (which is actually an ANSI standard: TIA/EIA-41).

The chip rate in CDMA2000 is not fixed as it is in UTRAN. It will be a multiple (up to 12) of 1.2288 Mcps, giving the maximum rate of 14.7456 Mcps. In the first phase of CDMA2000, the maximum rate will be three times 1.2288 Mcps – 3.6864 Mcps. As can be seen, this is quite close to the chip rate of UTRAN. However, it is unlikely that 3x rates will appear, because 1xEV-DO (IS-856) seems to satisfy the needs 3x is designed to address.

In CDMA2000 system specifications, the downlink is called the *forward link*, and the uplink is called the *reverse link*. The same naming convention is used in this section. The carrier composition of CDMA2000 can be different in the forward and reverse links. In the forward link the multicarrier configuration is always used (see Figure 1.5). In this configuration, several narrowband (1.25 MHz) carriers are bundled together. The original goal of CDMA2000 was to have a system with three such carriers (3x mode). These carriers have the same bandwidth as an IS-95 carrier and can be used in an overlay mode with IS-95 carriers. It is also possible to choose the spreading codes in CDMA2000 so that they are orthogonal with the codes in IS-95. In the reverse link the direct spread configuration will be employed. In this case the whole available reverse link bandwidth can be allocated to one direct spread wideband carrier. For example, a 5-MHz band could accommodate one 3.75-MHz carrier plus two 625-kHz guard bands. This option can be used in case the operator has 5 MHz of clear spectrum available. The CDMA2000 system does not use the time synchronized reverse link, and thus it cannot use mutual orthogonal codes with IS-95 systems. Therefore, splitting the wideband carrier into several narrowband carriers would not bring any benefits. Note, however, that in case of the 1x mode (the first stage in the CDMA2000 evolution path), there is only one 1.25-MHz carrier in the reverse link, and thus multicarrier and direct spread configurations would mean the same thing anyway. To the extent 1xEV-DO meets its expectations, the single carrier mode will likely be continued.

The evolutionary path from an IS-95A system into a full CDMA2000 system, that is, CDMA2000 3xRTT, can take many forms (see Figure 1.6). The first step could be IS-95B, which would increase the data rate from

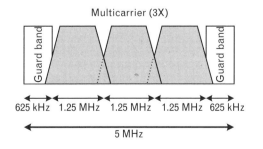
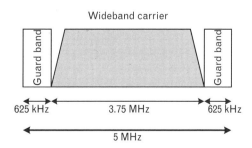

FIGURE 1.5   *CDMA2000 carrier types.*

14.4 Kbps to 64 Kbps. However, many IS-95 operators have decided to move straight into CDMA2000 1xRTT systems. Again, there are four levels of 1xRTT systems. The first one is known as 1xRTT release 0, or simply 1xRTT. This release can provide a 144-Kbps peak data rate. The next one is the 1xRTT release A, which can give 384-Kbps rates. The 1xEV-DO standard is the first system that can be regarded as a 3G system according to the ITU, the earlier ones being 2.5G systems. This system can provide 2+-Mbps data rates. The final phase (so far) under the 1xRTT banner is 1xEV-DV. This system is still under development, and it is comparable to the HSDPA upgrade in 3GPP systems. The peak data rate could be around 5 Mbps. Note that this number is already bigger than the planned peak data rate of the CDMA2000 3xRTT system. It remains to be seen whether CDMA2000 operators are actually interested in developing multicarrier (e.g., 3x) systems at all, if a single carrier system can provide comparable throughput. A 1xRTT system is easier to deploy because its carriers can be mapped one-to-one into IS-95 carriers. In any case, it is not necessary for an IS-95 operator to implement all of these evolution phases when upgrading its network; some of them could, and will be, skipped.

There are two kinds of channels in the CDMA2000 system. As in UTRAN, the physical channel exists in the air interface, and it is defined by a frequency and a spreading code. Logical channels exist just above physical channels. They define what kind of data will be transmitted on physical channels. Several logical channels can be mapped onto one physical channel. There is no transport channel concept in CDMA2000 and logical channels have taken their place.

The 3GPP2 membership includes ARIB, CWTS, TIA, TTA, and TTC. Although there are some common features in the 3GPP and 3GPP2 systems and they both belong under the common IMT-2000 umbrella, they are technically incompatible. The Operators' Harmonization Group (OHG) aims to coordinate these systems. The aim of this harmonization is not to produce one common specification for both systems; that would be a much too ambitious and impossible task. Merely, the harmonization work aims to make the life of the telecommunications industry and operators a little bit easier. For example, if certain operational parameters in these systems are close enough to each other, it could be possible to use same components for devices in both systems.

The CDMA2000 system is further discussed in [20].

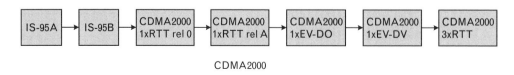

CDMA2000

FIGURE 1.6   *CDMA2000 evolution phases.*

1.6   3G Evolution Paths

Figure 1.7 describes a few possible evolution paths into 3G systems. Even though there are several IMT-2000 compatible systems, it seems that only two of them will survive in the end. WCDMA (IMT-DS), or UTRAN, is the most important one, and CDMA2000 (IMT-MC) will also gain a substantial but secondary market share. There will not be any IMT-SC systems (UWC-136), as the UWCC made a decision to join the WCDMA camp. As of this writing, the biggest question is the future of the IMT-TC, that is, the TDD mode of WCDMA. No operator has so far made orders for TDD mode equipment, and everybody seems to start their 3G deployments with FDD mode equipment. In China the TD-SCDMA systems may or may not become operational; the outcome of this is still too early to say in mid-2002. In any case, TDD mode systems will be deployed only after FDD mode systems if at all. There are some developments in the 3GPP FDD

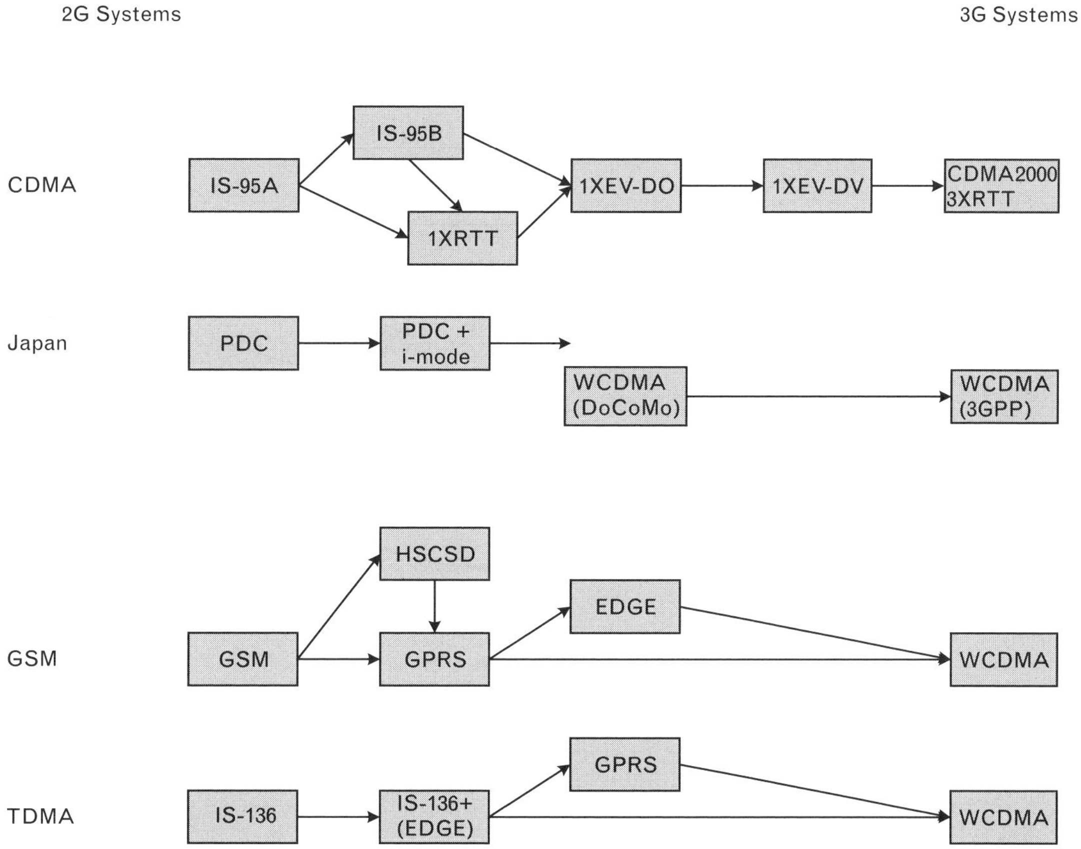

2G Systems   3G Systems

FIGURE 1.7   *3G evolution paths.*

mode standards, which could threaten the future of TDD mode systems (see Chapter 15).

REFERENCES

[1]   Walke, B., *Mobile Radio Networks*, New York: Wiley, 1999, pp. 4–7.

[2]   Redl, S., M. Weber, and M. Oliphant, *An Introduction to GSM*, Norwood, MA: Artech House, 1995, pp. 5–6.

[3]   Mehrotha, A., *GSM System Engineering*, Norwood, MA: Artech House, 1997, pp. 1–3.

[4]   Walke, B., *Mobile Radio Networks*, New York: Wiley, 1999, pp. 459–582.

[5]   Walke, B., *Mobile Radio Networks*, New York: Wiley, 1999, pp. 591–616.

[6]   "Enhanced Data Rates for GSM Evolution (EDGE)," Nokia Telecommunications Oy, Nokia white paper [on-line], March 1999, http://www.nokia.com/press/ background/pdf/edge_wp.pdf.

[7]   ITU press release ITU/99–22, November 5, 1999, accessible at http://www.itu.int/ newsarchive/press/releases/1999/99–22.htm1.

[8]   Prasad, R., W. Mohr, and W. Konhauser, *Third Generation Mobile Communication Systems*, Norwood, MA: Artech House, 2000.

[9]   Esmailzadeh, R., and M. Nakagawa, *TDD-CDMA for Wireless Communications*, Norwood, MA: Artech House, 2002.

[10] 3GPP TS 25.221, v 5.0.0, Physical Channels and Mapping of Transport Channels onto Physical Channels (TDD), 2002.

[11] 3GPP TS 25.222, v 5.0.0, Multiplexing and Channel Coding (TDD), 2002.

[12] 3GPP TS 25.223, v 5.0.0, Spreading and Modulation (TDD), 2002.

[13] 3GPP TS 25.224, v 5.0.0, Physical Layer Procedures (TDD), 2002.

[14] 3GPP TS 25.225, v 5.0.0, Physical Layer—Measurements (TDD), 2002.

[15] TS-SCDMA Forum home page, http://www.tdscdma-forum.org/.

[16] TS-SCDMA Standards, http://www.cwts.org/.

[17] TS-SCDMA Standards, English mirror site: http://www.geocities.com/tdscdma3g/.

[18] 3GPP TR 25.928, v 4.0.1, 1.28Mcps Functionality for UTRA TDD Physical Layer, 2001.

[19] 3GPP TR 25.834, v 4.1.0, UTRA TDD Low Chip Rate Option; Radio Protocol Aspects, 2001.

[20] Smith, C., and D. Collins, *3G Wireless Networks*, New York: McGraw-Hill, 2002.

CHAPTER 2

Principles of CDMA

In this chapter some basic concepts of CDMA are discussed. These concepts are CDMA specific, and often not used in other technologies, so some explanation may be necessary. An understanding of these concepts will make reading this book much easier. The examples in this section are WCDMA-specific.

2.1   Radio-Channel Access Schemes

The radio spectrum is a scarce resource. Its usage must be carefully controlled. Mobile cellular systems use various techniques to allow multiple users to access the same radio spectrum at the same time. In fact, many systems employ several techniques simultaneously. This section introduces four such techniques:

- *Frequency-division multiple access* (FDMA);

- TDMA;

- CDMA;

- *Space-division multiple access* (SDMA).

An FDMA system divides the spectrum available into several frequency channels (Figure 2.1). Each user is allocated two channels, one for uplink and another for downlink communication. This allocation is exclusive; no other user is allocated the same channels at the same time. In a TDMA system (Figure 2.2), the entire available bandwidth is used by one user, but only for short periods at a time. The frequency channel is divided into time slots, and these are periodically allocated to the same user so that other users can use other time slots. Separate time slots are needed for the uplink and the downlink.

GSM is based on TDMA technology. In GSM, each frequency channel is divided into several time slots (eight per radio frame), and each user is allocated one (or more) slot(s). In a TDMA system, the used system bandwidth is usually divided into smaller frequency channels. So in that sense GSM is actually a hybrid FDMA/TDMA system (as that shown in Figure 2.3), as are most other 2G systems. In a CDMA system all users occupy the same

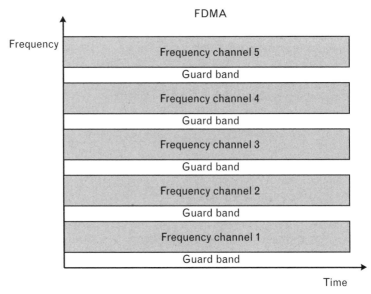

FDMA

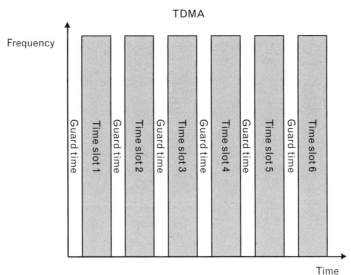

TDMA

frequency at the same time, no time scheduling is applied, and their signals are separated from each other by means of special codes (Figure 2.4). Each user is assigned a code applied as a secondary modulation, which is used to transform a user's signals into a spread-spectrum-coded version of the user's data stream. The receiver then uses the same spreading code to transform the spread-spectrum signal back into the original user's data stream. These codes are chosen so that they have low cross-correlation with other codes. This means that correlating the received spread-spectrum signal with the assigned code despreads only the signal that was spread using the same code. All other signals remain spread over a large bandwidth. That is, only the

FIGURE 2.3
Hybrid FDMA/TDMA.

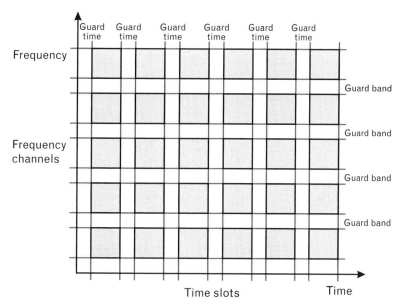

FIGURE 2.4   *CDMA.*

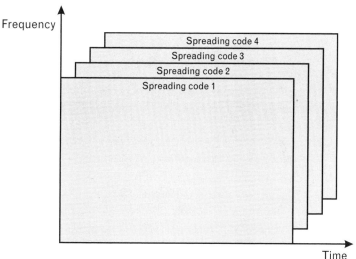

receiver knowing the right spreading code can extract the original signal from the received spread-spectrum signal.

In addition, as in TDMA systems, the total allocated bandwidth can be divided into several smaller frequency channels. The CDMA spread-spectrum scheme is employed within each frequency channel. This scheme is used in the UMTS Terrestrial Radio Access Network (UTRAN) frequency-division duplex (FDD) mode. The TDD mode uses a combination of CDMA, FDMA, and TDMA methods, because each radio frame is further divided into 15 time slots.

There are several methods used to modulate CDMA signals. The example in Figure 2.4 depicts *direct-sequence spread spectrum* (DS-SS) modulation.

With this method, the modulated signal occupies the whole carrier bandwidth all the time. Other modulation schemes include *frequency-hopping spread spectrum* (FH-SS), *time-hopping spread spectrum* (TH-SS), and various combinations of these. All these methods have their own advantageous properties. The 3GPP UTRAN system uses DS-SS modulation. Other CDMA modulation schemes are discussed in Section 4.1.

An SDMA system reuses the transmission frequency at suitable intervals of distance. If the distance between two base stations using the same frequency is large enough, the interference they inflict on each other is tolerable. The smaller this distance, the larger the system capacity. Therefore various techniques have been developed to take advantage of this phenomenon. Sectorization divides a cell into smaller "subcells," some of which can reuse the same frequency. A sector provides a fixed coverage area. Intelligent antennas can form narrow spot beams in desired directions, which increases the system capacity even further. Most digital 2G systems use some form of SDMA in addition to other above-mentioned techniques to improve the system capacity.

2.2   Spread Spectrum

Spread-spectrum transmission is a technique in which the user's original signal is transformed into another form that occupies a larger bandwidth than the original signal would normally need. This transformation is known as spreading. The original data sequence is binary multiplied with a spreading code that typically has a much larger bandwidth than the original signal. This procedure is depicted in Figure 2.5. The bits in the spreading code are called chips to differentiate them from the bits in the data sequence, which are called symbols. The term "chip" describes how the spreading operation chops up the original data stream into smaller parts, or chips.

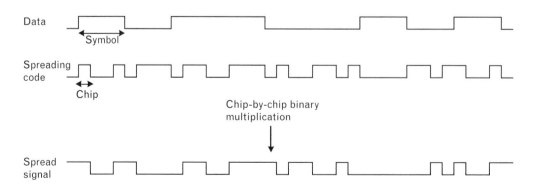

FIGURE 2.5   *Spreading.*

Each user has its own spreading code. The identical code is used in both transformations on each end of the radio channel, spreading the original signal to produce a wideband signal, and despreading the wideband signal back to the original narrowband signal (see Figure 2.6). The ratio between the transmission bandwidth and the original bandwidth is called the processing gain (also known as the spreading factor). Note that this ratio simply means how many chips are used to spread one data symbol. In the UTRAN, the spreading-factor values can be between 4 and 512 (however, in the TDD mode also SF=1 is allowed). The lower the spreading factor, the more payload data a signal can convey on the radio interface.

The spreading codes are unique, at least at the cell level. This means that once a user despreads the received wideband signal, the only component to despread is the one that had been spread with the same code in the transmitter. Two types of spreading codes are used in the UTRAN: orthogonal codes and pseudo-noise codes. These are further discussed in Chapter 5.

Spreading codes have low cross-correlation with other spreading codes. In the case of fully synchronized orthogonal codes, the cross-correlation is actually zero. This implies that several wideband signals can coexist on the same frequency without severe mutual interference. The energy of a wideband signal is spread over so large a bandwidth that it is just like background noise compared with the original signal; that is, its power spectral density is small. When the combined wideband signal is correlated with the particular spreading code, only the original signal with the corresponding spreading code is despread, while all the other component original signals remain spread (see Figure 2.7). Thus the original signal can be recovered in the receiver as long as the power of the despread signal is a few decibels higher than the interfering noise power; that is, the *carrier-to-interference ratio* (C/I) has to be large enough. Note that the power density of a spread signal can be much lower than the power density of the composite wideband signal, and the recovery of the original signal is still possible if the spreading factor is high enough, but if there are too many users in the cell generating too much

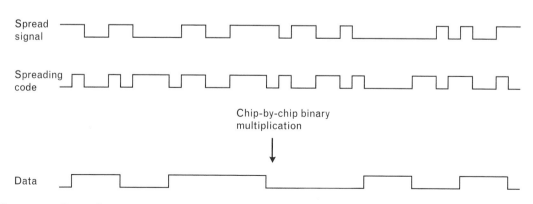

FIGURE 2.6   *Despreading.*

FIGURE 2.7
Recovery of despread signal.

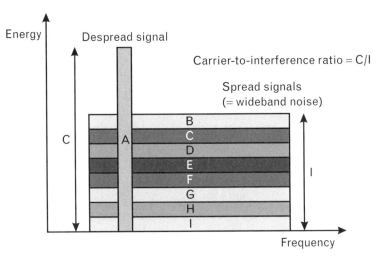

interference, then the signal may get blocked and the communication becomes impossible, as depicted in Figure 2.8.

Note that a wideband carrier does not increase the capacity of the allocated bandwidth as such. In principle, a set of narrowband carriers occupying the same bandwidth would be able to convey as much data as the wideband signal. However, in a wideband system the signals are more resistant to intercell interference, and thus it is possible to reuse the same frequency in adjacent cells. This means that the frequency reuse factor is one, while in typical GSM systems the value is at least four; that is, the same frequency can be reused at every fourth cell at most. This fact alone provides a substantial capacity gain over narrowband systems, although the capacity increase is not simply directly proportional to the reuse factor.

FIGURE 2.8
Unrecoverable signal.

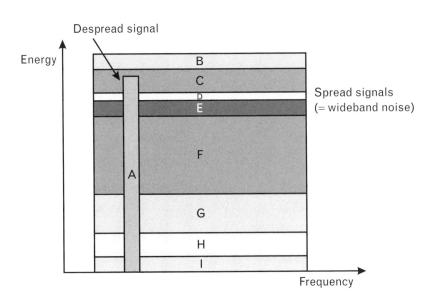

These wideband signals also have several other good properties (from [1]), which justify the consumption of a large chunk of expensive radio bandwidth:

1. *Multiple access capability.* Because all users have different spreading codes, which do not ideally cross-correlate too much, several users can coexist in the same frequency band. Despreading the received composite signal with a code signal of a certain user yields the original data signal of the user. However, increasing the number of users in the cell increases the interfering background noise. If this interference becomes too large, the signal recovery in the receiver will no longer succeed.

2. *Protection against multipath interference.* Multipath interference is a result of reflections and diffractions in the signal path. The various component signals may be interference to each other. A spread-spectrum CDMA signal can resist the multipath interference if the spreading codes used have good autocorrelation properties.

3. *Good jamming resistance.* Because the power spectral density of the signal is so low and resembles background noise, it is difficult to detect and jam on purpose. Therefore, CDMA communication systems are popular with the military.

4. *Privacy.* An intruder cannot recover the original signal unless he knows the right spreading code and is synchronized to it. However, this property only gives protection against an intruder with limited resources. To the extent the code-generating algorithms are known, a resourceful intruder can always record the intercepted wideband signal and then demodulate it with all possible spreading codes in his supercomputer. Therefore, WCDMA systems also need dedicated encryption procedures to be able to resist attacks from all kinds of intruders. Note, however, that implementing ciphering is optional for the UTRAN, but mandatory for *user equipment* (UE); a handset has to be able to support ciphering if the network so requires.

5. *Narrowband interference resistance.* A wideband signal can resist narrowband interference especially well. While the demodulation process will despread the original signal, it will also spread the interfering signal at the same time (see Figure 2.9). Thus the interference is spread over a wide spectrum. Demodulation will be successful if the spread interference is weak enough in the narrow despread signal bandwidth.

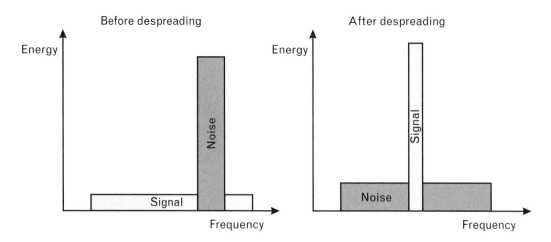

FIGURE 2.9   *Narrowband interference resistance.*

2.3   RAKE Receiver

In a multipath channel, the original transmitted signal reflects off obstacles in its journey to the receiver, and the receiver receives several copies of the original signal with different delays. These multipath signals can be received and combined using a RAKE receiver. A RAKE receiver is made of correlators, also known as RAKE fingers, each receiving a multipath signal. After despreading by correlators with a local copy of the appropriately delayed version of the transmitter's spreading code, the signals are combined. Since the received multipath signals are fading independently, this method improves the overall combined signal quality and performance.

It is called a RAKE receiver for two reasons. One is that most block diagrams of the device resemble a garden rake; each tine of the rake is one of the fingers. The other reason is that a common garden rake can illustrate the RAKE receiver's operation. The manner in which a garden rake eventually picks up debris off a patch of grass resembles the way the RAKE's fingers work together to recover multiple versions of a transmitter's signal (Figure 2.10). An individual signal received by a RAKE finger may be too weak to produce a correct result. However, combining several composite signals in a RAKE receiver increases the likelihood of reproducing the right signal. The RAKE receiver was patented in 1956, so the patent expired a long time ago.

2.4   Power Control

Efficient power control is very important for CDMA network performance. It is needed to minimize the interference in the system, and given the nature

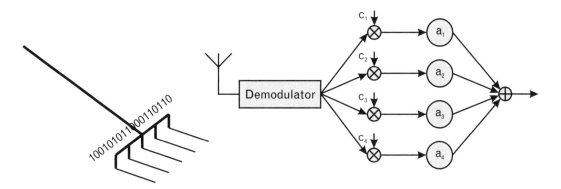

c_n = time-aligned spreading codes
a_n = weight factors, such as path gain (the higher the gain, the bigger the issued weight factor)

FIGURE 2.10   *RAKE receiver.*

of the DS-CDMA (all signals are transmitted using the same frequency at the same time), a good power control algorithm is essential. In this section the reasons for and the principles of CDMA power control are explained. Power control is needed both in the uplink and in the downlink, although for different reasons.

In the uplink direction, all signals should arrive at the base station's receiver with the same signal power. The mobile stations cannot transmit using fixed power levels, because the cells would be dominated by users closest to the base station and faraway users couldn't get their signals heard in the base station. The phenomenon is called the near-far effect (Figure 2.11). This problem calls for uplink power control. The mobile stations far away from the base station should transmit with considerably higher power than mobiles close to the base station.

The situation is different in the downlink direction. The downlink signals transmitted by one base station are orthogonal. Signals that are mutually orthogonal do not interfere with each other (the concept of orthogonality is discussed further in Section 5.1). However, it is impossible to achieve full orthogonality in typical usage environments. Signal reflections cause nonorthogonal interference even if only one base station is considered. Moreover, signals sent from other base stations are, of course, nonorthogonal and thus they increase the interference level. We must also remember that in a CDMA system the neighbor cells use the same downlink frequency carrier.

Therefore, power control is also needed in the downlink. The signals should be transmitted with the lowest possible power level, which maintains the required signal quality. Note that a mobile station close to the base station would not suffer if the signals it receives have been sent using too much power. But other users, especially those in other cells, could receive this

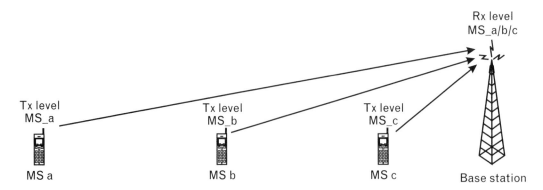

Without power control: Tx level MS_a =Tx level MS_b =Tx level MS_c
–> Rx level MS_a < Rx level MS_b < Rx level MS_c

With power control:Tx level MS_a >Tx level MS_b >Tx level MS_c
   –> Rx level MS_a = Rx level MS_b = Rx level MS_c

FIGURE 2.11   *Near-far effect in the uplink direction. (MS: mobile station.)*

signal as nonorthogonal noise, and therefore unnecessary high power levels should be avoided.

There are two basic types of power control: open loop and closed loop. The open-loop power control technique requires that the transmitting entity measures the channel interference and adjusts its transmission power accordingly. This can be done quickly, but the problem is that the interference estimation is done on the received signal, and the transmitted signal probably uses a different frequency, which differs from the received frequency by the system's duplex offset. As uplink and downlink fast fading (on different frequency carriers) do not correlate, this method gives the right power values only on average.

However, if a UMTS terrestrial radio access (UTRA) *time-division duplex* (TDD) mode is employed, then both the uplink and downlink use the same frequency and thus their fading processes are strongly correlated. This means that open-loop power control gives quite good results with the TDD mode. If the downlink transmission power is constant and known, then the mobile station can simply measure the received power level and adjust its own transmission level accordingly. The higher the reception level, the lower should be the transmission level. The sum of the reception and transmission levels should be a constant value (see Figure 2.12).

If the base station transmission power is not constant for a given channel, then the MS must either monitor some other channel (a downlink control channel may work for this technique) or it must receive the used transmission power from the base station somehow; for example, it could be sent via a synchronization channel. It is common practice to use the

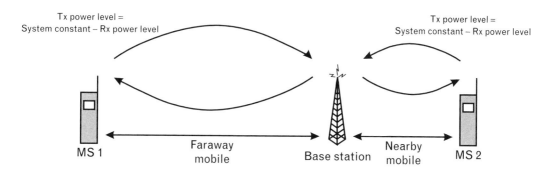

Tx power level =
System constant – Rx power level

Tx power level =
System constant – Rx power level

MS 1   Faraway mobile   Base station   Nearby mobile   MS 2

FIGURE 2.12   *Uplink TDD open-loop power control. (MS: mobile station.)*

received transmission value to estimate the current path loss. We assumed that broadcast channels do not use power control; that is, they are sent using a constant power level, and dedicated channels do use power control.

In the closed-loop power control technique, the quality measurements are done on the other end of the connection in the base station, and the results are then sent back to the mobile's transmitter so that it can adjust its transmission power. This method gives much better results than the open-loop method, but it cannot react to quick changes in channel conditions. This technique can also be applied in the opposite sense, in which the roles of the base station and the mobile are reversed.

The UTRAN FDD uses a fast closed-loop power control technique both in the uplink and downlink. In this method the received *signal-to-interference ratio* (SIR) is measured over a 667-microsecond period (i.e., one time slot), and based on that value, a decision is made about whether to increase or decrease the transmission power in the other end of the connection. Note that the delay inherent in this closed-loop method is compensated for by making the measurements over a very short period of time. The *transmit power control* (TPC) bits are sent in every time slot within the uplink and the downlink. There is not a neutral signal; all power control signals contain either an increase or decrease command. The SHO procedure adds additional complexity to power control, as then several base stations are transmitting to the UE at the same time and on the same frequency resulting in several, possibly contradictory, power commands to be received simultaneously by the mobile station. Two different algorithms have been defined for the TPC commands and the network indicates which one the UE should use.

The fast closed-loop power control is also called the inner loop power control. The uplink closed mechanism also contains another loop: the outer loop. The outer loop power control functions within the base station system, and adjusts the required SIR value (SIR_{target}), Which is then used in the inner loop control. Different channel types, which can be characterized by, for example, different coding and interleaving methods, constitute a channel's parameters. Different channel parameters may require different SIR_{target} values. The final result of the transmission process can only be known after the decoding process, and the resulting quality parameter is then used to adjust the required SIR value. If the used SIR value still gives a low quality bit stream, then the outer loop power control must increase the SIR_{target} value. This change in the outer loop will trigger the inner loop power control to increase the mobile station transmission power accordingly (see Figure 2.13).

The UTRAN FDD mode also uses the fast closed-loop power control technique for the downlink DPCCH/DPDCH.[1] The principle is similar to uplink fast closed-loop power control. The received signal SIR value is kept as close as possible to the SIR_{target}, value by sending power control commands up to the network. This is the inner loop control. There is also the outer loop control in the UE, which adjusts the SIR_{target} value. In all, this is a mirror image of the uplink fast closed-loop power control technique (see Figure 2.14).

It is possible that the network may transmit downlink signals only from one base station, even in the case of an SHO situation. The UE's signal is received and processed by several base stations, but only one of them is chosen to transmit the downlink. This scheme is known as *site-selection diversity transmission* (SSDT), which can be employed to reduce the overall system interference level.

Note that downlink power control is done separately for each mobile station. Each power control command affects only one channel component in the spread-spectrum signal. The total power level contained in the signal also depends on all the other channels in the wideband signal. Thus, one

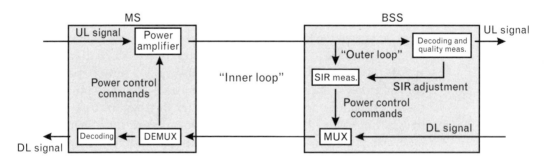

FIGURE 2.13   *Uplink closed-loop power control. (MS: mobile station; UL: uplink; and DL: downlink.)*

1   The *dedicated physical control channel* (DPCCH) and the *dedicated physical data channel* (DPDCH) are two of many physical channels within the WCDMA radio interface. These are fully described in Chapter 3. Dedicated channels are those associated with a particular connection between a base station and a mobile station. Dedicated connections need to be maintained for optimum performance with several channel maintenance tasks such as power control.

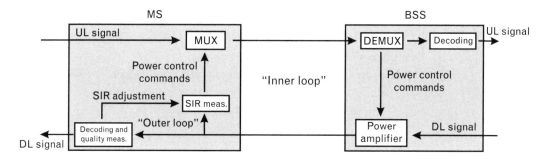

FIGURE 2.14 *Downlink closed-loop power control. (MS: mobile station; UL: uplink; and DL: downlink.)*

power control command probably doesn't have much effect on the total transmission power, even as it may have the desired effect on one of the component channels.

For the *downlink shared channel* (DSCH), the network may use the fast closed-loop power control of the associated DPCCH, or it may apply slow closed-loop power control. In any case, there must be some uplink channel available, otherwise the UE cannot send its power control commands to the network.

Closed-loop power control is called slow closed-loop power control if it is based on *frame error rate* (FER) results. The procedure reacts more slowly to changes in channel conditions, but on the other hand the FER is a more reliable measurement than the SIR.

In general, the lowest possible transmission power should always be used. This reduces the system interference level and also decreases the power consumption of mobile stations.

In the worst case, a defective power control results in pulsating coverage areas of cells ("breathing cells"). In this scenario the interference level in a cell increases gradually until the mobile stations farthest from the base station cannot be served anymore. Thus the effective range of the cell gets smaller. The reduction in coverage means fewer users and lower interference levels in the cell. This revised condition means that mobile stations far away from the base station can find the service again. The result is a pulsating cell radius, which was a curse on several early CDMA networks.

UTRAN FDD power control is further discussed in Chapter 5 of [2], Section 7.2.4.8 of [3], and [4].

2.5 Handovers

Mobile phones can maintain their connections in cellular networks when they move from one cell area to another. The procedure, which switches a connection from one base station to another, is called a *handover (HO)* or a handoff. It is possible that an HO does not involve a change of the base station but only a change of radio resources.

HOs in CDMA are fundamentally different from HOs in TDMA systems. While an HO in a TDMA system is a short procedure, and the normal state of affairs is a non–HO situation, the situation in a CDMA system is dramatically different. A UE communicating with its serving network can spend a large part of the connection time in a Soft Handover (SHO) state.

The various HO types are discussed at a rather high level in this section. The exact implementation and the specific HO procedures are explained more thoroughly in Section 11.4 for the UTRAN case of 3G.

2.5.1   Soft Handover

In an SHO, a UE is connected simultaneously to more than one base station (see Figure 2.15). The UE receives the downlink transmissions of two or more base stations. For this purpose it has to employ one of its RAKE receiver fingers for each received base station. Note that each received multipath component requires a RAKE finger of its own. Each separate link from a base station is called a *soft-handover branch*. Indeed, from the point of view of the UE, there is not much difference between being connected to one base station or several ones; even in the case of one base station, a UE has to be prepared to receive several multipath components of the same signal using its RAKE receiver. As all base stations use the same frequency in an

FIGURE 2.15
*SHO. (RNC: radio
network controller.)*

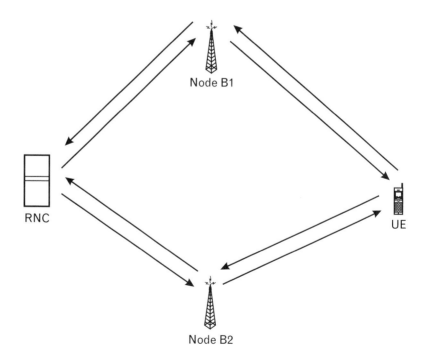

SHO, a UE can consider their signals as just additional multipath components. An important difference between a multipath component and an SHO branch is that each SHO branch is coded with a different spreading code, whereas multipath components are just time-delayed versions of the same signal.

An SHO is typically employed in cell boundary areas where cells overlap. It has many desirable properties. In the cell edges, a UE can collect more signal energy if it is in SHO than if it has only a single link to a base station. Without SHO, a communicating base station would have to transmit at a higher power level to reach the UE, which would probably increase the overall system interference level. Additionally, if a UE is in SHO, the connection is not lost altogether if one branch gets shadowed.

The SHO procedure should not be used without constraints. More transmitted signals may mean more energy in the air, which means more interference to the radio environment in the downlink direction. The control procedure in the UTRAN has to be very clever indeed to meet the conflicting demands of mobility and low interference levels. SHO branches should be added to a connection only when the estimated resulting total interference level is less than it would be without the SHO.

A softer HO is an HO between two sectors of a cell. From a UE's point of view, it is just another SHO. The difference is only meaningful to the network, as a softer HO is an internal procedure for a Node B (a UTRAN base station has the curious name Node B), which saves the transmission capacity between Node Bs and the RNC (a UTRAN base station controller). The uplink softer HO branches can be combined within the Node B, which is a faster procedure, and uses less of the fixed infrastructure's transport resources than most other types of HOss in CDMA systems.

As in GSM, all HOs are managed by the network. For this purpose the network measures the uplink connection(s) and receives measurement results from downlink connection(s) made by the UE. The cells to be measured are divided into three sets: *active, monitored,* and *detected.* Each set has its own requirements on how to perform measurements in the cells.

The active set contains all those base stations involved in an SHO with a UE. When the signal strength of a base station transmission exceeds the addition threshold in the UE, this base station is added to the active set and the UE enters into an SHO state if it is not already there. This threshold value, the addition threshold, is an important network performance parameter, and thus it can be set dynamically by the network. The UE does not add or remove base stations to or from its active set on its own initiative; these modifications are requested by the network through signaling mechanisms (see Section 7.5.2.13).

Another threshold parameter set by the network, the drop threshold, prevents the premature removal of base stations from the active set. The value of the drop threshold is always lower than the add threshold, but the exact value is again a system performance parameter and it can be set dynamically. When the signal strength value drops below the set threshold value, a drop timer is started in the network. If the value stays below the drop threshold until the timer expires, the base station in question is finally removed from the active set. This timer must be long enough to prevent a Ping-Pong effect, that is, the same base station is repeatedly added and removed from the active set. However, the drop timer must be short enough so that unusable base stations are not used for communication unnecessarily.

Both the add and the drop thresholds are used by the UTRAN to determine when it is necessary to update the active set. These thresholds are applied to UE measurements, so the UE must use the current thresholds to trigger the sending of measurement reports to the UTRAN. When a monitored cell exceeds the UTRAN-defined add threshold, a measurement report containing the latest results is sent to the network. The network may then send an active set update message to the UE, if the control algorithm decides to do so. There are also other parameters and considerations in the control algorithm besides the add threshold. For example, a cell may be so overloaded that no new connections can be allowed in the cell.

The monitored set includes cells that have been identified as possible candidates for HO but have not yet been added to the active set. These are indicated to the UE by the UTRAN in the neighbor cell list. The UE has to monitor these cells according to given rules. If a cell in the monitored set exceeds the add threshold, a measurement report will be triggered.

The detected set contains all the other cells that the UE has found while monitoring the radio environment and that are not included in the neighbor cell list. The UE may be requested by the UTRAN to report unlisted cells that it has detected. The triggering event that causes the UE to send a measurement report message is when a detected cell exceeds an absolute threshold.

SSDT power control (Figure 2.16) is a form of power control for the downlink that can be applied while a UE is in SHO. In a normal SHO, a UE has downlink connections with more than one cell, but in SSDT it has the downlink connection with only one base station at a time. Every downlink radio connection increases the system interference level. SSDT is a power control method that reduces the downlink interference generated while the UE is in an SHO. The principle of SSDT is that the best cell of the active set is dynamically chosen as the only transmitting site, and the other cells involved turn down their DPDCHs addressed to the UE in question. The DPCCH is transmitted in a normal fashion via all base stations in the active set.

The UE selects one of the cells from its active set to be the primary cell. All other cells are classed as nonprimary. The main objective is to transmit on the downlink only from the primary cell, thus reducing the interference

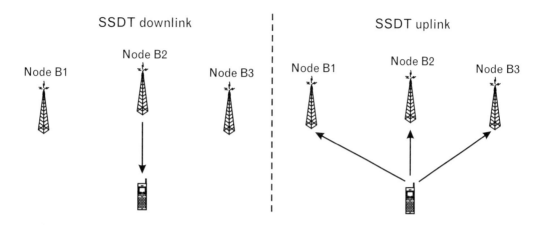

FIGURE 2.16 *SSDT.*

caused by multiple transmissions in an SHO mode. A second objective is to achieve faster site selection without network intervention, thus maintaining the advantage of the SHO. In order to select a primary cell, each cell is assigned a temporary *identification* (ID) and the UE periodically shares the primary cell ID with all the connecting cells. The nonprimary cells selected by the UE switch off their transmissions to this UE. The primary cell ID is delivered by the UE to the active cells via the uplink *feedback information* (FBI) field. The SSDT activation, SSDT termination, and ID assignment are carried out by the *radio resource control* (RRC) active set update procedure.

Note that successive SHOs may trigger a relocation procedure in the UTRAN. This is a kind of HO, although it does not take place in the radio interface. Thus, the UE does not directly know about it. Relocation is explained in the next section.

2.5.2 Relocation

Serving radio network subsystem (SRNS) relocation is a procedure in which the routing of a UE connection in the UTRAN changes. This procedure is best explained in a series of figures.

If a UE is in an SHO and all participating Node Bs belong to the same *radio network controller* (RNC), then the signals will be combined in this RNC and sent further to the serving *mobile services switching center* (MSC). If the SHO exists between sectors of the same Node B (softer handover), then the combining will be performed in Node B. In the downlink direction the splitting of the signal is done in corresponding places. (See Figure 2.17.)

If the UE moves to a position where it is in SHO with Node Bs belonging to different RNCs, then the signals will be relayed to the anchor RNC

FIGURE 2.17
Relocation, part 1.

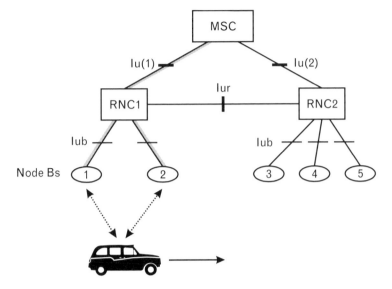

(RNC1), which combines the signals and sends them to the MSC. RNC1 is called the *serving RNC* (SRNC). There is always only one SRNC for each UE that has a connection to the UTRAN. The SRNC is in charge of the RRC connection between the UE and the UTRAN. (See Figure 2.18.)

The *relaying RNC* (RNC2) is called the *drift RNC* (DRNC). It provides its radio resources for the SRNC when the connection between the UTRAN and the UE needs to use cells controlled by the DRNC. In this example the combining of signals from cells 3 and 4 will be done in the DRNC by default, although the SRNC can override this and request all signals to be relayed to it without combining. Note that the combining process in the DRNC saves transport capacity in the Iur interface. There may be several DRNCs for one UE connection with the network.

FIGURE 2.18
Relocation, part 2.

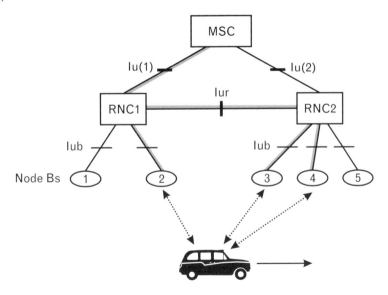

If the UE continues moving and leaves all cells controlled by RNC1, then the situation depicted in Figure 2.19 occurs. RNC1 is still the serving RNC (SRNC) and all traffic between the core network and the UE travels through it. This is clearly a waste of RNC1 resources and it loads up the Iur interface unnecessarily. Relocation is a process in which the SRNC status is moved from RNC1 to RNC2, as shown in Figure 2.20.

Although the relocation procedure itself is transparent to the UE, it may trigger the transmission of certain information to the UE. For example, the new SRNC can allocate a new *UTRAN radio network temporary identity* (U-RNTI) to the UE. The SRNC identity actually forms a part of the

Figure 2.19
Relocation, part 3.

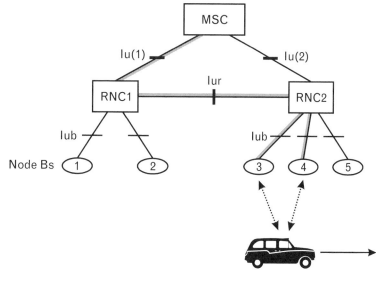

FIGURE 2.20
Relocation, part 4.

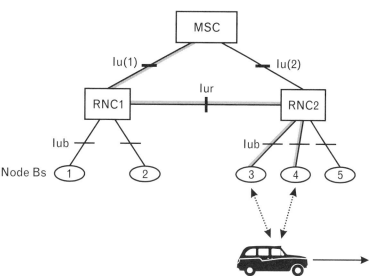

U-RNTI, so when a SRNC changes, the U-RNTI also changes (even though U-RNTI as such is unique within an UTRAN).

Note that the relocation procedure takes place in the UTRAN because of two special system characteristics. First, the UTRAN air interface makes it possible for a UE to perform a series of SHOs. Second, the core network should not know anything about the used radio access network technology. It does not know the concept of SHOs, and thus the combining of macrodiversity must be done in the UTRAN. The UTRAN is not allowed to route signals from the same UE connection to the core network via two Iu-interface instances. These signals must be combined in the RNC (if not already combined in Node B).

2.5.3   Hard Handover

Hard handover is also known as an interfrequency handover. During an HHO the used *radio frequency* (RF) of the UE changes. In principle, an HHO is a so-called break-before-make handover; that is, it is not seamless. The UE stops transmission on one frequency before it moves to another frequency and starts transmitting again. However, it is possible to make an HHO more seamless, for example, with the use of a compressed mode. If both frequencies use overlapping compressed mode gaps, and the switch is done during such a gap, then an HHO could be seamless.

HHOs are difficult for a mobile station in a CDMA system, as it is receiving and transmitting continuously and there are no free time slots for interfrequency measurements. But preliminary measurements before an HO are always necessary. The network has to get these results so that it can estimate which cell would be the most suitable for the UE. One possible solution could be to use another receiver just to measure other frequencies, but this would result in more expensive hardware. Thus, the 3GPP has specified a method called *compressed mode*. In compressed mode, not all time slots in downlink are used for data transmission. As we saw earlier in this chapter, RF transmissions occur in a pattern of time slots. The network defines a pattern of idle slots among these: the number of successive idle slots (i.e., the transmission gap), the location of the transmission gap, the repetition pattern of the transmission gap, and the duration of the compressed mode period. The network informs the UE about this pattern. The "unsent" data from idle slots will be sent during the normal slots using a lower spreading ratio and higher power. It is also possible to reduce the amount of data to be sent or puncture it (i.e., reduce redundancy).

The use of compressed mode is mandatory for UEs that do not have dual receivers. It makes the measurements for intersystem and interfrequency HOs possible. However, compressed mode also results in poorer performance, as it means either less data over the air interface (reduced data or puncturing) or more interference and higher code usage (lower spreading ratio). Compressed mode patterns are specified in [5].

Once the necessary measurements have been completed and reported to the network, the UE starts the HO procedure, if necessary, with messages from the HHO procedure. There is not a defined *protocol data unit* (PDU) called a *hard handover command* in the air interface specifications. Instead, the HHO procedure can be performed by several other procedures, such as physical channel reconfiguration, radio bearer establishment, radio bearer reconfiguration, radio bearer release, or transport channel reconfiguration. An HHO is just a series of normal radio link reconfigurations. If the reconfiguration happens to include a new frequency, then an HHO will take place.

Note that a special case of an HHO in the UTRAN is the *intermode handover* HO , which means an HO between FDD and TDD modes. For this procedure, a special dual-mode terminal is needed. Note that in this book the term *dual mode* refers to a terminal that can operate in both FDD and TDD modes. A UMTS/GSM-capable terminal is called a *dual-system terminal*. Generally, the literature does not seem to have any consistent view regarding this issue; a dual-mode terminal can mean a UMTS/GSM terminal in some sources and a FDD/TDD terminal in others.

2.5.4 Intersystem Handovers

Intersystem HOs are HOs between two different *radio access technologies* (RATs). 3GPP has specified intersystem HOs between GSM and UTRAN systems. Intersystem HOs are especially difficult procedures. There are plenty of problems that must be solved before such an HO is possible. A prerequisite for this procedure is that we have a dual-system 3G-GSM mobile phone capable of communicating with both systems. The first problem deals with measurements. Before a UE can start any HO, it must measure the quality of the new cell/carrier. Since it is busy communicating with the old channel, doing any measurements in another system is problematic.

First the UE must know the frequency (and in case of an HO to the UTRAN, the spreading code as well) in which the new cell in the other system is transmitting. This information must be relayed to the UE via the old cell. This information is typically sent within some kind of measurement control message.

Second, the UE must be able to measure the signal strength of the new carrier, or some other parameter on which the HO algorithm is based. This operation must be accomplished simultaneously with the operations of the old channel. In the case of a UTRAN-to-GSM HO this is difficult because typically a UTRAN's UE is receiving all the time and there are no idle slots in which to take measurements on the other frequency.

There are two alternatives to solving this problem:

1. Dual receiver;

2. Compressed mode.

If the UE has two receivers, then one receiver can perform interfrequency measurements, while another receives the normal UTRAN transmission. However, another receiver might be too expensive, at least for the mass-market handsets. Moreover, if the used GSM band is 1,800/1,900 MHz, it may be so close to the UMTS band used that the intercarrier interference becomes a problem.

Therefore, compressed mode is employed for intersystem measurements. Compressed mode was discussed in Section 2.5.3. This mode creates transmission gaps through which the UE can measure other systems. The length of one gap in the case of GSM measurements or decoding can be 3, 4, 7, 10, or 14 time slots (although exact values have been removed from the latest specifications, as this is an implementation issue for operators). Different gap lengths are used for different purposes. The preparation for an intersystem HO includes power level measurements (3 slots), initial synchronization to GSM's *frequency correction channel* (FCCH) and *synchronization channel* (SCH) on 7, 10, or 14 slots, tracking of the FCCH/SCH (4 slots) and *base station identity code* (BSIC) decoding (any gap length).

Once the required measurements have been made and reported back to the network, it may command an intersystem HO to be performed. Intrafrequency HOs in a UTRAN are typically seamless; that is, from the quality of the call, the user cannot notice the HO occurring. This requires that the connection is maintained simultaneously (at least for a moment) with both the old and the new base stations. However, in the case of a UTRAN/GSM HO (as well as with other interfrequency HOs), this is not possible if only one transmitter/receiver pair (transceiver) is available in the UE. The UE must stop transmitting in one system before it can start transmitting in another. The switching and routing delays in the network will cause additional delays in the procedure. This situation would be different if the UE could transmit and receive simultaneously in both systems. In this case, the old channel could be released only after the new channel is working nicely, resulting in a seamless HO.

An additional problem with the UTRAN-to-GSM HO is the different maximum data rates of these systems. This procedure must cope with a situation in which the UTRAN connection was using close to 2-Mbps data rates and after the HO the new connection can only get a small part of this rate.

In the GSM-to-UTRAN direction the HO procedure is probably technically easier, as GSM provides idle time slots in which it is possible to measure other frequencies, and also GSM's maximum data rates are lower than 3G maximum data rates.

A special problem with the inter-RAT HO to UTRAN is that synchronization to a UTRAN requires a large amount of information about the cell and the system, and relaying that information to the UE using an extended (and thus segmented) GSM HO command would be impractical. A 3GPP technical report [6] explains the use of predefined UMTS radio configurations. The UE should download up to 16 predefined radio configurations via UTRAN system information broadcast (SIB 16) message. Once the HO takes place, the network indicates only the identity of the preconfiguration to be used and possibly some additional parameters in the GSM handover command message. Note that this approach requires that the UE has been in the coverage area of the UMTS network prior to accessing the GSM network, as otherwise it could not have downloaded the configuration information. Also, this configuration information is naturally different for each public land mobile network (PLMN).

Finally, the case of a GSM-to-UTRAN intersystem HO can be divided into two different scenarios depending on whether the HO is between the circuit-switched or packet-switched domains. Actually, the procedure between packet-switched domains is not called a handover at all but an intersystem change, according to 3GPP jargon. It must be noted that the term "handover" in fact refers to a change in a circuit-switched connection, and is therefore not really applicable to packet-switched virtual connections.

It is possible that some other intersystem HOs will be defined in the future. These could include HOs between two different UMTS RATs, such as *broadband radio access network* (BRAN) to-UTRAN HOs, or HOs between a UTRAN and some other 2G cellular networks. The 3GPP specifications also mention the *UMTS satellite radio access network* (USRAN). These are not yet specified, but if and when they are implemented, an intersystem HO between the UTRAN and the USRAN must also be specified. A USRAN system could provide global coverage and therefore it could be an important supplement to terrestrial UTRAN coverage. However, as USRAN technologies have not been chosen yet, there are no specifications for these HOs either.

The intersystem HO procedures are discussed in more detail in Chapter 11. They are also discussed in [6].

2.6   Multiuser Detection

The capacity of a DS-CDMA system using RAKE receivers is interference limited. This means that when a new user (an interferer) enters into the network, the service quality of other users will degrade. The more the network can resist interference, the more users it can serve.

Multiuser detection (MUD) (also known as *joint detection and interference cancellation*) reduces the effect of interference and hence increases the system capacity. The idea behind MUD is that an optimum receiver would detect and receive all signals simultaneously, and then other signals would be subtracted from the desired signal. However, optimal MUD algorithms are too complex to be used in practice, and thus suboptimum multiuser receivers have been developed. These are divided into two main categories: linear detectors and interference cancellation.

Linear detectors apply a linear transform into the outputs of the matched filters that are trying to remove the multiple access interference (i.e., the interference due to correlations between user codes). Examples of linear detectors are decorrelators and *linear minimum mean square error* (LMMSE) detectors.

Interference cancellation is done by first estimating the multiple access interference and then subtracting it from the signal received. Interference cancellation methods include *parallel interference cancellation* (PIC) and *serial interference cancellation* (SIC).

MUD is currently a popular topic in telecommunications science, and it is studied widely. There are a wealth of academic literature and many research papers available on this subject. MUD is not a purely CDMA-specific issue. In principle it could also be used in GSM and other TDMA systems to improve performance.

References

[1]   Ojanperä, T., and R. Prasad, *Wideband CDMA for Third Generation Mobile Communications*, Norwood, MA: Artech House, 1998, pp. 34–36.

[2]   3GPP TS 25.214, v 5.0.0, Physical Layer Procedures (FDD), 2002.

[3]   3GPP TS 25.401, v 5.2.0, UTRAN Overall Description, 2002.

[4]   Holma, H., and A. Toskala, (eds.), *WCDMA for UMTS: Radio Access For Third Generation Mobile Communications*, New York: Wiley, 2000, pp. 109–110.

[5]   3GPP TS 25.215, v 5.0.0, Physical Layer-Measurements (FDD), 2002.

[6]   3GPP TR 25.922, v 5.0.0, Radio Resource Management Strategies, 2002.

CHAPTER 3

WCDMA Air Interface: Physical Layer

3.1 General

We begin our examination of the WCDMA air interface by looking at its organization and some of its characteristics. Following the OSI protocol model, radio interface protocols in the UTRAN system can be described by using a layered three-level protocol model. The lowest layer in this interface is the *physical layer* (listed as PHY in Figure 3.1). Layer 2 consists of the *medium access control* (MAC), the *radio link control* (RLC), the *broadcast multicast control* (BMC), and the *Packet Data Convergence Protocol* (PDCP) sublayers. Layer 3 includes the following sublayers: RRC, *mobility management* (MM), *GPRS mobility management* (GMM), *call control* (CC), *supplementary services* (SS), *short message service* (SMS), *session management* (SM), and *GPRS short message service support* (GSMS). These protocols are depicted in Figure 3.1.

As the physical layer is such an important part of the UTRAN system, it is discussed in four chapters. This chapter gives a general physical layer presentation. Chapters 4–6 include a more detailed discussion about some specific physical layer issues (modulation techniques, spreading codes, and channel coding). The other tasks (i.e., layers 2 and 3) in this protocol model are discussed in Chapter 7.

The physical layer is the lowest layer in the WCDMA air interface protocol model. It has to handle slightly different tasks depending on whether it is in the UE or in Node B, but the basic principles presented here are the same regardless of the location.

The physical layer has logical interfaces to both the MAC and RRC sublayers (see Figure 3.2). The interface to the MAC is named PHY, and it is used to transfer data (transport channels). The *control PHY* (CPHY) interface lies between physical layer and RRC, and is used for control and measurement information transfer. It is only used for layer 1 management, and the information the physical layer obtains through it is not meant to be sent further across the air interface.

The UTRAN can operate in two modes, FDD and TDD, and these modes set slightly different requirements for layer 1 functionality. In the

49

FIGURE 3.1
Air interface protocol model.

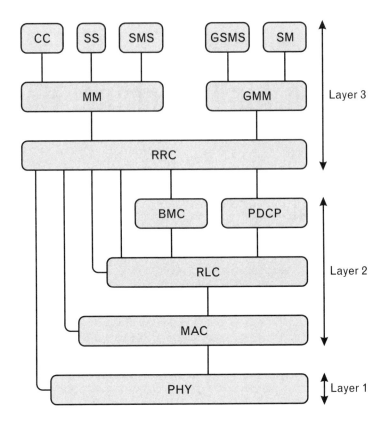

FDD mode, the uplink and downlink transmissions use different frequency bands. In the TDD mode, the uplink and downlink transmissions are on the same frequency but in different time slots. Thus, a WCDMA-TDD system is actually a CDMA/TDMA system because of this time slicing component. A physical channel in the FDD mode is defined as a frequency and a code, and is also defined in the TDD mode as a sequence of time slots.

The chip rate in the standard UTRAN air interface is 3.84 Mcps. One 10-ms radio frame is divided into 15 slots, which makes 2,560 chips per slot. This means that one time slot could transfer 2,560 symbols.[1] However, this is the total data rate available, and the rate one user gets depends on the *spreading factor* (SF) used in the channel. In FDD the spreading factors are from 4 to 256 for the uplink and from 4 to 512 for the downlink. In TDD they are from 1 to 16 in both directions. This gives channel rates from 7,500

---

1. Note, that here we are using the term "symbol" instead of "bit." Even though it may sound strange at this stage of the book, it is possible to transfer more than one bit of data within one chip if higher-order modulation schemes are employed. For example, Release'99 3GPP systems employ QPSK modulation, and with that modulation it is possible to transfer 2 bits of raw data within each chip. Thus a 3.84 Mcps QPSK system could in fact transfer 7.68 Mbps of raw data. But because the highest allowed spreading factor in FDD systems is 4, one user can only get 7.68 Mbps/4 = 1.92 Mbps via one physical channel. Thus, to ignore the effects of modulation process at this stage, we will simply assume that one chip can accommodate one symbol. The number of bits one symbol can contain will then be explained in Chapter 4.

FIGURE 3.2
Physical layer interfaces.

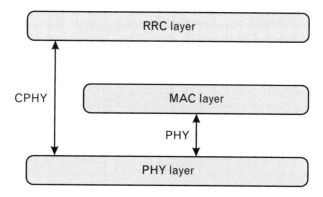

symbols/second to 960 ksymbols/second for FDD and from 240 ksymbols to 3.84 Msymbols/second for TDD. Again it must be noted that here we are discussing single channels. One user can have several channels (i.e., codes) simultaneously.

The physical layer must perform the following functions [1]:

1. FEC encoding/decoding of transport channels;

2. Radio measurements and indications to higher layers;

3. Macrodiversity distribution/combining and soft handover execution;

4. Error detection on transport channels;

5. Multiplexing of transport channels and demultiplexing of coded composite transport channels (CCTrCHs);

6. Rate matching;

7. Mapping of CCTrCHs on physical channels;

8. Modulation, spreading/demodulation, and despreading of physical channels;

9. Frequency and time synchronization;

10. Closed-loop power control;

11. Power weighting and combining of physical channels;

12. RF processing;

13. Timing advance on uplink channels (TDD only);

14. Support of uplink synchronization (TDD only).

These are discussed in detail in the following sections. The various channel types, spreading and scrambling, diversity schemes, and transport formats are introduced in the remaining sections of the chapter. Finally, in Section 3.6, we see how these functions work together in the physical layer.

3.1.1   Forward Error Correction Encoding/Decoding

Forward error correction (FEC) schemes aim to reduce transmission errors. Error correction coding is also known as channel coding. The idea is to add redundancy to the transmitted bit stream, such that occasional bit errors can be corrected in the receiving entity. There are numerous error-control schemes available, each having different capabilities. The choice of the channel-coding scheme depends on the requirements of the channel in question.

The UTRAN employs two FEC schemes: convolutional codes and turbo codes. Convolutional coding can be used for low data rates, and turbo coding for higher rates. Indeed, turbo coding is the most efficient with high bit rates. It is not suitable for low rates, as it does not perform well with short blocks of data. This is because low bit rates mean less bits in the turbo code internal interleaver. This results in a weaker performance. Also, turbo codes make blind rate detection more difficult. Blind transport format detection[2] can be used in the receiving entity when the transport format used is not signaled via a physical control channel. Note that the use of turbo codes in the UE is optional. The UTRAN learns from the UE's capability information whether the UE supports turbo codes, so it knows which codes to use with a particular UE.

Table 3.1 describes the channel coding algorithms used in the UTRAN air interface.

The code rate indicates the ratio between the number of input bits and the number of output bits of the channel coding function. In convolutional and turbo coding, it is typically either 1/2 or 1/3; twice as many bits or

TABLE 3.1 CHANNEL-CODING SCHEMES USED IN UTRAN AIR INTERFACE [2]

TrCH Type	Coding Scheme	Code Rate
BCH	Convolutional coding	1/2
PCH		
RACH		
CPCH, DCH, DSCH, FACH		1/2 or 1/3
	Turbo coding	1/3
HS-DSCH		

TrCH: transport channel; BCH: broadcast channel; PCH: paging channel; RACH: random access channel; CPCH: common packet channel; DPCH: dedicated physical channel; DCH: dedicated channel; DSCH: downlink shared channel; FACH: forward access channel; and HS-DSCH: high-speed downlink shared channel.

2.   The receiving entity has to establish the transport format for example by using the CRC. Once the decoding process produces the right CRC, the receiving entity knows that it has used the right transport format. The details of blind transport format detection can be read from Annex A of [2].

three times as many bits respectively emerge from the channel coder as enter it.

In the UTRAN the channel coding is combined with the CRC error-detection function (see Section 3.1.4) to form a hybrid ARQ scheme. This means that the channel coding aims to fix as many errors as possible, and then the error-detection function checks whether the result was correct. Erroneous packets are detected and indicated to higher layers for retransmission. More precisely, the retransmission of missing or corrupted packets belongs to RLC layer functionality.

Channel coding and the theory behind it are further discussed in Chapter 6.

3.1.2 Radio Measurements and Indications to Higher Layers

The radio measurements to be performed in the UTRAN air interface are specified in [3, 4]. Some measurement types are specific to either the UE or the Node B, but there are also measurements that are applicable to both. The measurement results are reported to higher layers (in the UE this is the RRC layer), and from thereon to the peer entity in Node B in some cases.

Radio measurements are typically controlled by the RRC layer in the UE. The RRC receives the necessary control information from the UTRAN in measurement control messages. In idle mode and in connected-mode states CELL_FACH, CELL_PCH, and URA_PCH (i.e., when there is no dedicated connection), the messages are broadcast in *system information blocks* (SIBs). In the dedicated state CELL_DCH, a special measurement control message is used. The RRC substates (CELL_DCH, CELL_FACH, CELL_PCH, and URA_PCH) are discussed in Section 7.6. The RRC may explicitly ask the physical layer to perform a certain measurement, or it can set certain conditions and the fulfillment of these then triggers the measurement process. Some measurements are continuous and are performed periodically when the physical layer is in a certain state.

Possible measurement types for the UE physical layer include:

- Received signal code power (RSCP);
- Received signal strength indicator (RSSI);
- Received energy per chip divided by the power density in the band (Ec/No);
- *Block error rate* (BLER);
- UE transmitted power;
- Connection frame number-system frame number (CFN-SFN) observed time difference;

- SFN-SFN observed time difference;

- UE Rx-Tx time difference;

- Observed time difference to GSM cell;

- UE GPS timing of cell frames for UE positioning;

- UE GPS code phase.

Correspondingly, the measurement types for the UTRAN include:

- Received total wide band power;

- Signal-to-interference ratio (SIR);

- SIR error;

- Transmitted carrier power;

- Transmitted code power;

- Bit error rate (BER);

- Round-trip time (RTT);

- UTRAN GPS timing of cell frames for UE positioning;

- PRACH/PCPCH propagation delay;

- Acknowledged PRACH preambles;

- Detected PCPCH access preambles;

- Acknowledged PCPCH access preambles;

- SFN-SFN observed time difference.

Although the list of possible measurements is quite long, it must be noted that not all of these measurements are performed continuously, but only when certain conditions are true. The long list is like a large toolbox; it is useful to have, but most of the tools will remain unused most of the time. For the exact definition of these measurements and their usage, consult [3, 4].

The purpose of the measurements is rather different in idle and in connected modes. In idle mode the purpose of the measurements is to help the UE in the cell-reselection process; that is, to make sure that it is camped on the best available (or at least good enough) cell. This is also for the most part true in the connected-mode states CELL_FACH, CELL_PCH, and URA_PCH. In these states the UE also performs measurements to gain information for the cell-reselection procedure. However, in the CELL_FACH state the UE may also have to report back some results. In the dedicated state (CELL_DCH) the measurements are typically done to help the UTRAN maintain the optimal radio connection. Thus the CELL_DCH state includes extensive reporting by the UE.

In the FDD mode the interfrequency measurements introduce a problem for the UE. In dedicated FDD mode the UE is normally receiving and transmitting all the time, so interfrequency measurements would be, in principle, impossible if only one receiver is used. Extra receivers are expensive, and thus a better solution has been specified. This is called the compressed mode. This means that pauses are created to signal transmissions, and interfrequency measurements can be performed during these intervals or pauses.

Uplink compressed mode must be used if the frequency to be measured is close to the uplink frequency used by the UTRAN air interface (i.e., frequencies in TDD mode/GSM-1,800/1,900). Otherwise interfrequency interference may affect the results.

Downlink compressed mode is not necessary if the UE has dual receivers. In this case one receiver can perform interfrequency measurements while the other handles the normal reception. Note, however, that double receivers in the UE do not remove the need for uplink compressed mode. If the uplink frequency is close enough to the downlink frequency to be measured, then compressed mode must be employed in the uplink to prevent interfrequency interference.

Compressed mode is discussed further in Section 2.5.3. See also [3].

3.1.3 Macrodiversity Distribution/Combining and Soft Handover Execution

Macrodiversity (i.e., soft handover) is a situation in which a receiver receives the same signal from different sources. This happens if a UE receives the same transmission from several base stations. Similarly, an RNC may combine the same signal sent by the UE and received by several base stations. The more energy the receiver is able to collect from transmissions, the more likely it can construct the original signal from the components.

The usage of this phenomenon is essential in a WCDMA system, because all base stations use the same frequency (frequency reuse = 1) and fast power control. Without macrodiversity combining, the system interference level would be increased and the capacity decreased by a considerable amount.

In the downlink the UE can receive, at most, as many macrodiversity components as it has fingers in its RAKE receiver. Thus the more RAKE fingers the UE has, the better performance it has, providing that all fingers find a separate diversity component. However, from the system point of view this case is not so clear. Each new transmission may also increase the system interference. If too many base stations are used in an SHO, the system interference level increases instead of decreasing and preserving the usefulness of an SHO.

In the uplink the effects of macrodiversity are only positive, as the more base stations that can receive the signal from a UE, the better the probability that some of them will receive it successfully. This does not generate more transmissions or interference. Indeed, the opposite is true, as the UE transmission power level can probably be lower if macrodiversity is used.

Because of the increased interference the downlink macrodiversity may generate, it is also possible to use site-selection diversity transmission (SSDT) power control. In this method the macrodiversity only exists in the uplink direction. The UE selects one cell from its set of active cells to be a primary cell. This selection is based on the downlink reception level measurements of the common pilot channel (CPICH) of each cell. The identification of the chosen cell is signaled to the network, and the UTRAN sends the downlink transmission only via this cell. Thus several Node Bs participate in reception, but only one in transmission.

There are two algorithms for combining the transmit power control (TPC) bits in an SHO situation. These are explained in [5].

Note that macrodiversity is not the only form of diversity that can be used beneficially in a WCDMA system. Diversity is further discussed in Section 3.4.

3.1.4   Error Detection on Transport Channels

The purpose of error detection is to find out whether a received block of data was recovered correctly. This is done on transport blocks using a cyclic redundancy check (CRC) method. There are five CRC polynomial lengths in use (0, 8, 12, 16, and 24 bits), and higher layers will indicate which one should be used for a given transport channel.

The sending entity calculates the CRC checksum over the whole message and attaches it to the end of the message. The receiving entity checks whether the CRC of the received message matches with the received CRC. The CRC calculation in the UTRAN is discussed with block codes in Section 6.3 and also in [2].

An erroneous CRC result must be indicated to layer 2 (L2). If the RLC PDUs are mapped one-to-one onto the transport blocks, then the error detection facility in layer 1 (L1) can be used by the retransmission protocol in L2.

In the UTRAN, the error detection is combined with the channel coding scheme (see Section 3.1.1) to form a hybrid ARQ scheme. The idea behind this scheme is that the channel coding reduces the amount of faulty packets before they get detected in the error-detection function. Channel coding aims to fix as many errors as possible, and then the error-detection checks whether the result was correct. Erroneous packets are detected, and indicated to higher layers for retransmission.

3.1.5 Multiplexing of Transport Channels and Demultiplexing of CCTrCHs

Each UE can have several transport channels in use simultaneously. Every 10 ms, one radio frame from each transport channel is multiplexed into a coded composite transport channel (CCTrCH). This multiplexing is done serially; that is, the frames are simply concatenated together.

There can be more than one CCTrCH per connection (although not yet with Release 99 of the specifications). In the FDD mode each UE can have only one CCTrCH on the uplink. In TDD the uplink can accommodate several CCTrCHs. On the downlink both modes can have several CCTrCHs per UE. The different CCTrCHs can have different C/I requirements to provide different quality of service (QoS) on the mapped transport channels. See [1] for further information about CCTrCHs.

3.1.6 Rate Matching

The number of bits on a transport channel can vary with every transmission time interval. However, the physical channel radio frames must be completely filled. This means that some sort of adjusting must be done to match the two given rates.

In the uplink the total bit rate after transport channel multiplexing must match the total physical channel bit rate. This is done by either repeating or puncturing bits. There are special rules for what bits can be punctured and what bits cannot. Puncturing means that bits are deleted from the output stream according to a predefined scheme. It is possible to puncture some bits, as this process will be done after channel coding, which had already added redundancy to the code. Thus puncturing can be seen as a deletion of some redundant bits. However, this weakens the resulting code, and there are limits on how many bits can be punctured.

In the downlink, the network can interrupt the transmission if the number of bits to be sent is lower than the maximum available. This is called the discontinuous transmission (DTX) mode, and it is done to reduce the overall interference in the radio path. Rate matching is needed in the downlink to determine how many DTX bits need to be transmitted. This is done by calculating the possible peak data rate and comparing it with the offered data rate.

Rate matching is explained in [2].

3.1.7 Mapping of CCTrCHs on Physical Channels

If there is more than one physical channel in use, then the bits in the CCTrCH must be divided among them. This is simply done by segmenting the input bits evenly for each physical channel. Remember that rate matching is already done in an earlier phase, so the bits should fit nicely into physical channels.

After the physical channel segmentation, the next phase is the 2nd interleaving (1st interleaving is discussed in Section 3.6). This is a block interleaving, where the bits are written into a matrix row by row, and read from it column by column. Before reading the bits out an intercolumn permutation is performed. The intercolumn permutation means that the order of columns is changed according to a predefined pattern.

The last phase of mapping is the actual filling of radio frames with bits. In the uplink all frames are completely filled if they are used (except in compressed mode). In the downlink the frames are also logically completely filled, but the DTX bits are not actually sent. They are placeholders to tell the transmitter that there is nothing to be sent, and the transmitter can be turned off for the duration of these bits.

The mapping of transport and physical channels is shown in Figure 3.3. This diagram is for the FDD mode only, and only shows the mapping relationship between these channels. There are many physical channels not shown here, because they do not map into any transport channel; the data they carry is generated and consumed by physical layer peer entities. The channels are discussed further in Section 3.2.

3.1.8   Modulation, Spreading/Demodulation, and Despreading of Physical Channels

Spreading is an important and quite complex issue. It is discussed more thoroughly in a separate chapter in this book (Chapter 5). This paragraph contains only a very brief summary on modulation, spreading, and spreading codes.

There are two families of spreading codes: orthogonal codes and pseudorandom (also called pseudo-noise [PN]) codes. These have different properties and both types of codes are used in the UTRAN system.

Spreading means increasing the bandwidth of the signal beyond the bandwidth normally required to accommodate the information. The spreading process in UTRAN consists of two separate operations or steps: channelization and scrambling. The uses of spreading codes are somewhat different in the uplink and in the downlink.

Channelization transforms each data symbol into several chips. The ratio (number of chips/symbol) is called the spreading factor. Data symbols on I- and Q-branches are multiplied with a channelization code. Channelization codes are orthogonal codes, meaning that in an ideal environment they do not interfere with each other. However, orthogonality can only be achieved if the codes are time synchronized. Thus it can be used in the downlink to separate different users within one cell, but in the uplink only to separate the physical channels of one user. It cannot be used by the base

FIGURE 3.3
Transport and physical channel mapping.

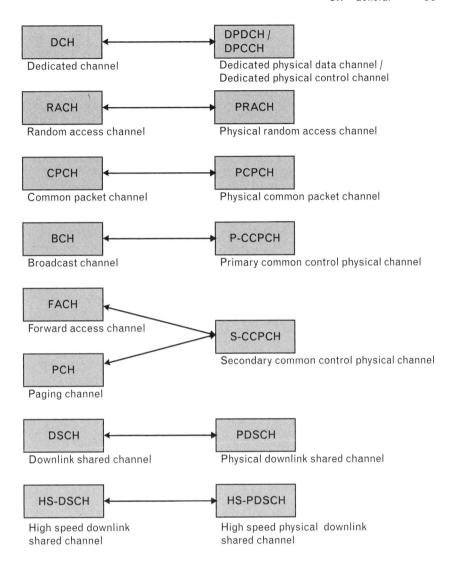

stations to separate different users, as all mobiles are unsynchronized in time, and thus their codes cannot be orthogonal. However, note that in the TDD mode it is possible to have a finely time-synchronized uplink; see Sections 3.1.13 and 3.1.14. Note that in the TDD mode the uplink is in any case crudely time synchronised, as UE transmissions must fit into their own time slots, but this is still not the same thing as chip level synchronisation that is required for code orthogonality.

In the scrambling procedure, the I- and Q-phases are further (after channelization) multiplied by a scrambling code. These scrambling codes have good autocorrelation properties. In the uplink, different users have different long code offsets, and the network can recognize different users from their offsets. And once the synchronization is achieved, various services of the user can be separated using orthogonal codes.

In the downlink, pseudorandom scrambling codes are used to reduce inter-base-station interference. Each Node B has only one unique primary scrambling code, and this is used to separate various base stations.

The modulation scheme in the UTRAN is quadrature phase shift keying (QPSK), and also 16 QAM on the HS-PDSCH channel. Modulation is a process where the transmitted symbols are multiplied with the carrier signal. The modulating symbols are called chips, and their modulating rate is 3.84 Mcps. Modulation and spreading in UTRAN are specified in [6] (FDD) and [7] (TDD).

3.1.9   Frequency and Time Synchronization

In this section we will discuss the frequency and time synchronisation procedure. This procedure takes place when the power is turned on in the UE. The synchronization procedure starts with downlink SCH synchronization. The UE knows the SCH primary synchronization code, which is common to all cells. The slot timing of the cell can be obtained by receiving the primary synchronization channel (P-SCH) and detecting peaks in the output of a filter that is matched to this universal synchronization code. The slot synchronization takes advantage of the fact that the P-SCH is only sent during the first 256 chips of each slot. The whole slot is 2,560 chips long. This is depicted in Figure 3.4. Thus the UE can determine when a slot starts, but it does not know the slot number yet (there are 15 slots in each frame), and thus it does not know where the radio frame boundary may be.

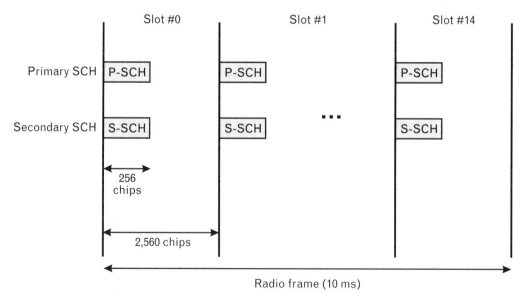

FIGURE 3.4   *Structure of synchronization channels.*

Thereafter the UE correlates the received signal from the secondary synchronization channel (S-SCH) with all secondary synchronization codes (SSC), and identifies the maximum correlation value. The S-SCH is also only sent during the first 256 chips of every slot. One SSC is sent in every time slot. There are 16 different SSCs, and they can form 64 unique secondary SCH sequences. One sequence consists of 15 SSCs, and these sequences are arranged in such a way that in any nonzero cyclic shift less than 15 of any of the 64 sequences is not equivalent to some other sequence. This means that once the UE has identified 15 successive SSCs, it can determine the code group used as well as the frame boundaries (i.e., frame synchronization).

An example of frame synchronization is as follows: if the UE receives a sequence of SSCs (Figure 3.5) from S-SCH, it must compare it with all SSC sequences from Table 3.2 (the table is taken from [6]), and once a match is found, it knows the used code group for the Node B sending it and the frame boundaries.

In this example the matching code group is 30, and the frame boundary is before the time slot #0, that is, before SSC 2 in code group 30 (see Figure 3.6).

Each code group identifies eight possible primary scrambling codes, and the correct one is found by correlating each candidate in turn over the CPICH of that cell. Once the correct primary scrambling code has been identified, it can be used to decode BCH information from the primary common control physical channel (P-CCPCH), which is covered with the cell's unique primary scrambling code. The CPICH also acts as a timing reference for the P-CCPCH. Note that the P-CCPCH doesn't use the first 256 chips of each slot, whereas the P-SCH and S-SCH use only these chips.

Note the important difference between the two primary codes. The primary synchronization code is common to all cells, and it is used to gain slot synchronization from the P-SCH. The primary scrambling code is unique to a cell; it is gained from the CPICH and used to demodulate common control channels.

Dedicated channel synchronization is skipped here, as it is simpler than the initial synchronization. Curious readers can study it in [5].

3.1.10   Inner-Loop Power Control

In general, power control comes in two forms, open- and closed-loop control. The basic difference between these methods is that the closed-loop control is based on the explicit power control commands received from the

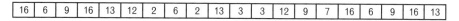

| 16 | 6 | 9 | 16 | 13 | 12 | 2 | 6 | 2 | 13 | 3 | 3 | 12 | 9 | 7 | 16 | 6 | 9 | 16 | 13 |

FIGURE 3.5   *Received sequence of SSCs.*

TABLE 3.2   SSCs and Scrambling Code Groups

Scrambling Code Group	Slot Number														
	#0	#1	#2	#3	#4	#5	#6	#7	#8	#9	#10	#11	#12	#13	#14
Group 0	1	1	2	8	9	10	15	8	10	16	2	7	15	7	16
Group 1	1	1	5	16	7	3	14	16	3	10	5	12	14	12	10
Group 2	1	2	1	15	5	5	12	16	6	11	2	16	11	15	12
Group 3	1	2	3	1	8	6	5	2	5	8	4	4	6	3	7
Group 4	1	2	16	6	6	11	15	5	12	1	15	12	16	11	2
Group 5	1	3	4	7	4	1	5	5	3	6	2	8	7	6	8
Group 6	1	4	11	3	4	10	9	2	11	2	10	12	12	9	3
Group 7	1	5	6	6	14	9	10	2	13	9	2	5	14	1	13
Group 8	1	6	10	10	4	11	7	13	16	11	13	6	4	1	16
Group 9	1	6	13	2	14	2	6	5	5	13	10	9	1	14	10
Group 10	1	7	8	5	7	2	4	3	8	3	2	6	6	4	5
Group 11	1	7	10	9	16	7	9	15	1	8	16	8	15	2	2
Group 12	1	8	12	9	9	4	13	16	5	1	13	5	12	4	8
Group 13	1	8	14	10	14	1	15	15	8	5	11	4	10	5	4
Group 14	1	9	2	15	15	16	10	7	8	1	10	8	2	16	9
Group 15	1	9	15	6	16	2	13	14	10	11	7	4	5	12	3
Group 16	1	10	9	11	15	7	6	4	16	5	2	12	13	3	14
Group 17	1	11	14	4	13	2	9	10	12	16	8	5	3	15	6
Group 18	1	12	12	13	14	7	2	8	14	2	1	13	11	8	11
Group 19	1	12	15	5	4	14	3	16	7	8	6	2	10	11	13
Group 20	1	15	4	3	7	6	10	13	12	5	14	16	8	2	11
Group 21	1	16	3	12	11	9	13	5	8	2	14	7	4	10	15
Group 22	2	2	5	10	16	11	3	10	11	8	5	13	3	13	8
Group 23	2	2	12	3	15	5	8	3	5	14	12	9	8	9	14
Group 24	2	3	6	16	12	16	3	13	13	6	7	9	2	12	7

TABLE 3.2 (CONTINUED)

SCRAMBLING CODE GROUP	SLOT NUMBER														
	#0	#1	#2	#3	#4	#5	#6	#7	#8	#9	#10	#11	#12	#13	#14
Group 25	2	3	8	2	9	15	14	3	14	9	5	5	15	8	12
Group 26	2	4	7	9	5	4	9	11	2	14	5	14	11	16	16
Group 27	2	4	13	12	12	7	15	10	5	2	15	5	13	7	4
Group 28	2	5	9	9	3	12	8	14	15	12	14	5	3	2	15
Group 29	2	5	11	7	2	11	9	4	16	7	16	9	14	14	4
Group 30	2	6	2	13	3	3	12	9	7	16	6	9	16	13	12
Group 31	2	6	9	7	7	16	13	3	12	2	13	12	9	16	6
Group 32	2	7	12	15	2	12	4	10	13	15	13	4	5	5	10
Group 33	2	7	14	16	5	9	2	9	16	11	11	5	7	4	14
Group 34	2	8	5	12	5	2	14	14	8	15	3	9	12	15	9
Group 35	2	9	13	4	2	13	8	11	6	4	6	8	15	15	11
Group 36	2	10	3	2	13	16	8	10	8	13	11	11	16	3	5
Group 37	2	11	15	3	11	6	14	10	15	10	6	7	7	14	3
Group 38	2	16	4	5	16	14	7	11	4	11	14	9	9	7	5
Group 39	3	3	4	6	11	12	13	6	12	14	4	5	13	5	14
Group 40	3	3	6	5	16	9	15	5	9	10	6	4	15	4	10
Group 41	3	4	5	14	4	6	12	13	5	13	6	11	11	12	14
Group 42	3	4	9	16	10	4	16	15	3	5	10	5	15	6	6
Group 43	3	4	16	10	5	10	4	9	9	16	15	6	3	5	15
Group 44	3	5	12	11	14	5	11	13	3	6	14	6	13	4	4
Group 45	3	6	4	10	6	5	9	15	4	15	5	16	16	9	10
Group 46	3	7	8	8	16	11	12	4	15	11	4	7	16	3	15
Group 47	3	7	16	11	4	15	3	15	11	12	12	4	7	8	16
Group 48	3	8	7	15	4	8	15	12	3	16	4	16	12	11	11
Group 49	3	8	15	4	16	4	8	7	7	15	12	11	3	16	12

TABLE 3.2   (CONTINUED)

SCRAMBLING CODE GROUP	SLOT NUMBER														
	#0	#1	#2	#3	#4	#5	#6	#7	#8	#9	#10	#11	#12	#13	#14
Group 50	3	10	10	15	16	5	4	6	16	4	3	15	9	6	9
Group 51	3	13	11	5	4	12	4	11	6	6	5	3	14	13	12
Group 52	3	14	7	9	14	10	13	8	7	8	10	4	4	13	9
Group 53	5	5	8	14	16	13	6	14	13	7	8	15	6	15	7
Group 54	5	6	11	7	10	8	5	8	7	12	12	10	6	9	11
Group 55	5	6	13	8	13	5	7	7	6	16	14	15	8	16	15
Group 56	5	7	9	10	7	11	6	12	9	12	11	8	8	6	10
Group 57	5	9	6	8	10	9	8	12	5	11	10	11	12	7	7
Group 58	5	10	10	12	8	11	9	7	8	9	5	12	6	7	6
Group 59	5	10	12	6	5	12	8	9	7	6	7	8	11	11	9
Group 60	5	13	15	15	14	8	6	7	16	8	7	13	14	5	16
Group 61	9	10	13	10	11	15	15	9	16	12	14	13	16	14	11
Group 62	9	11	12	15	12	9	13	13	11	14	10	16	15	14	16
Group 63	9	12	10	15	13	14	9	14	15	11	11	13	12	16	10

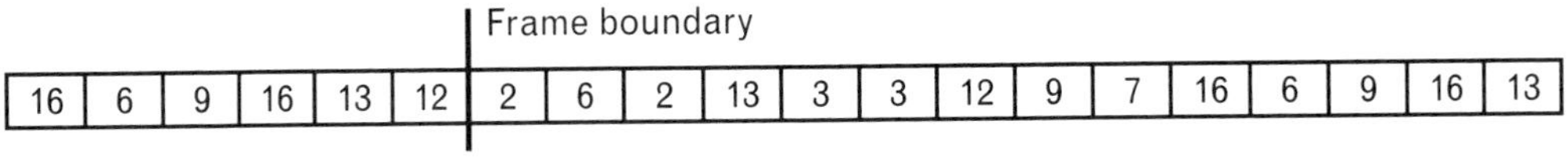

Frame boundary

16	6	9	16	13	12	2	6	2	13	3	3	12	9	7	16	6	9	16	13

Figure 3.6   Frame synchronization obtained from a sequence of SSCs.

peer entity, whereas in the open-loop control the transmitting entity estimates the required power level by itself from the received signal. Both of these methods are used in the UTRAN.

The closed-loop power control in the UTRAN can be further divided into two processes: inner-loop and outer-loop power control. The outer-loop power control sets the signal-to-interference ratio (SIR_{target}) and the inner-loop power control in layer 1 adjusts the peer entity transmit power so that the measured SIR fulfills the SIR_{target} requirement.

This adjustment is done with transmit power control (TPC) commands. The receiving entity (UE layer 1 or Node B layer 1) measures the SIR and compares it to the SIR_{target}. If SIR_{est} SIR_{target} then the TPC bit is set to 0 (reduce power) in the peer entity. Otherwise it is set to 1 (increase power). This TPC bit is transmitted to the peer entity once every time slot. There is no neutral TPC command; it is always either an increase or a decrease command. As there are 1,500 time slots in one second, this makes the inner loop power control a very fast method to adjust transmission power. Therefore inner-loop power control is also known as fast power control. It is performed entirely in layer 1. The basic principles of inner-loop power control are similar in both the UE and in the UTRAN.

Outer-loop power control is handled by the RRC in layer 3. Power control as a whole is a very important issue for CDMA systems, and it is further discussed in Section 2.4.

3.1.11 Power Weighting and Combining of Physical Channels

In the uplink, one UE can transmit simultaneously one DPCCH and up to six DPDCHs. This is depicted in Figure 3.7. The control channel (DPCCH) will be sent in the Q-plane, and the data channels (DPDCH) in both planes. However if there is only one data channel, then it is sent only in the I-plane.

The channelization codes are orthogonal codes, and the scrambling code is a pseudo-noise sequence. The gain factors β_d and β_c can be used to set different quality of service requirements for different channels (i.e., channels with higher QoS requirements can be sent using relatively high power levels). As seen, data and control channels may have different gain factors, but all data channels have an equal factor. These values can be signaled to layer 1 from the higher layers, or they can be computed autonomously in layer 1. The largest gain factor should be set to 1. This is explained in [5]. The gain factors may vary from frame to frame based on the current transport format combination (TFC).

The power weighting in the downlink is depicted in Figure 3.8. All channels have their own power weight factor G. Thus it is possible for the UTRAN to set different weights on different channels according to their Quality of Service (QoS) requirements.

Note that all physical channels except the SCH are processed in the same way as a DPDCH. Other physical channel types are, however, left out of Figure 3.8 to keep the size of the diagram from growing too large. Note that all channels (except the SCH) are scrambled with the same scrambling code. (SCH carries fixed code sequences. There is only one universal primary synchronisation code for P-SCH, and only 64 secondary synchronization codes for S-SCHs).

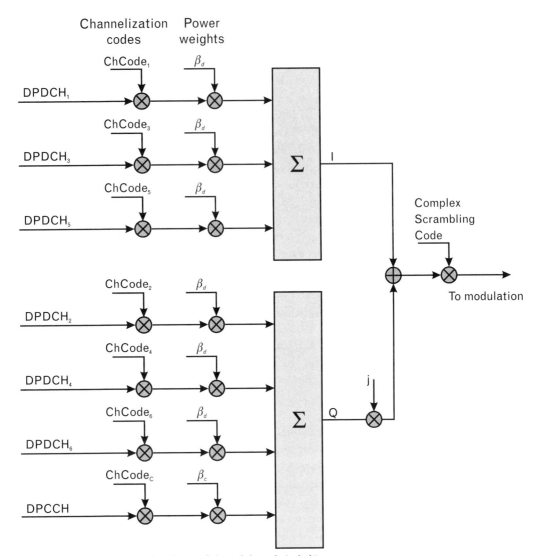

FIGURE 3.7   *Power weighting and combining of physical channels (uplink).*

3.1.12   RF Processing

The RF characteristics in the UTRAN are defined in the 3G TS 25.1XX series of specifications. This section only mentions the main topics of RF processing very briefly.

3.1.12.1   UE Power Classes

There will be four power classes for UEs. Table 3.3 gives the maximum output power for each class in the FDD mode.

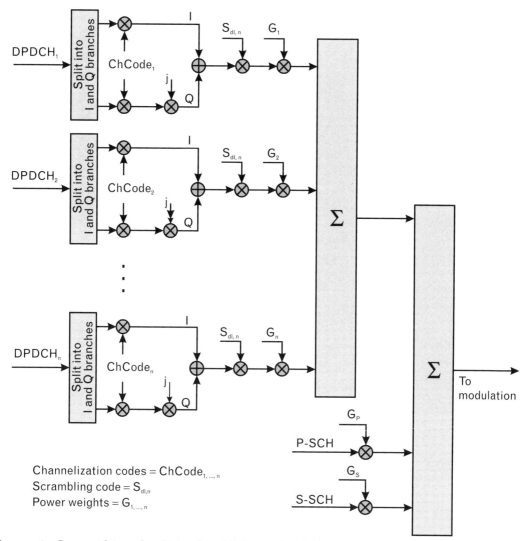

FIGURE 3.8   *Power weighting and combining of physical channels (downlink).*

3.1.12.2   Frequency Bands

The total bandwidth for 3G is divided between the FDD and the TDD modes. The paired band is allocated for the FDD mode and the smaller unpaired band for the TDD mode. UTRA/FDD is designed to operate in the following paired bands[3]:

---

3.   There are three ITU regions. ITU Region 1 includes Africa, Europe, and Northern Asia. ITU Region 2 includes North and South America. ITU Region 3 includes Southern Asia (including Japan and China), Australia, including most of the Pacific Islands.

TABLE 3.3   UE POWER CLASSES

POWER CLASS	MAXIMUM OUTPUT POWER
1	+33 dBm
2	+27 dBm
3	+24 dBm
4	+21 dBm

Note that power classes 1 and 2 are only allowed in ITU-Region 1 operating bands.

The allocations for the FDD mode are:
1,920–1,980 MHz: Uplink;
2,110–2,170 MHz: Downlink.

In ITU-Region 2 the FDD allocations are:
1,850–1,910 MHz: Uplink;
1,930–1,990 MHz: Downlink.

In ITU-Region 3 they are:
1,710–1,785 MHz: Uplink;
1,805–1,880 MHz: Downlink.

The allocations for the TDD mode are:
1,900–1,920 MHz;
2,010–2,025 MHz.

For Region 2 they are:
1,850–1,910 MHz;
1,910–1,930 MHz;
1,930–1,990 MHz;

In TDD mode, the same frequency carriers are used both in the uplink and in the downlink. Note that the bands described are taken from the 3GPP specifications, but each country may define its own frequency bands for 3G. ITU's worldwide spectrum-allocation recommendations for the future are discussed in Section 15.1.

The nominal channel spacing is 5 MHz.

3.1.12.3   RF Specifications

Consult the following specifications for further information:

25.101	UE Radio transmission and reception (FDD)
25.102	UE Radio transmission and reception (TDD)
25.104	UTRA (BS) FDD; Radio transmission and reception
25.105	UTRA (BS) TDD; Radio transmission and reception
25.106	UTRA Repeater; Radio transmission and reception
25.113	Base Station EMC
25.123	Requirements for support of radio resource management (TDD)
25.133	Requirements for support of radio resource management (FDD)
25.141	Base Station conformance testing (FDD)
25.142	Base Station conformance testing (TDD)
25.143	UTRA Repeater; Conformance testing

These specifications define UE power classes, base station classes, frequency bands, transmitter and receiver characteristics, and performance requirements.

3.1.13 Timing Advance on Uplink Channels

In the TDD mode, the radio frame in the frequency channel is divided into 15 time slots. In the uplink direction, it will be necessary to have guard periods between these time slots to prevent them from interfering with each other in the base station if the cell is large. In a large cell a transmission from a user close to the base station arrives there much earlier than a transmission from another user close to the cell boundary. It may even happen that the near-user's time slot n + 1 overlaps with the far-user's time slot n in the base station receiver. This is called the "TDMA effect"; the effect is evident when a receiver has to be a certain location to recover time slots in their intended and correct order.

This can be prevented, if the transmission from the far-user is advanced in time. The UTRAN measures the timing of the received burst in the Node B, and calculates the timing difference to the optimal arrival time. This value is used as the required timing advance value, and signaled to the UE by means of higher-layer messages (e.g., uplink physical channel control). The required timing advance will be given as a 6-bit number (0–63) being the multiple of 4 chips, which is nearest to the required timing advance. Upon receiving this command, the UE knows the right timing advance for its uplink transmissions. Note that the timing advance procedure is not a one-off process, but a continuous one. The UTRAN has to measure the timing of uplink bursts continuously as the user may be moving.

The initial value for timing advance will be derived from the timing of the received PRACH in the UTRAN. In the case of a handover, the timing advance of the new cell is sent relative to the old cell timing advance. Timing advance is used in the uplink, both in case of uplink dedicated physical channels (DPCHs) as well as for physical uplink shared channels (PUSCHs).

However, this fine procedure may not be used at all in real TDD cells. As discussed elsewhere in this book, the TDD system will most probably be used to provide high data rates in traffic hot spots. This means that typical TDD cells will be micro- and picocells; that is, quite small in size. A maximum range of a microcell is a few hundreds of meters. This would generate only relatively small timing differences between users. The guard period around the transmission burst is 96 chips, which translates into 25 μs. Within that time, a radio signal can travel 3.75 km [8]. This clearly shows that in micro- and picocells, the uplink timing advance is not necessary, and thus the UTRAN most probably does not use it.

The timing advance is a TDD-mode-only concept. It is discussed in the TDD physical layer procedures specification [9] and also in the RRC specification [10].

3.1.14   Support of Uplink Synchronization

Uplink synchronization is also a TDD-mode-only concept. If UL (uplink) synchronization is used, then the timing advance needs to be much more accurate. The value of the timing advance parameter is given as a multiple of ¼ chips. This accuracy enables the usage of synchronous CDMA in the UL. The UTRAN will continuously measure the timing of a transmission from the UE and send back the calculated timing advance value. On receipt of this value the UE will adjust the timing of its transmissions accordingly in steps of ± ¼ chips. A synchronous uplink would be advantageous in that it would enable the usage of orthogonal codes, and thus reduce the amount of interference.

Support of UL synchronization is optional for a UE.

3.2   Channels

There are three separate channel concepts in the UTRAN: logical, transport, and physical channels (see Figure 3.9).

Logical channels define what type of data is transferred. These channels define the data-transfer services offered by the MAC layer; that is, the concept of logical channels is used in the interface above the MAC.

Transport channels define how and with which type of characteristics the data is transferred by the physical layer. These channels are used in the interface between the MAC and the PHY layers. The transport channel is a new concept if WCDMA is compared to the GSM system.

FIGURE 3.9
Channel concepts.

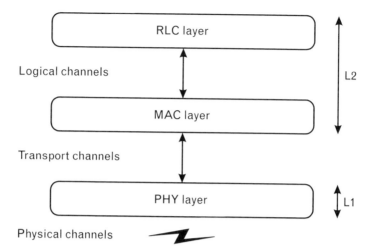

Physical channels define the exact physical characteristics of the radio channels. These are the channels used below the PHY layer; that is, in the radio interface.

Note that the choice of abbreviations for the channel names might be misleading at first. For example, the letters SCH may indicate shared channel, synchronization channel, or signaling channel, depending on the case. Be alert.

Physical and transport channels are discussed in [11] and [12]. Definitions for logical channels can be found in [13].

3.2.1 Logical Channels

Note that logical channels are not actually a layer 1 concept, but logically they exist in the interface between the MAC and the RLC protocol layers; that is, clearly within layer 2. However, they are described in this chapter for clarity.

Logical channels can be divided into control channels and traffic channels. A control channel can be either common or dedicated. A common channel is a point-to-multipoint channel; that is, common to all users in a cell, and a dedicated channel is a point-to-point channel; that is, used by only one user. Control channels transfer Control plane (C-plane) information and traffic channels User plane (U-plane) information.

The defined logical control channels are:

• Broadcast control channel (BCCH)

    Downlink common channel;

    Broadcasts system and cell-specific information.

• Paging control channel (PCCH)

Downlink channel;

Transfers paging information and some other notifications.

• Dedicated control channel (DCCH)

Bidirectional point-to-point channel;

Transfers dedicated control information.

• Common control channel (CCCH)

Bidirectional point-to-multipoint channel;

Transfers control information.

• Shared channel control channel (SHCCH)

Bidirectional;

Transfers control information for uplink and downlink shared channels;

Only in TDD mode.

The used logical traffic channels are:

• Dedicated traffic channel (DTCH)

Bidirectional point-to-point channel;

Transfers user info.

• Common traffic channel (CTCH)

Downlink point-to-multipoint channel;

Transfers dedicated user information for a group of users.

3.2.2   Transport Channels

The transport channels define how and with which type of characteristics the data is transferred by the physical layer. Transport channels are divided into common channels and dedicated channels. They are all unidirectional.
Common transport channels include:

• Broadcast channel (BCH)

A downlink channel for broadcast of system and cell-specific info.

• Paging channel (PCH)

A downlink channel used for transmission of paging and notification messages;

Transmission associated with transmission of paging indicator in PICH physical channel.

- Random access channel (RACH)

  A contention-based uplink channel;

  Used for initial access or non-real-time dedicated control or traffic data;

  A limited-size data field.

- Common packet channel (CPCH)

  A contention-based channel used for transmission of bursty data traffic;

  An uplink channel;

  Only in FDD mode.

- Forward access channel (FACH)

  A common downlink channel;

  May carry small amounts of user data.

- Downlink shared channel (DSCH)

  A downlink channel shared by several UEs;

  Used for dedicated control or traffic data;

  Associated with a DCH (does not exist alone).

- High-speed downlink shared channel (HS-DSCH)

  A downlink channel shared by several UEs;

  Optimized for very high speed data transfer;

  Employs efficient link adaptation scheme;

  Very quick rate changes (HSDPA frame is only 2 ms versus 10 ms of other channels);

  Associated with a DCH, and up to 4 HS-SCCHs (does not exist alone);

  Available only from Release 5 onward.

- Uplink shared channel (USCH)

  An uplink channel shared by several UEs;

  Carries dedicated control or traffic data;

  Only in TDD mode.

The only dedicated transport channel type is:

- Dedicated channel (DCH)

  For one UE only;

  Either uplink or downlink.

3.2.3 Physical Channels

There are two ways to use the allotted spectrum in UTRAN. In frequency-division duplex (FDD) mode, both the uplink and downlink bands have their own frequency channels. In time-division duplex (TDD) mode, there is only one frequency channel, which is then dynamically time-divided for both uplink and downlink slots. It seems that the FDD mode will be the more preferred technology choice by the operators (at least during the first years of the UTRAN deployment), and the TDD mode will be the less used choice. TDD is also given smaller frequency allocations. In practice, the FDD mode will be given the symmetric part of the UMTS spectrum allocation, and TDD will get the asymmetric "leftovers."

Depending on the operating mode, the physical channels will be somewhat different.

3.2.3.1 FDD Physical Channels

The physical channels in the FDD mode are described below.

Downlink

- Synchronization channel (SCH)

  Used for cell search;

  Two subchannels, the primary and secondary SCH.

  Transmitted only during the first 256 chips (i.e., one-tenth) of each timeslot.

- Common pilot channel (CPICH)

  Fixed rate of 30 Kbps;

  Carries a predefined bit sequence;

  Two types: primary and secondary CPICH:

  Primary CPICH (P-CPICH) is the phase reference for SCH, primary CCPCH, AICH, and PICH, and the default reference for other downlink physical channels;

  Secondary CPICH (S-SPICH) may be the reference for the downlink DPCH, and for the associated PDSCH. The presence of S-CPICH in a cell is optional.

- Primary common control physical channel (P-CCPCH)

  Fixed rate of 30 Kbps;

  Carries BCH;

  Not transmitted during the first 256 chips of each timeslot.

- Secondary common control physical channel (S-CCPCH)

  Variable rate;

  Carries FACH and PCH;

  FACH and PCH can be mapped to the same or separate channels;

  Transmitted only when there is data available.

- Physical downlink shared channel (PDSCH)

  Carries DSCH (downlink shared channel);

  Always associated with a downlink DPCH, which carriers its control information.

- Paging indicator channel (PICH)

  Carries page indicators to indicate the presence of a page message on the PCH.

- Acquisition indicator channel (AICH)

  Carries acquisition indicators (= signatures for the random access procedure).

- CPCH Access preamble acquisition indicator channel (AP-AICH)

  Carries AP acquisition indicators of the associated CPCH.

- CPCH status indicator channel (CSICH)

  Carries CPCH status information.

- CPCH Collision-detection/channel-assignment indicator channel (CD/CA-ICH)

  Carries CD (collision detection) indicators only if the CA (channel assignment) is not active, or both CD indicators and CA indicators at the same time if the CA is active.

Note that the three previous channels are control channels for uplink PCPCH. If CPCH functionality is not supported by the network, then these channels do not exist.

- High-speed physical downlink shared channel (HS-PDSCH)

Carries HS-DSCH;

Can employ either QPSK or 16 QAM modulation;

A HS-PDSCH frame is 2 ms, consisting of 3 time slots;

Uses always the spreading factor of 16;

One UE may receive several HS-PDSCH simultaneously.

· Shared control channel for HS-DSCH (HS-SCCH)

Carries downlink signaling related to HS-DSCH transmission;

Indicates when there is data to be received on HS-DSCH for this UE;

A fixed rate channel (SF = 128, i.e., 60 kbps);

There can be up to 4 HS-SCCHs a UE has to monitor.

Downlink and Uplink

· Dedicated physical data channel (DPDCH)

Carries DCH (dedicated channel);

Carries data generated at layer 2 and above.

· Dedicated physical control channel (DPCCH)

Carries control information generated at layer 1.

Note that in the uplink these two channels are I/Q code multiplexed, but in the downlink they are time multiplexed. Sometimes the DPDCH and DPCCH together is an entity known as a dedicated physical channel (DPCH).

Uplink

· Physical random access channel (PRACH)

Carries RACH;

Uses slotted ALOHA technique with fast acquisition indicators.

· Physical common packet channel (PCPCH)

Carries CPCH (common packet channel);

Uses DSMA-CD technique with fast acquisition indication.

· Uplink dedicated control channel for HS-DSCH (HS-DPCCH)

Carries HSDPA feedback information (HARQ acknow-ledgements and channel quality indications);

Multiplexed with a DPCCH;

Uses SF=256.

3.2.3.2 TDD Physical Channels

The physical channels in the TDD mode are as follows.

Downlink

- Primary common control physical channel (P–CCPCH)

    Carries BCH.

- Secondary common control physical channel (S-CCPCH)

    Carries PCH and FACH;

    One or more instances per cell.

- Synchronization channel (SCH)

    Gives the code group of a cell;

    Indicates the position (timeslot and code) of P–CCPCH.

- Paging indicator channel (PICH)

    Carries page indicators to indicate the presence of a page mes-sage on the PCH.

- Physical downlink shared channel (PDSCH)

    Carries DSCH;

- Physical Node B synchronisation channel (PNBSCH)

    Carries Node-B synchronisation bursts;

    Used by the network to gain time synchronisation among Node-Bs;

    No meaning for UE, so in a way this channel is neither a downlink nor an uplink channel; as it is only used between Node-Bs. See [9].

- High speed physical downlink shared channel (HS-PDSCH)

    Carries HS-DSCH;

    Can employ either QPSK or 16 QAM modulation;

    One UE may receive several HS-PDSCHs simultaneously.

- Shared control channel for HS-DSCH (HS-SCCH)

Carries downlink signalling related to HS-DSCH transmission;

Indicates when there is data to be received on HS-DSCH for this UE;

There can be up to 4 HS-SCCHs a UE has to monitor.

Downlink and Uplink

• Dedicated physical channel (DPCH)

Carries DCH;

Uplink

• Physical random access channel (PRACH)

Carries RACH;

One or more instances per cell.

• Physical uplink shared channel (PUSCH)

Carries USCH.

• Shared information channel for HS-DSCH (HS-SICH)

Carries HSDPA feedback information to Node-B

All TDD mode physical channels in the downlink use SF=16, or in special cases also SF=1 (i.e., no spreading). In the uplink the spreading factor can vary between 1 and 16.

3.2.4   Shared Channels

As seen from the previous sections, the UTRAN specifications contain shared channels in the radio interface. This is a new concept when compared to GSM, so these are discussed further in this chapter.

The idea behind shared channels is a more efficient usage of spectrum capacity. To save capacity, the network does not assign a dedicated channel for every user. If the expected/measured data traffic is of low or medium volume, or it is a bursty type, then shared channels could be used instead of dedicated channels. On a shared channel the resource is open to be used by everybody, and each user can request a temporary allocation for a short time using a special resource-reservation procedure. The capacity is granted by the scheduling function in the controlling radio network controller (CRNC). The temporary reservation is typically quite short, and once it has expired, the allocation procedure must be performed again in case new data has to be sent. Thus a shared channel can be used by only one active user at a time, but that user may change frequently. The shared channels support fast power control, but they do not support soft handovers.

In the uplink direction there are two shared channels, the common packet channel (CPCH) in the FDD mode and the uplink shared channel (USCH) in the TDD mode. Although their implementation is different, they both perform the same function, which is the transfer of bursty or low volume (non-real-time) packet traffic. Both uplink shared channels must be combined with other channel types for data transfer, which is a RACH/FACH pair. The RACH/FACH is used for the shared channel allocation and the FACH also for relaying the downlink acknowledgements. Acknowledgements can also be sent via a downlink shared channel (DSCH), if such is allocated.

In the FDD mode the CPCH data transmission is based on the digital sense multiple access–collision detection (DSMA-CD) approach with fast acquisition indication. First the UE sends access preambles on the CPCH with increasing power levels until an acknowledgement is received via the access preamble-acquisition indicator channel (AP-AICH). A positive acknowledgement indicates that access has been granted. The UE must then send a CD preamble and wait for a response via collision detection/channel assignment indicator channel (CD/CA-ICH). This is the contention-resolution procedure. The response will define the allocated channel with, for example, the scrambling and channelization codes to be used with the message part. Next the UE sends the power control preamble and immediately after that it follows up with the actual message part.

In brief, the CPCH is very much like the RACH channel in UTRAN. The RACH can be used to send small amounts of data (one or two frames at a time), but without a power control loop. The CPCH includes fast power control, as the downlink power control bits are conveyed via the associated DPCCH. See Section 11.3 for the description of the CPCH data-transfer procedure.

The USCH in the TDD mode is used by only one UE at a time, but this allocation can be changed on a frame-by-frame basis. The basic working principle is that a UE asks for the USCH allocation using a RACH (a slotted ALOHA approach with a fast acquisition indication), and the allocation is granted using a FACH. The UE then confines its transmissions to the allocated frames. The acknowledgement messages are sent via the DSCH. Interleaving for the USCH may be applied over multiple radio frames.

Both uplink shared channels can use fast power control, and their data rates can also be modified rapidly.

The downlink shared channel is defined for both the FDD and the TDD modes. A DSCH does not use SHO either; that is, it is transmitted only in a single cell.

In the FDD mode a DSCH can be allocated on a radio frame basis to different UEs. It is also possible that a single UE has been allocated several parallel DSCHs at the same time. These channels may have different spreading factors. Each DSCH is associated with a dedicated downlink channel, which carries all the necessary control information for the associated DSCH. The UTRAN can tell the UE that there is data to be decoded on the DSCH by using the TFCI field of the associated DPCH.

In the TDD mode a DSCH is associated with another downlink channel, a DCH. This associated channel carries the control information for the DSCH. However, the shared channel may carry its own TFCI. In the TDD mode there are three ways the UTRAN can tell the UE that there is data to be decoded on the DSCH:

1. By using the TFCI field of the associated channel or PDSCH;

2. By using a user-specific midamble on the DSCH (i.e., the UE will decode the PDSCH if the PDSCH was transmitted with the midamble assigned to the UE by the UTRAN);

3. By using higher-layer signaling.

In both modes the uplink acknowledgements and the power control bits are sent via uplink dedicated channels. Interleaving for the DSCH may be applied over a multiple of radio frames.

From Release 5 onwards there is also a new special high speed DSCH, HS-DSCH. That channel is used in a new high speed data transmission service, High Speed Downlink Packet Access (HSDPA). This is further discussed in Chapter 12. The DSCH data transmission procedure described earlier does not apply to HS-DSCH.

Note that the RACH and FACH are not considered shared channels but common channels. The difference between these concepts is that common channels cannot be allocated to one user; they are common to everybody. The RACH is a contention-based uplink channel where the permission to send each burst must be acquired separately. The FACH is a common downlink channel. A UE must receive all data packets from the FACH assigned to it, and then check the address information to find out whether the packet was addressed to it. Typically these channels are used to relay control information, but it is also possible to send small amounts of user data in them.

3.2.5 Channel Mapping

The following diagrams (Figures 3.10–3.13) depict the mapping of channel types to each other. There are many physical channels, especially in the downlink FDD diagram, that do not map into any transport channel at all. This is because they are some type of indication channel, which indicates

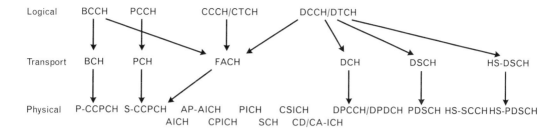

FIGURE 3.10 *FDD mode channels, downlink.*

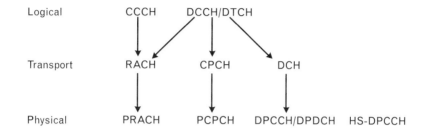

FIGURE 3.11 *FDD mode channels, uplink.*

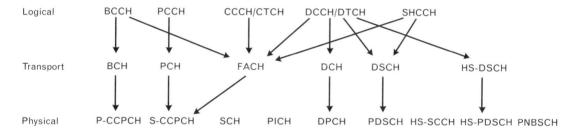

FIGURE 3.12 *TDD mode channels, downlink.*

FIGURE 3.13
*TDD mode channels,
uplink.*

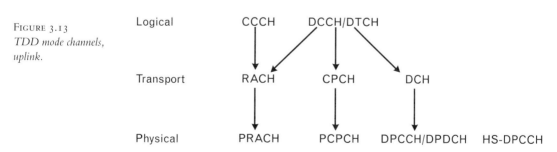

something to the receiving physical channel in a unidirectional scheme. The information is of no interest to the higher layers, and thus it is not necessary to map these channels to any kind of transport channels. Consult [1] and [11–14] for further information on channel types and their mapping. The acronyms used in the channel diagrams are explained in Table 3.4.

TABLE 3.4 CHANNEL NAMES AND ACRONYMS

AICH	Acquisition Indicator Channel	HS-DSCH	High-Speed Downlink Shared Channel
AP-AICH	Access Preamble—Acquisition Indicator Channel	HS-PDSCH	High Speed Physical Downlink Shared Channel
BCCH	Broadcast Control Channel	HS-SICH	Shared information channel for HS-DSCH
BCH	Broadcast Channel	HS-SCCH	Shared control channel for HS-DSCH
CCCH	Common Control Channel	PCCH	Paging Control Channel
CD/CA-ICH	Collision Detection/Channel Assignment Indicator Channel	P-CCPCH	Primary Common Control Physical Channel
CPCH	Common Packet Channel	PCH	Paging Channel
CPICH	Common Pilot Channel	PCPCH	Physical Common Packet Channel
CSICH	CPCH Status Indicator Channel	PDSCH	Physical Downlink Shared Channel
CTCH	Common Traffic Channel	PICH	Paging Indicator Channel
DCCH	Dedicated Control Channel	PNBSCH	Physical Node B synchronisation channel
DCH	Dedicated Channel	PRACH	Physical Random Access Channel
DPCCH	Dedicated Physical Control Channel	PUSCH	Physical Uplink Shared Channel
DPCH	Dedicated Physical Channel	RACH	Random Access Channel
DPDCH	Dedicated Physical Data Channel	S-CCPCH	Secondary Common Control Physical Channel
DSCH	Downlink Shared Channel	SCH	Synchronization Channel
DTCH	Dedicated Traffic Channel	SHCCH	Shared Channel Control Channel
FACH	Forward Access Channel	USCH	Uplink Shared Channel
HS-DPCCH	Uplink dedicated control channel for HS-DSCH		

3.3   Spreading and Scrambling Codes

In a DS–CDMA transmitter the information signal is modulated by a spreading code (to make it a wideband signal) and in the receiver it is correlated with a replica of the same code. The spreading process actually consists of two phases, spreading and scrambling, and both of them use different types of codes with different characteristics (Figure 3.14).

The spreading phase is also known as channelization. Channelization increases the bandwidth of the signal. The codes used in this phase are orthogonal codes. The UTRAN uses orthogonal variable spreading factor (OVSF) codes. In an ideal orthogonal system the cross-correlation between the desired and the interfering orthogonal signals is zero. However, in a real

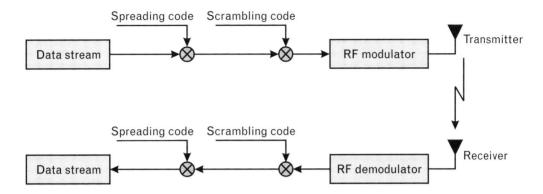

FIGURE 3.14   *Spreading and scrambling.*

system there are always some multipath components (reflections and distractions) of the same signal. These will distort the orthogonality. Moreover, the number of codes is finite and thus they have to be reused in every cell. Therefore, the same code can be allocated to different users in adjacent cells. A UE cannot normally know which of the downlink signals is addressed to it without some help. Also, because the UEs are not time synchronized, their uplink transmissions are asynchronous and not orthogonal.

This all means that in addition to channelization something else is needed. The solution is scrambling. Scrambling is done after the spreading in the transmitter. In the scrambling process the code sequence is multiplied with a pseudorandom sequence of bits (i.e., the scrambling code). In the downlink direction each base station has a unique scrambling code, and in the uplink it is different for each UE. These are codes that are generated with good autocorrelation properties. The autocorrelation and cross-correlation functions are connected in such a way that it is generally not possible to achieve good autocorrelation and cross-correlation values at the same time.

A scrambling code can be either short or long. Short codes span over one symbol period, while long codes span over several symbol periods.

This subject is thoroughly discussed in Chapter 5. The use of spreading codes in the UTRAN is specified in [1].

3.4   Diversity

Diversity is defined in the *Oxford English Dictionary* as "the state of being varied."

Diversity as a concept is used in many different ways in the UTRAN. A signal can be subjected to time diversity, multipath diversity, and antenna diversity. Furthermore, these classes of diversity contain further variations and subclasses. So one could say that the uses of diversity in the UTRAN are quite diverse.

The following forms of diversity are available in the FDD mode: time diversity, multipath diversity, macrodiversity, and antenna diversity. Even though it may sound strange, diversity is quite often a desired property in a CDMA system, and some forms of it may be generated artificially to the signal.

3.4.1 Time Diversity

Time diversity means that the signal is spread in the time domain. If there is a short period of time in which signals interfere with each other, which distorts part of the signal, time diversity may help to reconstruct the signal in the receiver despite the errors. The methods for achieving time diversity are channel coding, interleaving, and retransmission protocols.

Time diversity spreads the faulty bits over a longer period of time, and thus makes it easier to reconstruct the original data. If there are 4 successive erroneous bits in one byte, it is very difficult to recover the original data (see Figure 3.15). However if these 4 false bits from the radio interface are evenly spread over 4 bytes by means of interleaving, then it is much easier to recover the data, for example, by means of error correcting coding. The longer the interleaving period, the better the protection provided by the time diversity. However, longer interleaving increases transmission delays

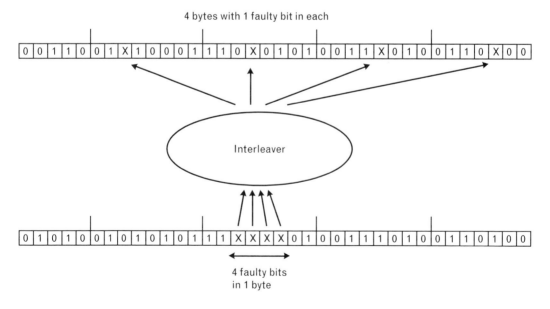

FIGURE 3.15   *Time diversity.*

and a balance must be found between the error resistance capabilities and the delay introduced.

3.4.2 Multipath Diversity

Multipath diversity is a phenomenon that happens when a signal arrives at the receiver via different paths (i.e., because of reflections). There is only one transmitter, but various obstacles in the signal path cause different versions of the signal to arrive at the receiver from different directions and possibly at different times (see Figure 3.16).

In second-generation GSM systems too much multipath diversity means trouble, as GSM receivers are not able to combine the different components, but typically they just have to use the strongest component. In a WCDMA system the receiver is typically able to track and receive several multipath components and combine them into a composite signal. The receiver is usually of the RAKE variety, which is well suited to the task. The more energy that can be collected from the multipath components, the better will be the signal estimation.

3.4.3 Macrodiversity

In a CDMA system the same signal can be transmitted over the air interface, on the same frequency, from several base stations separated by considerable distances. This scheme is called the soft handover (SHO). In a SHO all the participating base stations use the same frequency, and the result is a macrodiversity situation. Note the difference in these concepts: a SHO is a procedure. Once it is performed, the result is a macrodiversity situation.

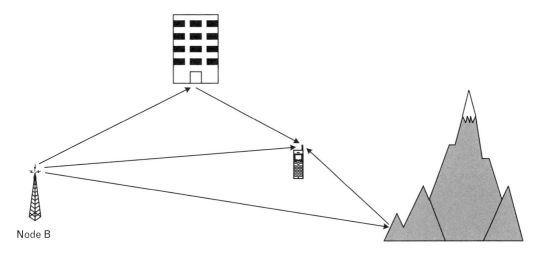

Node B

FIGURE 3.16   *Multipath diversity in the downlink.*

Macrodiversity is achieved through SHO. However, these concepts are often used without any distinction in the literature.

In macrodiversity the mobile's transmission is received by at least two base stations, and similarly the downlink signal is sent by at least two base stations. The gain from macrodiversity is highest when the path losses of the SHO branches are about equal. If one of the participating base stations is clearly stronger than the others, then macrodiversity cannot provide much gain.

Macrodiversity also provides protection against shadowing. Without macrodiversity (and multipath diversity) a UE can easily get shadowed if a large obstacle gets between the UE and the base station. In SHO the UE has at least one other path that can maintain the service if one radio link suffers from shadowing.

Macrodiversity components will be combined in the physical layer, and not in the protocol stack. The most suitable place to perform this is in the mobile station's RAKE receiver, as this provides the largest gain. There are also other receiver techniques that can perform the combining. Advanced receiver techniques are discussed in [15].

Macrodiversity is an especially suitable method for improving the gain of services with strict delay requirements. With non-real-time services the same effect could be achieved with time diversity; that is, with longer interleaving periods and retransmission protocols. Macrodiversity in the downlink can increase the overall interference level in the system, and thus it should only be used when necessary. Additionally, a SHO requires one channelization code per radio link.

Site-selection diversity transmission (SSDT) is a special case of SHO (see Figures 3.17 and 3.18). The principle of SSDT is that the best cell of the active set is dynamically chosen as the only transmitting site, and the other cells involved turn down their DPDCHs. The DPCCH is transmitted as it normally would be via all cells. Because only one base station is transmitting in downlink data channels, the interference level is lower than with normal SHO.

The working principle of the SSDT is that the UE selects one of the cells from its active set to be the primary cell, and the other cells are

FIGURE 3.17
SSDT uplink.

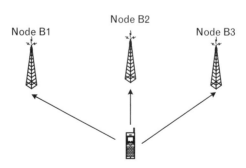

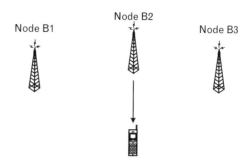

FIGURE 3.18
SSDT downlink.

Node B1   Node B2   Node B3

nonprimary. Only the primary cell transmits in the downlink data channels. Each cell in the active set is assigned a temporary identification. The UE periodically informs the UTRAN about the current primary cell ID. This information is delivered via the uplink feedback information (FBI) field.

In order for the UE to continuously perform measurements and to maintain synchronization, the nonprimary cells must continue to transmit pilot information on the DPCCH in case the situation in the air interface changes.

SSDT and the coding of FBI bits are discussed in [5].

3.4.4 Antenna Diversity

Antenna diversity means that the same signal is either transmitted or received (or both) via more than one antenna element in the same base station. Antenna diversity can sometimes also be applied to the UE. Transmission and receiver antenna diversities are not the same, and thus they are discussed separately in this section. Here only base station transmission/reception diversity is considered. Antenna diversity in mobile terminals is problematic: it is expensive and tends to increase the size of mobiles beyond what the market will accept.

3.4.4.1 TX Diversity (Base Station)

Transmitter-antenna diversity can be used to generate multipath diversity in places where it would not otherwise exist. Multipath diversity is a useful phenomenon, especially if it can be controlled. It can protect the UE against fading and shadowing. TX diversity is designed for downlink usage. Transmitter diversity needs two antennas, which would be an expensive solution for the UEs.

The UTRA specifications divide the transmitter diversity modes into two categories: (1) open-loop mode and (2) closed-loop mode. In the open-loop mode no feedback information from the UE to the Node B is available. Thus the UTRAN has to determine by itself the appropriate parameters for the TX diversity. In the closed-loop mode the UE sends feedback information up to the Node B in order to optimize the transmissions from the diversity antennas.

Thus it is quite natural that the open-loop mode is used for the common channels, as they typically do not provide an uplink return channel for the feedback information. Even if there was a feedback channel, the Node B cannot really optimize its common channel transmissions according to measurements made by one particular UE. Common channels are common for everyone; what is good for one UE may be bad for another. The closed-loop mode is used for dedicated physical channels, as they have an existing uplink channel for feedback information. Note that shared channels can also employ closed loop power control, as they are allocated for only one user at a time, and they also have a return channel in the uplink.

There are two specified methods to achieve the transmission diversity in the open-loop mode and two methods in closed-loop mode.

Open-Loop Mode

The TX diversity methods in the open-loop mode are (1) spacetime-block-coding-based transmit-antenna diversity (STTD) and (2) time-switched transmit diversity (TSTD).

In STTD the data to be transmitted is divided between two transmission antennas at the base station site and transmitted simultaneously. The channel-coded data is processed in blocks of four bits. The bits are time reversed and complex conjugated, as shown in Figure 3.19.

The STTD method, in fact, provides two brands of diversity. The physical separation of the antennas provides the space diversity, and the time difference derived from the bit-reversing process provides the time diversity. These features together make the decoding process in the receiver more reliable.

In addition to data signals, pilot signals are also transmitted via both antennas. The normal pilot is sent via the first antenna and the diversity pilot via the second antenna. The symbol sequence for the second pilot is given in [11]. The two pilot sequences are orthogonal, which enables the receiving UE to extract the phase information for both antennas.

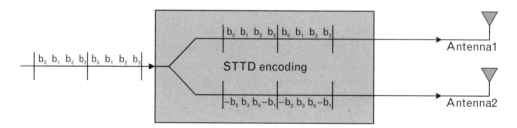

FIGURE 3.19   *STTD encoding.*

The STTD encoding is optional in the UTRAN, but its support is mandatory for the UE's receiver.

Time-switched transmit diversity (TSTD) can be applied to the SCH. Just as with STTD, the support of TSTD is optional in the UTRAN, but mandatory in the UE. The principle of TSTD is to transmit the synchronization channels via the two base station antennas in turn. In even-numbered time slots the SCHs are transmitted via antenna 1, and in odd-numbered slots via antenna 2. This is depicted in Figure 3.20. Note that SCH channels only use the first 256 chips of each time slot (i.e., one-tenth of each slot).

Open-loop transmit diversity is discussed in [11].

Closed-Loop Mode

The closed-loop-mode transmit diversity can only be applied to the downlink channel if there is an associated uplink channel. Thus this mode can only be used with dedicated channels (DPCH and PDSCH with an associated DPCH).

The chief operating principle of the closed loop mode is that the UE can control the transmit diversity in the base station by sending adjustment commands in FBI bits on the uplink DPCCH. This is depicted in Figure 3.21. The UE uses the base station's common pilot channels to estimate the channels separately. Based on this estimation, it generates the adjustment information and sends it to the UTRAN to maximize the UE's received power.

There are actually two modes in the closed-loop method. In mode 1 only the phase can be adjusted; in mode 2 the amplitude is adjustable as well as the phase. Each uplink time slot has one FBI bit for closed-loop-diversity control. In mode 1 each bit forms a separate adjustment command, but in mode 2 four bits are needed to compose a command.

Closed-loop transmit diversity is described in [5].

Transmit-Diversity-Mode Control Strategies

The transmit-diversity-mode to be used will be determined by the UTRAN. In fact, the whole system of transmit-antenna diversity is optional for the UTRAN, but the support for all TX diversity modes is mandatory for the UE. If the UTRAN employs any form of TX diversity, it informs

FIGURE 3.20   *TSTD.*

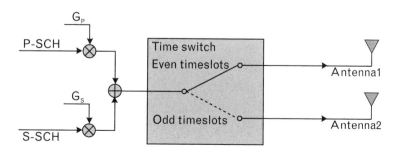

FIGURE 3.21
Closed-loop transmit diversity, UTRAN side.

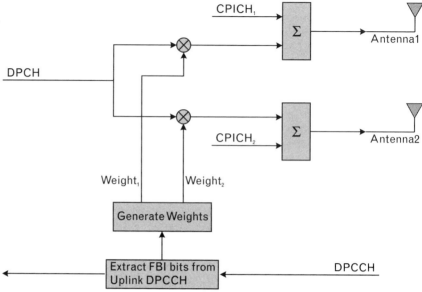

the UE about it. In the case of common channels, the UE is informed by the system information broadcast on the BCCH whether TX diversity is used. The use of TX diversity on dedicated channels is signaled in the call setup phase. The possible diversity modes are given in Table 3.5, which is taken from [11].

If TX diversity is used, then the information in Table 3.5 dictates the diversity mode. With most channels there is no choice, as there is only one mode available. Only the dedicated mode downlink channels, DPCH and PDSCH, make it possible to employ either the STTD or the closed-loop mode. The choice will be made by the UTRAN and the decision will be based on the radio channel conditions. Different modes will provide the best performance depending on the conditions in the radio channel. Generally, the closed loop tends to give better performance if the receiver moves at slow speed, but it loses this advantage if the receiver moves faster, because then the channel conditions are changing rapidly, and the feedback information required by the closed loop tends to be out-of-date by the time they arrive at the base station.

We are not allowed to employ both the STTD and the closed-loop mode on the same physical channel at the same time. But it is possible for the UTRAN to change the mode. Also, if TX diversity is used on any of the downlink physical channels it is also used on the P-CCPCH and the SCH of the same cell. Thus the UE can detect the usage of the transmit diversity from the SCH channel. If the UE receives synchronization bursts only in every other time slot, it knows that TX diversity is used and that the missing SCH bursts are actually being sent via the second diversity antenna.

3.4.4.2 Reception Diversity

Reception-antenna diversity is mostly used in the uplink direction, but it is possible to implement it in the UEs. However, note that if some form of TX antenna diversity is already used in the downlink, then the additional gain by using receiver-antenna diversity in the UE is small. The purpose of the various diversity schemes is to help the receiver collect more energy from the downlink signal. But if one diversity scheme has already succeeded in collecting plenty of energy, there is not much the other diversity schemes can do to increase the gain. Thus the reception diversity will most probably be implemented only in the base station. As the capacity in a typical UTRAN cell is uplink limited, reception diversity in the base station can increase the whole system capacity and cell ranges.

There are two ways to achieve receive-antenna diversity; space and polarization diversity. In space diversity the reception antennas are physically separated. This is a suitable solution for large outdoor base stations. In polarization diversity the antennas do not need to be physically separated, and thus this solution is suitable for small-sized receivers (i.e., indoor base stations or even mobile stations).

In both cases receiver diversity protects the receiver against signal fading and interference. The idea of receiver antenna diversity is that fading does not usually correlate between diversity antennas. Of course, this cannot be

TABLE 3.5 ANTENNA TX DIVERSITY MODES

CHANNEL TYPE	OPEN LOOP		CLOSED LOOP
	TSTD	STTD	
P-CCPCH		X	
SCH	X		
S-CCPCH		X	
DPCH		X	X
PICH		X	
PDSCH		X	X
AICH		X	
CISCH		X	
AP-AICH		X	
CD/CA-ICH		X	
DL-DPCCH for CPCH		X	X
HS-PDSCH		X	X
HS-SCCH		X	X

From: [11].

guaranteed in all cases, but having two receivers even with correlated fading provides some gain, because two receivers can usually collect more energy than one receiver.

There is also a new interesting antenna diversity scheme being studied; Multiple Input Multiple Output (MIMO) antennas. This topic is further discussed in Chapter 12. MIMO will eventually be part of 3GPP specifications, but not yet in Release 5.

3.5   Transport Formats

The transport channel is a concept applied to the interface between the physical layer and the MAC layer. These channels are used by the MAC layer to access physical layer services. All transport channels are unidirectional. Some types of transport channels can exist in both the uplink and downlink directions, but these entities are still separate resources. The physical layer operates in 10-ms time slices (radio frames) in the connected mode (however, HSDPA employs 2-ms frames). These frames will be filled with data sent from the MAC layer to the physical layer for processing and transmitting (and similarly something will be received from the physical layer). This data is sent using transport blocks. The MAC layer generates a new transport block every 10 ms (or a multiple of that), fills it with the necessary information, and sends it to the physical layer. The CRC is added to the transport block by the physical layer. It is possible to send several transport blocks via the same transport channel within one frame in parallel. A set of simultaneous transport blocks is called the transport block set.

The transport block size gives the size of the transport block in bits and similarly the transport block set size gives the size of the transport block set in bits. Because one transport block set always consists of similar-sized transport blocks, the size of a transport block set is a multiple of the transport block size.

The transmission time interval (TTI) is defined as the inter-arrival time of transport block sets. This is always a multiple of an L1 radio frame duration, the exact value being either 10, 20, 40, or 80 ms (again, in HSDPA this is 2 ms). Each transport channel can have its own TTI. Note that a TTI value does not tell you anything about the amount of data to be sent, but just how often the MAC layer sends data to the physical layer. The size of the individual data chunk is determined by the transport block size and the transport block set size parameters. This is depicted in Figure 3.22. The higher the block in the figure, the higher the data rate. We can see that the TTI indicates how often the transport channel data rate can be modified. With 10 ms TTI, the rate can be modified every 10 ms; with an 80-ms setting the modification can be done only every 80 ms. Note that it is also possible to have a zero size transport block in a TTI.

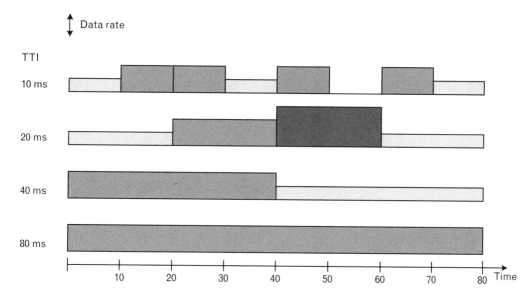

FIGURE 3.22   *Transmission time intervals.*

The transport format defines the data in a transport block set and how it should be handled by the physical layer. In effect the transport format defines the characteristics of a transport channel. The transport format consists of two parts, semistatic and dynamic. Note that the following discussion does not apply to the HS-DSCH channel in all its details as the link adaptation scheme changes the transport format contents slightly, see Chapter 12 for HSDPA.

The semistatic part definitions are common to all transport formats in a transport channel. These define the service attributes, such as quality and transfer delay, for the data transfer. These definitions include:

- Transmission time interval (TTI);
- Type of error protection scheme;
- Size of the CRC;
- Static rate matching parameter.

The dynamic part definitions can be different for every transport format:

- Transport block size;
- Transport block set size.

A transport format might look like the following:

Semistatic part: {10ms, turbo coding, static rate matching
   parameter = 1}
Dynamic part: {320 bits, 640 bits)

All transport formats associated with a transport channel form a transport format set. The semistatic parts of the transport formats are similar, so the only varying component within a transport format set is the dynamic part.
A transport format set might look like this:

Semistatic part: {10 ms, turbo coding, static rate matching
   parameter = 1}
Dynamic parts: {40 bits, 40 bits}; {40 bits, 80 bits}; {40 bits, 120 bits}

Each transport format within a transport format set has a unique identifier called the transport format identifier (TFI). It is used in the interlayer communication between the MAC layer and the physical layer to indicate the transport format. See Figure 3.23.
Several transport channels can exist simultaneously, each of them having different transport characteristics. These transport channels are multiplexed together in layer 1, and the composite is called the coded composite transport channel (CCTrCH). The collection of transport formats used in a CCTrCH is called the transport format combination. This combination can be different for each 10-ms frame.
Note that the transport format combination does not contain all possible dynamic parts of the corresponding transport formats, but only those that are currently used in a frame. Also, some combinations of transport formats are not allowed in the transport format sets because their combined high bit rates would exceed the capacity of the physical channel.
An example of a transport format combination could be:

DCH1:
Semistatic part: {10 ms, turbo coding, static rate matching
   parameter = 1}
Dynamic part: {40 bits, 40 bits}

DCH2:
Semistatic part: {10 ms, convolutional coding, static rate matching
   parameter = 3}
Dynamic part: {320 bits, 320 bits}

DCH3:
Semistatic part: 110 ms, turbo coding, static rate matching
   parameter = 2}
Dynamic part: {320 bits, 1,280 bits}

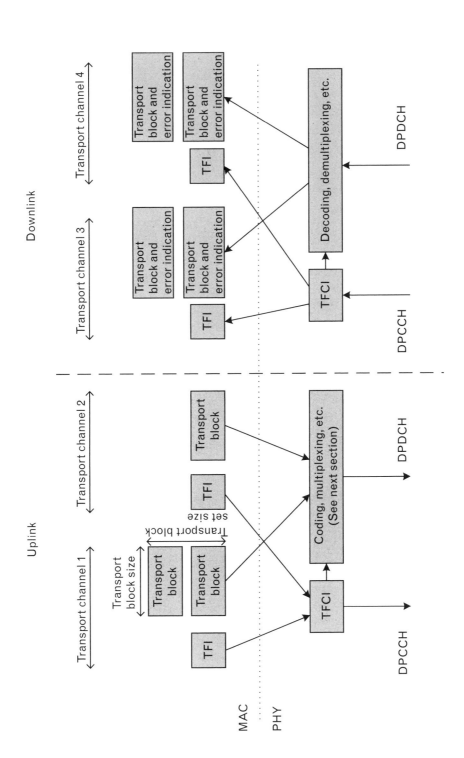

FIGURE 3.23 *Transport formats.*

The set of all transport format combinations is called the transport format combination set. This is what is given by the RRC to the MAC layer in control signals. Higher layers decide the transport format combination set to be used, but the MAC can select from that set the exact transport format combination to be used in any given frame. This means that the MAC layer can maintain a fast radio resource control, as it can adjust the bit rate on a frame-by-frame basis.

The transport format combination is identified from the transport format combination identifier (TFCI). This identifier is used in peer-to-peer communication to inform the receiving entity about the transport format combination. It can be signaled to the peer entity or it can be detected there blindly. It is not needed in interlayer communication, as the physical layer can determine it by itself from the received TFIs from the MAC layer. A transport format combination set might look like this:

Dynamic Part:
Combination1
DCH1: {40 bits, 40 bits}; DCH2: {320 bits, 640 bits}; DCH3: {320 bits, 320 bits}

Combination2:
DCH1: {40 bits, 80 bits}; DCH2: {320 bits, 320 bits}; DCH3: 320 bits, 1,280 bits}

Combination3:
DCH1: {40 bits, 160 bits}; DCH2: {320 bits, 320 bits}; DCH3: {320 bits, 320 bits}

Semistatic Part:
DCH1: {110 ms, turbo coding, static rate matching parameter = 1}
DCH2: {10 ms, convolutional coding, static rate matching parameter = 3}
DCH3: {10 ms, turbo coding, static rate matching parameter = 2}

Transport formats and the associated concepts are defined in [1]. See the MAC section from Chapter 7 to study how the physical layer, MAC, and RLC are cooperating when data is transmitted over the air interface. The physical layer knows the current channel conditions, the MAC knows the allowed transport formats and transport format combinations, and the RLC knows how much data is available for transmission. This information must be combined once every radio frame in a complex process the outcome of which are filled data blocks in the physical layer ready for transmission.

3.6 Data Through Layer 1

This section ties the previous sections of this chapter together and shows how the data is processed while it is going through layer 1. Figure 3.24 describes the data processing in the downlink direction and Figure 3.25 depicts the uplink case.

FIGURE 3.24
Downlink data path through physical channel.

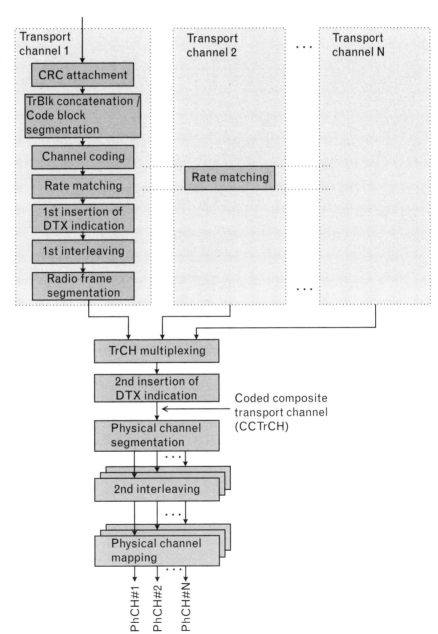

FIGURE 3.25
Uplink data path through physical channel.

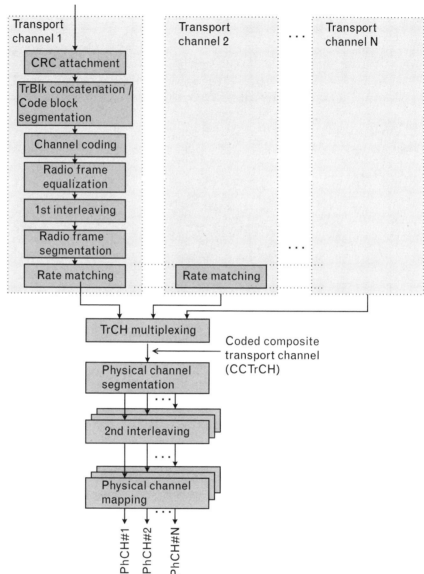

The topmost box in Figure 3.24 is the CRC attachment. It is discussed in Section 3.1.4 concerning error detection on transport channels. The second box, the transport block concatenation and code block segmentation, means that all transport blocks on a transport channel within a TTI are serially concatenated. If the resulting block size is larger than the maximum size of a code block, then additional code block segmentation is performed. The maximum size of a code block depends on the channel coding method to be used for the TrCH. Channel coding is discussed in Section 3.1.1, FEC encoding/decoding, and the rate matching box in Section 3.1.6, rate matching.

DTX bits are used to fill up the radio frames in the downlink. DTX bits are not transmitted over the air interface; they are simply placeholders to tell the transmitter when to turn the actual radio transmission off. They are not used in the uplink, where the radio frames are always completely filled if there is something to send.

The 1st insertion of DTX indication bits is needed only if the positions of the TrCHs in the radio frame are fixed. In this scheme a fixed number of bits is reserved for each TrCH in the radio frame. If a TrCH cannot fill its slot, then DTX bits are added. The 2nd insertion of DTX indication bits is done after the TrCH multiplexing. The purpose of the second insertion is to completely fill the radio frames, and these DTX bits are inserted to the end of the radio frame. However, the 2nd interleaving mixes the bits so that in the actual transmitted bit sequence, the DTX bits are no longer in the end of the frame.

The 1st interleaving is only used when the TTI is longer than one radio frame; that is, longer than 10 ms. This interleaving constitutes interframe interleaving and it is done over the whole TTI length.

Radio frame segmentation is also needed only if the transmission time interval is longer than 10 ms. In this case the input bit sequence must be segmented and mapped evenly into 2, 4, or 8 radio frames.

Transport channel multiplexing is explained in Section 3.1.5, multiplexing of transport channels and demultiplexing of CCTrCHs.

Physical channel segmentation, the 2nd interleaving, and physical channel mapping are discussed in Section 3.1.7.

In the uplink direction the functionality is similar to the downlink situation. Radio frame equalization is used only in the uplink. When the transmission time interval is longer than 10 ms (i.e., 20, 40, or 80 ms), the input bit sequence will be segmented and mapped onto consecutive (2, 4, or 8) radio frames. Radio frame equalization ensures that the contents of the input block can be evenly divided into equal-sized blocks.

FDD mode data processing in the physical layer is discussed in [2, 16]. In the TDD mode the processing is only slightly different; see [17] for the TDD mode description.

REFERENCES

[1]   3GPP TS 25.302, v 5.0.0, Services Provided by the Physical Layer, 2002.

[2]   3GPP TS 25.212, v 5.0.0, Multiplexing and Channel Coding (FDD), 2002.

[3]   3GPP TS 25.215, v 5.0.0, Physical Layer—Measurements (FDD), 2002.

[4]   3GPP TS 25.225, v 5.0.0, Physical Layer—Measurements (TDD), 2002.

[5]   3GPP TS 25.214, v 5.0.0, Physical Layer Procedures (FDD), 2002.

[6]   3GPP TS 25.213, v 5.0.0, Spreading and Modulation (FDD), 2002.

[7]   3GPP TS 25.223, v 5.0.0, Spreading and Modulation (TDD), 2002.

[8]   Holma, H., and A. Toskala, (eds.), *WCDMA for UMTS: Radio Access For Third Generation Mobile Communications*, New York: Wiley, 2000, p. 297.

[9]   3GPP TS 25.224, v 5.0.0, Physical Layer Procedures (TDD), 2002.

[10] 3GPP TS 25.331, v 5.0.1, RRC Protocol Specification, 2002.

[11] 3GPP TS 25.211, v 5.0.0, Physical Channels and Mapping of Transport Channels onto Physical Channels (FDD), 2002.

[12] 3GPP TS 25.221, v 5.0.0, Physical Channels and Mapping of Transport Channels onto Physical Channels (TDD), 2002.

[13] 3GPP TS 25.321, v 5.0.0, MAC Protocol Specification, 2002.

[14] 3GPP TS 25.301, v 5.0.0, Radio Interface Protocol Architecture, 2002.

[15] Prasad, R., W. Mohr, and W. Konhäuser, *Third Generation Mobile Communication Systems*, Norwood, MA: Artech House, 2000, Chapter 4.

[16] Holma, H., and A. Toskala, (eds.), *WCDMA for UMTS: Radio Access for Third Generation Mobile Communications*, New York: Wiley, 2000, Chapter 6.

[17] 3GPP TS 25.222, v 5.0.0, Multiplexing and Channel Coding (TDD), 2002.

Modulation Techniques and Spread Spectrum

4.1 Spreading Techniques

There are several techniques employed for spreading the information signal. The most important ones are discussed below, although these are by no means the only ones, and these techniques can be combined to form hybrid techniques. UTRAN uses the direct-sequence CDMA (DS-CDMA) modulation technique.

4.1.1 DS-CDMA

In DS-CDMA, the original signal is multiplied directly by a faster-rate spreading code (Figure 4.1). The resulting signal then modulates the digital wideband carrier. The chip rate of the code signal must be much higher than the bit rate of the information signal. The receiver despreads the signal using the same code. It has to be able to synchronize the received signal with the locally generated code; otherwise, the original signal cannot be recovered.

4.1.2 Frequency-Hopping CDMA

In frequency-hopping CDMA (FH-CDMA), the carrier frequency at which the signal is transmitted is changed rapidly according to the spreading code. Frequency-hopping (FH) systems use only a small part of the bandwidth at a time, but the location of this part changes according to the spreading code (Figure 4.2). The receiver uses the same code to convert the received signal back to the original. FH-CDMA systems can be further divided into slow- and fast-hopping systems. In a slow-hopping system, several symbols are transmitted on the same frequency, whereas in fast-hopping systems, the frequency changes several times during the transmission of one symbol. The GSM system is an example of a slow FH system because the transmitter's carrier frequency changes only with the time slot rate—217 hops per second—which is much slower than the symbol rate. Fast FH systems are very expensive with current technologies and are not at all common.

FIGURE 4.1
DS-CDMA principle.

FIGURE 4.2
FH-CDMA principle.

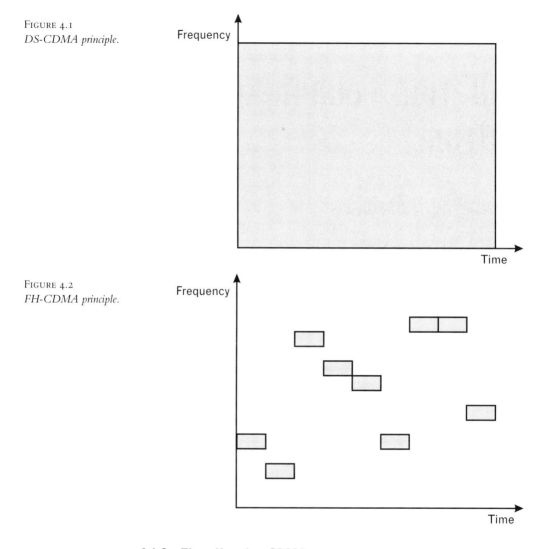

4.1.3   Time-Hopping CDMA

In time-hopping CDMA (TH-CDMA), the used spreading code modulates the transmission time of the signal. The transmission is not continuous, but the signal is sent in short bursts. The transmission time is determined by the code. Thus, the transmission uses the whole available bandwidth, but only for short periods at a time (see Figure 4.3).

4.1.4   Multicarrier CDMA

In multicarrier CDMA (MC-CDMA), each data symbol is transmitted simultaneously over N relatively narrowband subcarriers. Each subcarrier is encoded with a constant phase offset. Multiple access is achieved with different users transmitting at the same set of subcarriers, but with spreading

FIGURE 4.3
TH-CDMA principle.

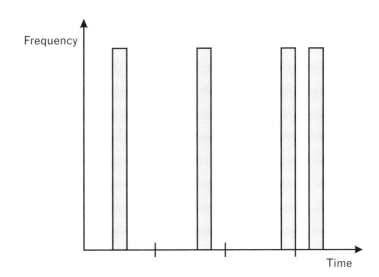

codes that are orthogonal to the codes of the other users. These codes are a set of frequency offsets in each subcarrier. It is unlikely that all of the subcarriers will be located in a deep fade and, consequently, frequency diversity is achieved (see Figure 4.4).

Note that one of the IMT-2000 families of protocols is based on MC-CDMA technology. The IMT-MC protocol (CDMA2000) uses MC-CDMA spreading in the downlink, although in the uplink direction, the IMT-MC uses DS-CDMA, just like the UTRAN FDD mode [1]. The first release of CDMA2000 will support only one downlink 1.2288-Mcps carrier (1xRTT), so it cannot be regarded as an MC-CDMA system. However, in later releases, the IMT-MC downlink should support three parallel subcarriers (3xRTT). See also [2–4].

FIGURE 4.4
MC-CDMA principle.

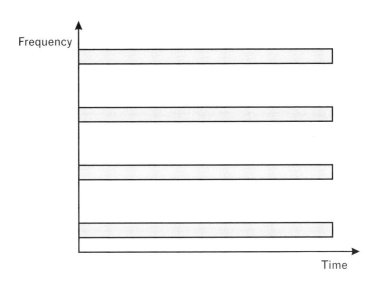

....

4.2   Data Modulation

A data–modulation scheme defines how the data bits are mixed with the carrier signal, which is always a sine wave. There are three basic ways to modulate a carrier signal in a digital sense: amplitude shift keying (ASK), frequency shift keying (FSK), and phase shift keying (PSK). The curious term *keying* comes from the communications practices of long ago, when a transmitter, for example, was turned on and off according to a code with a single-pole-single-throw switch called a *key*. The key had a special handle that made it particularly easy to use as an on-off switch, which could be manipulated by a trained operator familiar with the digital codes of the time, such as the Morse code.

In ASK the amplitude of the carrier signal is modified (multiplied) by the digital signal. The modulated signal can be given as

$$s(t) = f(t) \sin(2\pi f_c t + \phi) \tag{4.1}$$

where $s(t)$ is the modulated carrier signal and $f(t)$ is the digital signal. The phase of the signal remains constant.

In FSK the frequency of the carrier signal is modified by the digital signal. If the digital signal has only two symbols, 0 or 1, then in the basic FSK scheme, the transmission switches between two frequencies to account for bi-level FSK. The amplitude of the signal is constant. For mathematically minded people:

$$s(t) = f_1(t) \sin(2\pi f_{C1} t + \phi) + f_2(t) \sin(2\pi f_{C2} t + \phi) \tag{4.2}$$

A multilevel FSK employs multiple frequencies between which the transmission switches according to the modulating digital signal.

In PSK the phase of the carrier signal is modified by the digital signal. Mathematically:

$$s(t) = \sin\left[2\pi f_C + \phi(t)\right] \tag{4.3}$$

The PSK family is the most widely used modulation scheme in modern cellular systems. There are many variants in this family, and only a few of them are mentioned here. For a more thorough treatment of PSK modulation possibilities, see [5]. In binary phase shift keying (BPSK) modulation, each data bit is transformed into a separate data symbol. The mapping rule is 1 -> + 1 and 0 ->- 1. There are only two possible phase shifts in BPSK, 0 and π radians.

The quadrature phase shift keying (QPSK) modulation has four phases: 0, $1/2\pi$, π, and $3/2\pi$ radians. Two data bits are transformed into one complex data symbol; for example, $(00 -> +1 +j)$, $(01 -> -1 +j)$, $(11 ->+1 -j)$, $(10 -> +1 -j)$. A symbol is any change (keying) of the carrier.

Generally, M-ary PSK has M phases, given as $2\pi m/M$; m = 0, 1, ..., $M - 1$.

Minimum shift keying (MSK) is a modification of QPSK, in which the modulating pulses are sinusoidal instead of rectangular. The GSM system uses Gaussian minimum shift keying (GMSK) modulation, in which the rectangular modulating pulses are first Gaussian-shaped-filtered. With GMSK modulation one symbol carries (approximately) one bit only, but GMSK has other good properties, such as low power consumption and low adjacent channel interference. MSK and GMSK are well presented in [6].

The number of times the signal parameter (amplitude, frequency, or phase) is changed per second is called the signaling rate or the symbol rate. It is measured in baud: 1 baud = 1 change per second. With binary modulations, such as ASK, FSK, and BPSK, the signaling rate equals the bit rate. With QPSK and M-ary PSK, the bit rate exceeds the baud rate. In CDMA systems, the terminology is a bit different, as here the data modulation rate is given as the chip rate. The chipping process is the last modulation stage applied to the signal in the transmitter.

The UTRAN air interface uses QPSK modulation in the downlink, although HS-PDSCH may also employ 16 Quadrature Amplitude Modulation (16 QAM). 16 QAM requires good radio conditions to work well. The modulation chip rate is 3.84 Mcps. The original proposal called for a chip rate of 4.096 Mcps, but this was modified later to make it closer to the CDMA2000 chip rate. Figure 4.5 depicts a phasor diagram for QPSK and 16 QAM. As seen, with 16 QAM also the amplitude of the signal matters. As explained, in QPSK one symbol carries two data bits; in 16 QAM each symbol includes four bits. Thus, a QPSK system with a chip rate of 3.84 Mcps could theoretically transfer $2 \times 3.84 = 7.68$ Mbps, and a 16 QAM system could transfer 4×3.84 Mbps = 15.36 Mbps. In 3GPP also the usage of 64 QAM with HSDPA has been studied, but rejected so far. In this scheme

FIGURE 4.5
Modulation schemes in
UTRAN air interface.

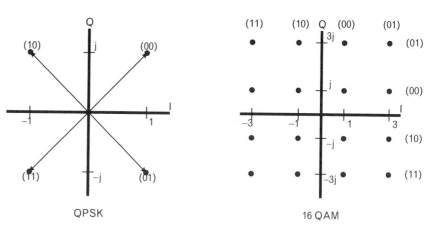

each symbol would contain six data bits, but the channel conditions should be very good indeed before it could be used.

There are two dedicated physical channels in the UTRAN air interface, namely the DPCCH (for control information) and the DPDCH (for user data). These are time-multiplexed in the downlink direction. The control channel must always be present because of the power control mechanisms, but the user data may be missing if there is nothing to be sent. With time multiplexing this results in a pulsed transmission, which is not a problem in the downlink, as common channels will be sent continuously. In the uplink, however, the discontinuous transmission with time multiplexing would cause severe electromagnetic problems, just as the bursts in GSM do today. This can be prevented by using code multiplexing instead to combine the control and data channels. The difference between time and code multi-plexing is shown in Figure 4.6.

Discontinuous transmission is not the only problem that needs solving in the WCDMA uplink. Uplink transmission is the biggest power consumer in a mobile handset. The modulation scheme used should be one that saves power as much as possible. The power amplifier in a handset is most effi-cient (and power saving) when it works close to its saturation point. To achieve this goal, the modulation scheme should produce signals with small peak-to-average power ratios; that is, if the difference between the peak power level and the average power level is small, then the power amplifier can be tuned to work closely to its saturation point.

FIGURE 4.6
Multiplexing schemes in
UTRAN air interface.

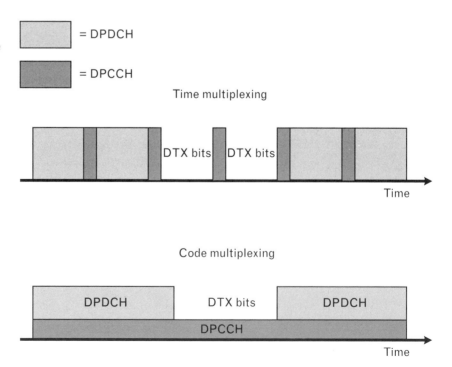

This problem is not easy to solve in WCDMA. A UE can transmit several physical channels at different power levels simultaneously. There is one dedicated control channel and one or more dedicated physical data channels for user data, including voice. If a traditional modulation technique, such as QPSK in the downlink, were used in a multichannel system, such as the UTRAN's uplink, it would result in a large number of zero crossings in the I/Q plane. This would yield a relatively high peak-to-average power ratio, diminished amplifier power efficiency, thus, shorter handset talk time.

Therefore, a complex scrambling scheme is used in the UTRAN uplink. This scheme has many names; the UTRAN generally uses the name dual-channel QPSK. Other names include hybrid phase shift keying (HPSK) and orthogonal complex quadrature phase shift keying (OCQPSK). In dual-channel QPSK, the physical channels are I/Q multiplexed (Figure 4.7). The control channel (DPCCH) will be sent via the Q path, and the first data channel (DPDCH) uses the I path. Additional data channels will be divided evenly between the I and Q paths. This was discussed in Section 3.1.11.

There are two main methods the dual-channel QPSK uses to achieve the goal, which is a low peak-to-average power ratio: selected orthogonal spreading codes and complex scrambling with a Walsh rotator. The complex scrambling principle is depicted in Figure 4.8. In this example we have the original data chip divided into its I and Q components (1,1) and a complex scrambling signal (−1,1). When complex scrambling takes place, the phases of these signals are added together ($45° + 135° = 180°$) and the resulting signal constellation is (−1,0).

The distance from the origin represents the power level of the signal. If the original data signal uses equal power levels for control and data channels, then the original data chips are always mapped into one of the constellation points [(1,1), (−1,1), (1,−1)]. When they are scrambled using a complex scrambling code, the result always lies on either the I or Q axis; that is, mapped into the constellation points (1,0), (0,1), (−1,0), and (0,−1). If the control and data channels have unequal power levels, then the result does not lie on the I/Q axes, but still, the resulting constellation points have a constant distance from the origin. This implies that no matter what the

FIGURE 4.7
I/Q code multiplexing.

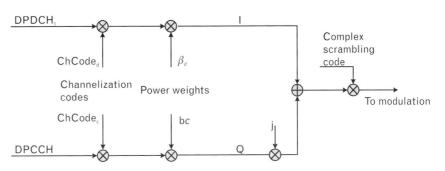

FIGURE 4.8
Complex scrambling.

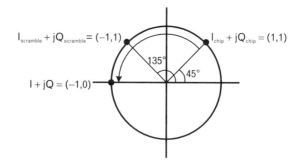

power difference is between the transmitted channels, complex scrambling can handle it and distribute the power evenly between the *I* and *Q* axes (see Figure 4.9).

The term *zero crossing* means that two successive resulting constellations are placed on opposite sides of the origin; that is, when these chips are transmitted, the transition must go via zero. In Figure 4.7 this would happen if the next resulting constellation would be placed on (1,0). This is bad for the peak-to-average power ratio. A UE can resist this by choosing suitable orthogonal scrambling codes to be used with a fixed repeating function, the Walsh rotator. This is defined as (1,1) for $I_{scramble}$ and (1,−1) for $Q_{scramble}$. If the Walsh rotator is used, two consecutive identical constellation points are scrambled in different ways. The first point is rotated by 45 degrees [i.e., constellation point (1,1)] and the second point by −45 degrees [i.e., constellation point (1,−1)] (see Figure 4.10). Thus, it is preferable to use such orthogonal codes, which have pairs of identical chips.

If the Walsh rotator and suitable orthogonal codes are used, then at least every other phase shift will be exactly 90 degrees. This means that these phase shifts cannot cause zero crossings. It also eliminates zero phase shifts from every other phase shift. A zero phase shift is equally harmful, as it causes overshooting trajectory (i.e., a power peak). This increases the peak-to-average power ratio. However, note that the scheme cannot altogether remove the zero crossings and zero phase shifts. The scheme discussed here only halves the number of their occurrences. They can still take place between the pairs of Walsh rotator sequences.

FIGURE 4.9
Constellation points for data channels with unequal power levels.

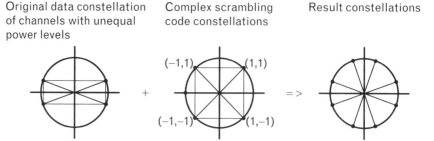

Original data constellation of channels with unequal power levels

Complex scrambling code constellations

Result constellations

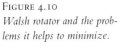

FIGURE 4.10
Walsh rotator and the problems it helps to minimize.

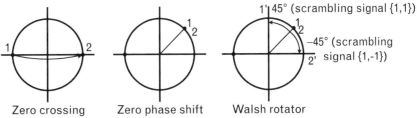

Zero crossing   Zero phase shift   Walsh rotator

Note that modulation is not the only issue that determines the system power efficiency, although it is a very important component of it. Another important factor is the power control scheme used.

Modulation is defined in the 3GPP specification [7]. The specification defines the subject, but does not explain it much. The modulation in UTRAN is briefly discussed in [8]. Dual-channel QPSK, or HPSK as it is also called, is well presented and explained in [9]. Note that CDMA2000 uses a similar modulation scheme in the reverse link.

REFERENCES

[1]   Holma, H., and A. Toskala, *WCDMA for UMTS: Radio Access for Third Generation Mobile Communications,* New York: Wiley, 2000, pp. 303–315.

[2]   Yee, N, J.P.M.G. Linnartz, and G. Fettweis, "Multi-Carrier CDMA in Indoor Wireless Networks," *IEICE Trans. on Communications [Japan],* Vol. E77-B, No. 7, July 1994, pp. 900–904.

[3]   Yee, N., and J.P.M.G. Linnartz, "Wiener Filtering for Multi-Carrier CDMA," *IEEE/ICCC Conference on Personal Indoor Mobile Radio Communications (PIMRC) and Wireless Computer Networks (WCN),* The Hague, The Netherlands, September 19–23, 1994, Vol. 4, pp. 1344–1347.

[4]   Yee, N., and J.P.M.G. Linnartz, "Multi-Carrier CDMA in an Indoor Wireless Radio Channel," Memorandum UCB/ERL M94/6, University of California at Berkeley, Electronics Research Lab, 1994. Available at http://cs-tr.cs.berkeley.edu/TR/UEB:ERL-94–6.

[5]   Xiong, F., *Digital Modulation Techniques,* Norwood, MA: Artech House, 2000.

[6]   Mehrotra, A., *GSM System Engineering,* Norwood, MA: Artech House, 1997, pp. 207–217.

[7]   3G TS 25.213, v 5.0.0, Spreading and Modulation (FDD), 2002.

[8]   Holma, H., and A. Toskala, *WCDMA for UMTS: Radio Access for Third Generation Mobile Communications,* New York: Wiley, 2000, pp. 80–88.

[9]   "HPSK Spreading for 3G," Agilent Technologies Application Note 1335 (on-line), December 1999, Agilent Technologies. Accessible at ttp://www.agilent.com/.

CHAPTER 5

Spreading Codes

This chapter continues with the subject discussed in Chapter 4, expanding on it further. Whereas the previous chapter handled the spread spectrum issue from a general point of view, presenting various spread-spectrum modulation techniques and their general principles, this chapter concentrates on the schemes specified by the 3GPP documents. Both the uplink and downlink cases are studied separately, and the differences between pseudorandom codes and orthogonal codes are explained. Understanding the contents of this chapter is a prerequisite for understanding the WCDMA air interface.

Spreading codes are also known as spreading sequences. There are two types of spreading codes in the UTRAN air interface: orthogonal codes and pseudorandom codes. Pseudorandom codes are also known as pseudonoise (PN) codes. Both kinds of codes are used together in the uplink and in the downlink. The same code is always used for both the spreading and despreading of a signal. This is possible because the spreading process is actually an XOR operation with the data stream and the spreading code. The reader should recall that two successive XOR operations will produce the original data. Another possible binary operation to combine the two bit streams could be N-XOR, as it is also reversible. The truth table for XOR is given in Table 5.1.

Spreading means increasing the bandwidth of the signal. At first hearing, this may not sound like a good idea because bandwidth is a scarce and expensive resource. There are some good reasons for doing this, however. The most important incentive for wideband spreading is the good interference resistance of a wideband signal. A wideband signal can survive in a very noisy environment. It is also difficult to jam because its energy is spread over so wide a spectrum that it is very difficult to locate. The low-energy-density property also means that emissions from the transmitter are very low. The characteristics of the wideband signal are further discussed in Section 2.2.

TABLE 5.1 XOR Truth Table

A	B	A XOR B
0	0	0
1	0	1
0	1	1
1	1	0

....

5.1   Orthogonal Codes

The spreading procedure in the UTRAN consists of two separate operations: channelization and scrambling. Channelization uses orthogonal codes and scrambling uses PN codes. Channelization occurs before scrambling in the transmitter both in the uplink and the downlink.

Channelization transforms each data symbol into multiple chips. This ratio (number of chips/symbol) is called the spreading factor (SF). Thus, it is this procedure that actually expands the signal bandwidth. Data symbols on the I and Q branches are combined with the channelization code. Channelization codes are orthogonal codes (more precisely, orthogonal variable spreading factor [OVSF] codes), meaning that in an ideal environment they don't interfere with each other. However, orthogonality requires that the codes be time synchronized. Therefore, it can be used in the downlink to separate different users within one cell, but in the uplink only to separate the different services of one user. It cannot be used to separate different uplink users in a base station, as all mobiles are unsynchronized in time; thus, their codes cannot be orthogonal (unless the system in question employs the TDD mode with uplink synchronization). Also, orthogonal signals cannot be used as such between base stations in the downlink because there is only a limited number of orthogonal codes. The orthogonal codes must be reused in every cell, and therefore it is quite possible that a UE in the cell boundary area receives the same orthogonal signal from two base stations, each directing their identical orthogonal codes to two different UEs. If only orthogonal spreading codes are used, these signals would interfere with each other very severely. However, in the uplink the transmissions from one user are, of course, time synchronous; thus, orthogonal codes can be used to separate the different channels of a user.

The generation method for channelization codes is defined in [1] and is illustrated in Figure 5.1.

This algorithm produces a tree of codes illustrated in Figure 5.2. This example shows only the root of the code tree. The UTRAN employs the spreading factors 4 through 512, where 4 to 256 appear in uplink, and SF 512 is added to the SF catalog in the downlink direction. This code tree also illustrates how the codes can be allocated. If, for example, the code $C_{8,2}$ is allocated, then no codes from its subtree can be used (i.e., $C_{16,4}$, $C_{16,5}$, $C_{32,8}$). These subtree codes would not be orthogonal with their parent code.

The use of orthogonal codes is depicted in Figure 5.3. A data sequence (1001) is combined with spreading code $C_{ch,4,1}$ (1,1,−1,−1).

This code has a spreading factor of 4, which means that for each data signal there are four chips in the spreading code. The resulting signal bandwidth is four times wider than the bandwidth of the original signal. We then look to see what happens when this spread signal is despread with two

FIGURE 5.1
*Generation of OVSF
codes.*

$$C_{ch,1,0} = 1$$

$$\begin{bmatrix} C_{ch,2,0} \\ C_{ch,2,1} \end{bmatrix} = \begin{bmatrix} C_{ch,1,0} & C_{ch,1,0} \\ C_{ch,1,0} & -C_{ch,1,0} \end{bmatrix} = \begin{bmatrix} 1 & 1 \\ 1 & -1 \end{bmatrix}$$

$$\begin{bmatrix} C_{ch,2^{(n+1)},0} \\ C_{ch,2^{(n+1)},1} \\ C_{ch,2^{(n+1)},2} \\ C_{ch,2^{(n+1)},3} \\ \vdots \\ C_{ch,2^{(n+1)},2^{(n+1)}-2} \\ C_{ch,2^{(n+1)},2^{(n+1)}-1} \end{bmatrix} = \begin{bmatrix} C_{ch,2^n,0} & C_{ch,2^n,0} \\ C_{ch,2^n,0} & -C_{ch,2^n,0} \\ C_{ch,2^n,1} & -C_{ch,2^n,1} \\ C_{ch,2^n,1} & -C_{ch,2^n,1} \\ \vdots & \vdots \\ C_{ch,2^n,2^n-1} & C_{ch,2^n,2^n-1} \\ C_{ch,2^n,2^n-1} & -C_{ch,2^n,2^n-1} \end{bmatrix}$$

codes, $C_{ch,4,1}$ and $C_{ch,4,2}$. Despreading with the correct code—the code which spread the signal—produces the original signal (1001) in the integrator, but if any of the wrong codes are used, then the result is noise. Note that an SF of 4 is a very low spreading factor, the lowest possible in the UTRAN.

In the uplink direction, these orthogonal codes are assigned on a per-UE basis, so the code management is quite straightforward. However, in the downlink direction, the same code tree is used by the base station for all mobiles in its cell area. Thus, careful code management is needed so that the base station does not run out of downlink channelization codes.

Note that when the signal was despread with a wrong code in Figure 5.3, the result in the integrator was exactly zero every time. This shows that in a fully orthogonal system, noise does not exist. Thus, in theory, power control would also be unnecessary in such a system. However, full orthogonality cannot be achieved in practice. There is always some noise in the system, and power control is needed to reduce it.

The previous example demonstrated how spreading works with one user. Typical practice in the downlink has the same composite signal spread using several orthogonal codes (one for each user). This is depicted in Figure 5.4. Note that the spreading codes must be time aligned; otherwise, the orthogonality is lost. This example shows how the original data streams can be resolved from the combined signal.

The downlink transmissions from separate base stations are not orthogonal. A UE must first identify the right base station transmission according to the scrambling code, and then from that signal extract its own data using the

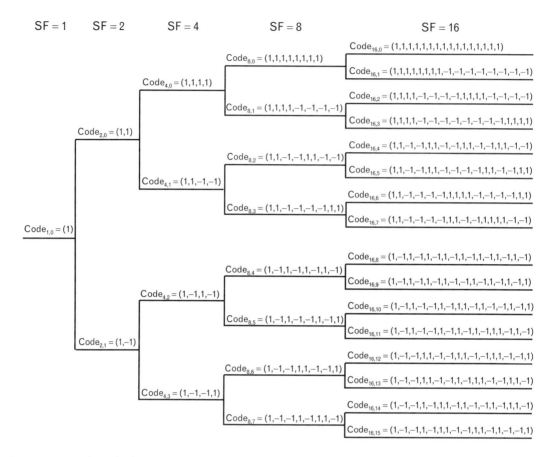

FIGURE 5.2   *Orthogonal codes.*

orthogonal channelization code. Thus, in the real world the downlink environment is never purely orthogonal and interference free. Intracell interference exists because of multipath reflections and intercell interference from asynchronous base stations. From the asynchronous nature of the system follows inter-base-station nonorthogonality. However, the intercell interference is not as serious a problem as it might at first seem because power control and soft handovers (SHOs) should keep the other base stations from interfering too much with the downlink signal to a particular UE.

5.2   PN Codes

The orthogonal codes alone cannot handle the spreading function in the UTRAN air interface. As explained earlier, they can only be used when the signals applying them are time synchronous. Clearly this is not the case

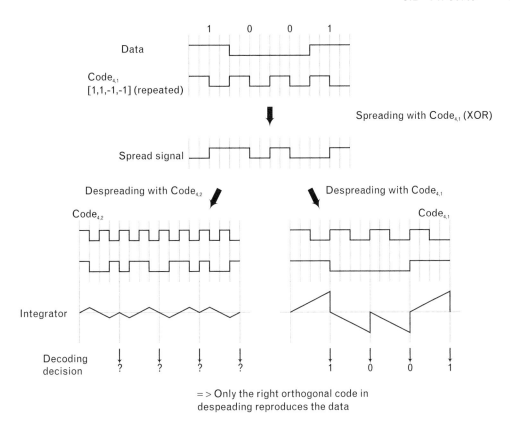

FIGURE 5.3 *Uses of orthogonal codes.*

between asynchronous users in the uplink direction. If orthogonal spreading codes alone were used in the uplink, then they could easily cancel each other. Moreover, the downlink signals are only orthogonal within one base station. But even in this case, orthogonality is partially lost with channel distortions. The base station's orthogonality decreases as we move out toward to the mobiles. Therefore, something else is needed.

To solve these problems, the system employs pseudorandom codes. They are used in the second part of the spreading procedure, which is called the scrambling stage. In the scrambling procedure, the signal, which is already spread to its full bandwidth with an orthogonal spreading code, is further combined (XORed) with a pseudorandom scrambling code. This scrambling code is either a long code (a Gold code with a 10-ms period) or a short code [S(2) code]. These pseudorandom codes have good autocorrelation properties (see Section 5.4). There are millions of scrambling codes available in the uplink, so no special code management is needed. A spreading code identifies the specific UE to the base station. And once the uplink synchronization is obtained, various services of this UE can be separated using orthogonal codes. Note, however, that there are proposals to

FIGURE 5.4
Multiuser spreading.

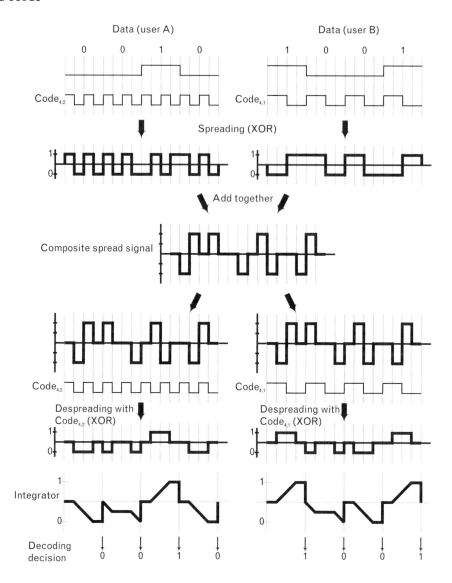

FIGURE 5.4
Multiuser spreading.

introduce orthogonality into the uplink in later releases of the 3GPP specifications. The work name of this item is Uplink Synchronous Transmission
Scheme (USTS), see [2]. Time-synchronization and orthogonal signals
would reduce the interference in the uplink direction because fully
orthogonal signals do not cause any interference with each other.

In the downlink direction, pseudorandom scrambling codes are used to
reduce the inter-base-station interference. Each Node B has only one primary scrambling code, and UEs can use this information to separate base stations. There are 512 different primary scrambling codes possible in the
downlink. This number should be enough for the cell planning purposes. A
bigger number would cause problems with the cell search procedure (see

Section 3.1.9). The primary scrambling codes are divided into 64 code groups, each consisting of 8 codes. Dividing the 512 possible primary scrambling codes into only 64 small groups of codes can speed up the synchronization procedure.

The specifications also define secondary scrambling codes. Each primary scrambling code has a set of 16 secondary scrambling codes. They can be employed while transmitting channels that do not need to be received by everyone in the cell. However, they should be used sparingly because channels transmitted with secondary scrambling codes are not orthogonal to channels that use the primary scrambling code. One possible application could be in sectored cells, where separate sectors do not have to be orthogonal to each other.

The calculation of different scrambling codes with their exact use, the types of scrambling codes used, and when they are used is specified in [1]. See also Figure 5.5.

5.3   Synchronization Codes

There is an exception to the general rule in the downlink direction that all physical channels are first combined with a channelization code and then with a scrambling code. Synchronization channels (both primary and secondary) are not subjected to either of these. Instead, they are combined with synchronization codes.

There are two types of synchronization codes, primary and secondary. Primary codes are used by the primary synchronization channels (P-SCH), and secondary codes by the secondary synchronization channels (S-SCH). The primary synchronization code is identical in all cells. This is a useful

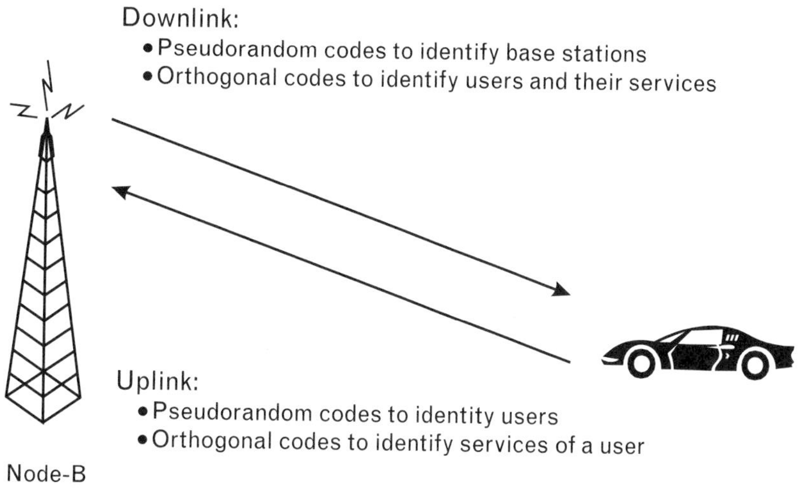

FIGURE 5.5
Code types in the air interface.

Downlink:
- Pseudorandom codes to identify base stations
- Orthogonal codes to identify users and their services

Uplink:
- Pseudorandom codes to identity users
- Orthogonal codes to identify services of a user

Node-B

property, as it can be used for downlink time slot synchronization in the UE's cell-search phase. This fixed sequence of bits is sent only during the first 256 chips of each slot (there are 2,560 chips in the whole time slot).

There are 16 different secondary synchronization codes. These are sent via the S-SCHs, but only during the first 256 chips of each time slot. These codes are fixed and known to all UEs, and form an alphabet of 16 symbols. The base station will change the transmitted code from slot to slot. There are thus 64 different secondary-synchronization-code transmit sequences. The UE can determine which of the 64 apply to a particular base station by reading all of the secondary synchronization codes that appear in a 10-ms radio frame. The particular code sequence tells the UE which of the 64 groups of the scrambling codes to search through in order to find the PN code for the Node B in a cell. Searching through only eight scrambling codes is much faster than a search of all 512 possible scrambling codes.

The use of synchronization codes in the downlink synchronization is further discussed in Section 3.1.9.

5.4   Autocorrelation and Cross-Correlation

We learned earlier that the pseudorandom codes should have good autocorrelation properties. *Autocorrelation* measures the correlation between the signal and a time-delayed version of itself. Thus, if the signal recovered in a receiver is combined with the pseudorandom code that made it, a good correlation should be found if the signal is the right one (i.e., the signal was modulated using the same pseudorandom spreading code in the transmitter). This property can be used in the initial synchronization sequence as well as to separate the multipath components of a signal.

Cross-correlation measures the correlation between a signal and some other (pseudorandom) code. This value should be low, as it indicates the level of interference caused by the other users of the system.

Autocorrelation and cross-correlation properties are connected, so it is generally not possible to achieve good properties from both of them at the same time; it is difficult to achieve a high autocorrelation and a low cross-correlation at the same time. Thus, a good compromise has to be found. This is a compromise between the fast acquisition performance of the good autocorrelation code and the low interference level of the good (low correlation) cross-correlation code. Various optimization criteria for the pseudorandom code design exist. For example, see [3], which is especially relevant to the UTRAN because it discusses code optimization in an asynchronous DS-CDMA system.

5.5   Intercell Interference

As stated earlier, downlink transmissions from different base stations are not orthogonal. Orthogonality requires that the signals originate from the same transmitter. The UE must first identify the right transmission using the scrambling code, and then from that signal extract its own data using the orthogonal channelization code. This is possible because all signals from one base station are orthogonal. But base stations are asynchronous and nonorthogonal to each other. This means that in the real world, where there are several base stations and some traffic in all of them, the downlink environment is never purely orthogonal and interference free.

Intracell interference exists because of multipath reflections. Intercell interference exists because of nonorthogonal transmissions from separate base stations. However, as we saw earlier, the intercell interference is not as serious a problem as it might seem at first because power control should keep the other base stations from interfering too much. If a UE is in the middle of a cell, the intercell interference should not be a serious problem, as the other base stations are far away. In the border areas, however, the transmissions from the serving base station may be too low compared with the overall interference level. This border-area problem can be eased with the SHO procedure in which the UE simultaneously receives the same transmission from two or more Node Bs. This helps the UE as long as it has enough RAKE fingers in its receiver for each received Node B, or multipath component, which requires a RAKE finger of its own. From the network point of view the situation is not purely beneficial as it has to assign a new orthogonal code in each base station that takes part in an SHO. Also, the more Node Bs that are enlisted to transmit a signal to a UE, the greater the overall interference level in the air interface may be, but the more Node Bs there are in an SHO, the less energy has to be sent via an individual Node B. Each SHO participant also requires fixed network data transmission resources.

Spreading codes are also discussed in the literature [4–7]. TDD-mode spreading is discussed in [8, 9].

REFERENCES

[1]   3GPP TS 25.213, v 5.0.0, Spreading and Modulation (FDD), 2002.

[2]   3GPP TS 25.854, v 5.0.0, Study Report for Uplink Synchronous Transmission Scheme, 2001.

[3]   Karkkainen, K.H.A., "Influence of Various PN Sequence Phase Optimization Criteria on the SNR Performance of an Asynchronous DS-CDMA System," *Proc. IEEE 1995 Military Communications Conference (MILCOM 95)*, San Diego, California, Nov. 1995, pp. 641–646.

[4]   Ojanpera, T., and R. Prasad, *Wideband CDMA for Third Generation Mobile Communications*, Norwood, MA: Artech House, 1998, pp. 108–115.

[5]   Holma, H., and A. Toskala (eds.), *WCDMA for UMTS: Radio Access for Third Generation Mobile Communications*, New York: Wiley, 2000, pp. 28–30; 79–80.

[6]   Walke, B., *Mobile Radio Networks*, New York: Wiley, 1999, pp. 67–72.

[7]   Prasad, R., W. Mohr, and W. Konhauser, *Third Generation Mobile Communication Systems*, Norwood, MA: Artech House, 2000, pp. 81–82.

[8]   3GPP TS 25.223, v 5.0.0, Spreading and Modulation (TDD), 2002.

[9]   Prasad, R., W. Mohr, and W. Konhauser, *Third Generation Mobile Communication Systems*, Norwood, MA: Artech House, 2000, pp. 33–42.

CHAPTER 6

Channel Coding

....

6.1   Coding Processes

An information stream going through a digital radio access network such as the UTRAN must undergo several coding processes, which are depicted in Figure 6.1. The information entering this system may already be in digital format (data) or it may be analog information (voice).

The source encoding function transforms the user's traffic into a digital format and, to the extent possible, compresses the data. The particular source encoder depends on the type of the information in need of encoding. Speech is encoded using a speech encoder [Adaptive Multi Rate (AMR) codec, in the UTRAN], video using a video encoder, and so forth. The source encoder tries to encode the information into the smallest possible number of bits from which the source decoder in the receiving entity can reconstruct the same original information if no errors were introduced to the data during its transmission. Various compression techniques may be used to accomplish the source-coding task. But source coding removes most of the redundancy inherent in the user's information. Another form of redundancy has to be added back into the data stream so that the data can be recovered in the receiver even after suffering the trials of the radio channel.

By inserting carefully contrived redundancy back into the user's data stream, channel coding is responsible for delivering the information bits without any errors over the radio interface. The transmission channel can introduce errors to transmitted bits, which must first be detected and then corrected, if possible. These tasks are not possible without adding some extra information (redundancy) to the data stream. The number of data bits is always increased in channel coding. How much they are increased depends on the specific coding scheme.

The modulator receives the bit sequence from the channel encoder, and converts the source and channel coded stream into a waveform suitable for the transmission channel. In the UTRAN the modulation scheme is QPSK (or 16 QAM) on the downlink and dual-channel QPSK (HPSK) on the uplink.

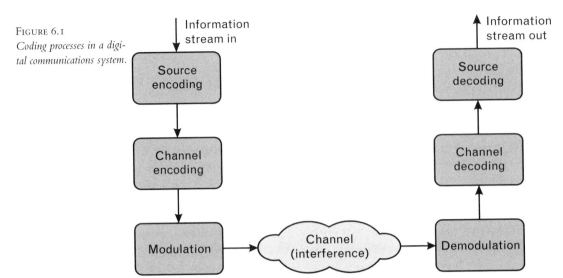

FIGURE 6.1
Coding processes in a digital communications system.

....

6.2   Coding Theory

In the last half of the 1940s, Claude Shannon developed a set of theories that are now commonly known as Shannon's Law [1]. Shannon studied the problem of maximizing the amount of information one can transmit over a noisy communication channel. Before Shannon it was expected that the higher the data rates, the more likely you are to get transmission errors. Shannon argued that this is not necessarily the case, but that to the extent it is possible to achieve error rates approaching zero, it is possible to achieve the maximum channel capacity or any data rate below the channel capacity.

The maximum channel capacity can be determined from the following equation:

$$C = W \log_2 \left(1 + S/N\right)^1 \tag{6.1}$$

where W is the channel bandwidth, S is the power of the signal, and N is the power of the noise. This law sets the absolute data transmission rate that can be achieved. One can easily see that an increase in either the bandwidth or the signal power can bring a higher maximum channel capacity. Another important observation is that the increase in noise level requires an increase in signal level so that the same data transmission rate can be supported as before the noise increased. The signal–to–noise ratio is usually given as the

---

1. Note that future MIMO systems (see Section 12.11.1) will be able to exceed the capacity limit set by Shannon. This is because these multiantenna systems can exploit the spatial characteristics of multipath propagation. In MIMO systems the hard capacity limit can be given as $C = WN \log2 (1 + S/N)$, where N is the number of independent radio paths.

ratio of energy level per bit (E_b) to the energy level per Hertz (N_0) of the noise, that is E_b/N_0.

This theory is important to remember, especially in a WCDMA system like the UTRAN, where the transmissions of users interfere with each other. A user can increase his data rate by increasing the signal power level, but this will also increase the system noise level and, thus, reduce the maximum data transmission capacity of other users.

Shannon's information theory is discussed in most communication books in some way or another as it is still a central theorem in today's practice. For example see [2] and [3].

The following sections consider three types of channel coding algorithms: block codes, convolutional codes, and turbo codes.

6.3 Block Codes

A block code manipulates the data one block at a time. The encoder adds some redundant bits to the block of bits and the decoder uses them to determine whether an error has occurred during the transmission (Figure 6.2). The output of the block coder is always larger than the input. The ratio between the block of information bits k (input) and the block of channel coded bits n (output) is called the code rate:

$$Rc = k/n \tag{6.2}$$

The corresponding code is referred to as (n, k) code. The added redundancy of a (n, k) code is thus ($n - k$)/n. The more redundancy there is, the more errors the channel decoder can fix. But the added redundancy also consumes channel bandwidth, so a good balance must be found.

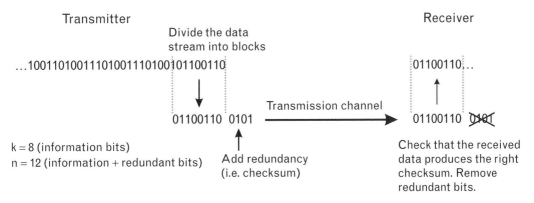

FIGURE 6.2   *Block codes.*

Depending on how the redundant bits are added to the code word n, the resulting code may be called a systematic or a nonsystematic code. In a systematic code, all redundant bits are added to the end of the code word. In a nonsystematic code, the redundant bits are mixed in with the information bits (see Figure 6.3).

There are 2^k possible information blocks, which can be mapped into 2^n possible code words. As we can see, most of the 2^n code words will be left unused. The set of code words to be brought into use is not chosen randomly, but in a way that maximizes the performance of the channel decoder.

The Hamming distance is the measure of the difference between two code words. For example, if there are two code words a and b:

$a = 100110001$

$b = 100101001$

The Hamming distance $d(a, b) = 2$ because the two code words differ in exactly two bit positions (bits 5 and 6). The smallest distance between all the code words is called the minimum distance d_{min}. It is an important measure as it indicates how good this code is for detecting errors. A minimum distance of i indicates that the channel decoder can detect up to $i - 1$ bit errors. If more bit errors than $i - 1$ are present, then the received code word will be similar to some other valid code word; thus, it could be accepted as correct.

Block codes are often used with an automatic repeat request (ARQ) method. Block codes are very efficient at finding errors, and when they are found, a retransmission of the block can be requested from the peer entity. This scheme requires that the data be block-oriented, the timing constraints with the data not be very tight, and that the user's data be tolerant of delays. Every retransmission adds to the overall transmission delay.

The cyclic redundancy check (CRC) is a common method of block coding. CRC bits are also used in WCDMA. Adding the CRC bits is done before the channel encoding and they are checked after the channel decoding. The size of the CRC field to be added to a transport block can be 0, 8, 12, 16, or 24 bits in WCDMA. The corresponding generator polynomials are given in Table 6.1. The generator polynomial can be explained with an

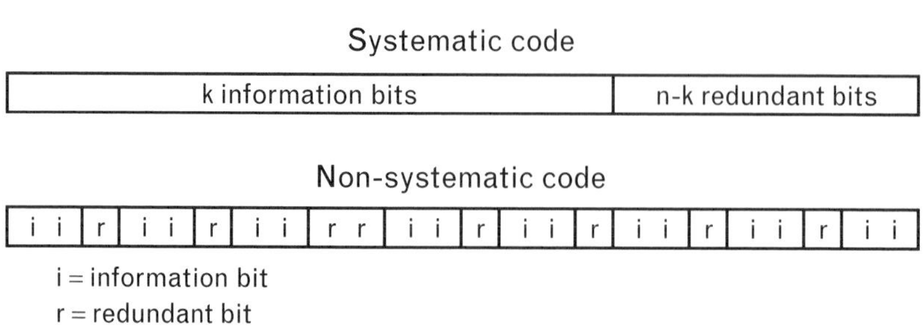

FIGURE 6.3

Systematic and nonsystematic codes.

Systematic code

k information bits	n-k redundant bits

Non-systematic code

i	i	r	i	i	r	i	i	r	r	i	i	r	i	i	r	i	i	r	i	i	r	i	i

i = information bit
r = redundant bit

TABLE 6.1   CRC GENERATOR POLYNOMIALS

No of CRC bits	Polynomial
24	$D^{24} + D^{23} + D^6 + D^5 + D + 1$
16	$D^{16} + D^{12} + D^5 + 1$
12	$D^{12} + D^{11} + D^3 + D^2 + D + 1$
08	$D^8 + D^7 + D^4 + D^2 + D + 1$

example from this table. The CRC generator polynomial $D^8 + D^7 + D^4 + D^2 + D + 1$ means that the polynomial bit string is 110010111. The information block is divided modulo 2 by the generator polynomial, and the remainder becomes the checksum field. A detailed example of the CRC calculation is given, for example, in [4].

Note that certain types of block codes can also be used for error correction, although these are not used in WCDMA.

6.4 Convolutional Codes

Convolutional coding is another way to protect the information bits against errors. Where block codes are used to detect errors and ARQ schemes to fix them, convolutional codes combine both of these functions. Convolutional codes are typically used when the timing constraints are tight and intolerant of the ARQ schemes. The coded data must contain enough redundant information to make it possible to correct at least some of the detected errors that appear in the channel decoder without having to ask for repeats.

This scheme is known as forward error correction (FEC). The receiver does not ask for a retransmission when an error is detected, but it attempts to fix the errors by itself.

Convolutional codes are different from block codes in that they operate continuously on streams of data. They also have a memory, which means that the output bits do not only depend on the current input bits, but also on several preceding input bits. A convolutional code can therefore be described using the format (n, k, m), where n is the number of output bits per data word, k is the number of input bits, and m is the length of the coder memory. The code rate of the convolutional code is, thus, similar to block codes:

$$R_c = k/n$$

(6.3)

A convolutional coder $(3, 1, 9)$ is shown in Figure 6.4. It is a combination of shift registers (D) and XOR functional units. In the end of the data sequence

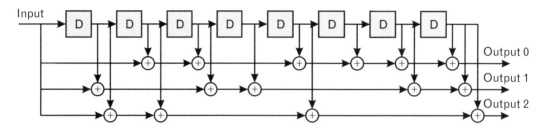

Generator polynomials: $G_0 = 557$ (octal)
$G_1 = 663$ (octal)
$G_2 = 711$ (octal)

FIGURE 6.4   *Convolutional 1/3 rate encoder.*

to be encoded, the convolutional coder adds $m - 1$ zeros to the output sequence. This is done periodically to force the encoder back to the initial state. Once a convolutional coder becomes overwhelmed with channel errors, it is impossible for it to recover from its confusion. Periodic resets solve the problem. The structure of the encoder is rather simple and its operation straightforward. The decoder, however, is something completely different.

The optimal convolutional decoder is the maximum likelihood sequence estimator (MLSE), which is based on the idea that for a finite sequence of bits, the receiver generates all possible sequences the encoder could have possibly sent. Next, the receiver compares the actual received bit sequence with each of the possible generated sequences and calculates the Hamming distance for each pair. The minimum Hamming distance should identify the most likely transmitted sequence.

The MLSE method provides the most efficient convolutional code decoder, but the problem is that once the size of the transmitted bit sequences increases, the complexity of the MLSE algorithm becomes unmanageable. A solution to this problem is to use the Viterbi algorithm, which estimates the MLSE algorithm well enough to still be efficient.

The actual theories behind the MLSE and Viterbi algorithms are outside the scope of this book. However, convolutional codes and maximum likelihood decoders are explained well in [3].

Convolutional decoders can be either hard or soft decision decoders. This distinction refers to the way the decoders receive the bit information from the demodulator. In the hard decision method the demodulator output is either a 0 or 1. In the soft decision method the demodulator returns not only the received bit (0 or 1), but also an estimation of the reliability of this decision. The estimation can be based on such things as the current received signal level. The decoder can then use the estimation data as one parameter in its maximum likelihood decoding algorithm.

Convolutional decoders work well against random errors, but they are quite vulnerable to bursts of errors, which are typical in mobile radio systems. The especially fast moving UEs in CDMA systems can cause bursty errors if the power control is not fast enough to manage the interference. This problem can be eased with interleaving, which spreads the erroneous bits over a longer period of time and, thus, makes the convolutional decoder more efficient.

6.5   Turbo Codes

Turbo codes are a relatively new invention, first discussed in 1993 [5]. They are found to be very efficient because they can perform close to the theoretical limit set by the Shannon's Law (see Section 6.2). Their efficiency is best with high data rate services; they are over-kill and not very efficient on low rate services.

In turbo coding the output of the decoding process is used to readjust the input data. This iterative process improves the quality of the decoder output, although the returns from the process diminish with every iterative loop. The original turbo code paper [5] used 18 loops.

The turbo encoder specified in the UTRAN is shown in Figure 6.5. This encoder is a parallel concatenated convolutional code (PCCC). It consists of two convolutional encoders in parallel separated by an interleaver. The encoders are recursive and systematic. This is not by any means the only

FIGURE 6.5
Turbo encoder.

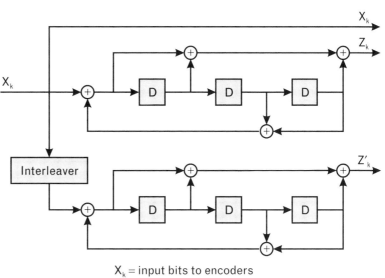

X_k = input bits to encoders
Z_k = output from 1st encoder
Z'_k = output from 2nd encoder
D = shift register

possible turbo coder implementation. The encoders can also be connected serially (serial concatenated convolutional coder [SCCC]), where the interleaver is between the encoders. The component encoders can also be block coders instead of convolutional encoders, in which case the turbo encoder type is a parallel concatenated block coder (PCBC) or a serial concatenated block coder (SCBC).

The task of the interleaver is to randomize the data before it enters the second encoder. The interleaver consists of a rectangular matrix. It performs both intra-row and inter-row permutations for the input bits, and the output bit sequence is pruned by deleting these bits, which were not part of the input bit sequence.

The turbo decoder is depicted in Figure 6.6. It consists of two soft-input-soft-output (SISO) decoders connected by interleavers and a deinterleaver. The extrinsic information is relayed back from the output of the second decoder to the input of the first decoder. Each iteration improves the estimate of the extrinsic information, which again improves the estimate for the decoded data. The data is processed in the iterative loop on a block-by-block basis. The size of the block can vary between 40 and 5,114 bits, inclusive of these values. Smaller amounts of data must be padded with dummy bits to achieve the minimum block size.

Turbo codes are discussed, for example, in [5–8]. The principles of turbo codes are rather complex; thus, most of the technical papers about the subject may be overwhelming for newcomers to the field. Probably for the same reason, the subject is also a popular research subject. As a starting point, there is an overview of turbo codes available from the Internet [9]; this is an unpublished paper by Professor William Ryan at New Mexico State University. An excellent "database" of turbo code research can be found on the Turbo Codes home page at the University of Virginia [10].

FIGURE 6.6
Turbo decoder.

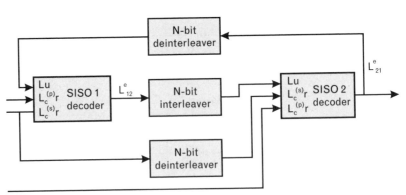

Lu = a priori values for all information bits u
L^e_{12} = extrinsic info from 1st to 2nd decoder
L^e_{21} = extrinsic info from 2nd to 1st decoder
$L^{(p)}_c r$ = parity information
$L^{(s)}_c r$ = systematic information

6.6   Channel Coding in UTRAN

Beyond the error checking performed by the block coding function, the UTRAN employs two FEC schemes: convolutional codes and turbo codes. Actually there is also a third scheme, which is no FEC coding at all. This makes a total of four channel coding mechanisms: block coding, convolutional coding, turbo coding, and no channel coding at all. Convolutional coding can be used for low data rates, and turbo coding for higher rates. Toskala [8] estimates that the suitable threshold value between these schemes is about 300 bits per TTI. At higher bit rates, turbo coding is more efficient than convolutional coding. Turbo coding is not suitable for low rates as it does not perform well on short blocks of data. Turbo codes also make blind rate detection more difficult. Note that the use of turbo codes in the UE is optional. The UE informs the networks about its capabilities so the network knows which codes to use.

Channel coding also increases the amount of bits to be sent, so it should not be used unnecessarily. More bits mean more interference. The channel coding schemes in UTRAN are shown in Table 3.1.

Note that both block codes and convolutional codes are used in the UTRAN. The idea behind this arrangement is that the channel decoder (either a convolutional or turbo decoder) tries to correct as many errors as possible, and then the block decoder (CRC check) offers its judgment on whether the resulting information is good enough to be used in the higher layers.

Channel coding in general is discussed in [11, 12].

REFERENCES

[1]   Shannon, C. E., "A Mathematical Theory of Communication," *Bell System Technical Journal*, Vol. 27, 1948, pp. 379–423, 623–656.

[2]   Black, U. D., *Data Networks: Concepts, Theory, and Practice*, Englewood Cliffs, NJ: Prentice Hall International, 1989, pp. 57–59.

[3]   Viterbi, A. J., *CDMA: Principles of Spread Spectrum Communication*, Reading, MA: Addison-Wesley, 1995.

[4]   Black, U. D., *Data Networks: Concepts, Theory, and Practice*, Englewood Cliffs, NJ: Prentice Hall International, 1989, pp. 256–259.

[5]   Berrou, C., A. Glavieux, and P. Thitimajshima, "Near Shannon Limit Error Correcting Coding and Decoding: Turbo-Codes (1)," *Proc. IEE Int. Conf. on Communications*, Geneva, Switzerland, May 1993, pp. 1064–1070.

[6]   3GPP TS 25.212, v 5.0.0, Multiplexing and Channel Coding (FDD), 2002.

[7]   Hagenauer, J., E. Offer, and L. Papke, "Iterative Decoding of Binary Block and Convolutional Codes," *IEEE Trans. on Information Theory*, Vol. 42, No. 2, March 1996, pp. 429–445.

[8]   Holma, H., and A. Toskala (eds.), *WCDMA for UMTS: Radio Access for Third Generation Mobile Communications*, New York: Wiley, 2000, pp. 101–102.

[9]   Ryan, W. E., "A Turbo Code Tutorial," available at http://www.ee.virginia.edu/re-search/CCSP/turbo_codes/tcodes-bib/turbo2c.ps, accessed May 20, 2002.

[10]   http://www.ee.virginia.edu/research/CCSP/turbo_codes/home.html.

[11]   Walke, B., *Mobile Radio Networks*, New York: Wiley, 1999, pp. 75–80.

[12]   Mehrotra, A., *GSM System Engineering*, Norwood, MA: Artech House, 1997, pp. 187–200.

CHAPTER 7

Wideband CDMA Air Interface: Protocol Stack

7.1   General Points

The unifying principle in the UTRAN development work has been to keep the mobility management (MM) and connection management (CM) layers independent of the air interface radio technology. This idea has been realized as the access stratum (AS) and nonaccess stratum (NAS) concepts (Figure 7.1). The AS is a functional entity that includes radio access protocols[1] between the UE and the UTRAN. These protocols terminate in the UTRAN. The NAS includes core network (CN) protocols between the UE and the CN itself. These protocols are not terminated in the UTRAN, but in the CN; the UTRAN is transparent to the NAS. The MM and CM protocols are GSM CN protocols; GPRS Mobility Management (GMM) and Session Management (SM) are GPRS CN protocols. Just as the NAS tries to be independent of the underlying radio techniques, so also have the MM, CM, GMM, and SM protocols tried to remain independent of their underlying radio technologies. The apparent dependence of these higher-layer protocols on the radio access protocols will be clarified later in this chapter.

The NAS protocols can be kept the same, at least in theory, regardless of the radio access specification that carries them. Thus, it should be possible to connect any 3G radio access network (RAN) to any 3G CN. This is a nice principle and a worthy goal, but in practice, its implementation is not simple. True independence among the layers in a protocol stack is difficult and expensive to implement. Time and budget constraints usually conspire to allow short cuts to appear in signaling implementations, which causes some interdependence to work itself into the details. The idea of separate access strata is, nevertheless, helpful in understanding the mechanisms and reduces development and testing costs.

One practical result of this concept is that GSM's MM and CM resources are used almost unchanged in 3G NAS. More precisely, the NAS

---

1   Protocol: "System of rules governing formal occasions" (Oxford English Dictionary). This is in fact a rather good description of a communications protocol. The UE and the network entities must have strict rules in their communication so that both entities know exactly what should be done in each occasion.

FIGURE 7.1
Stratum model.

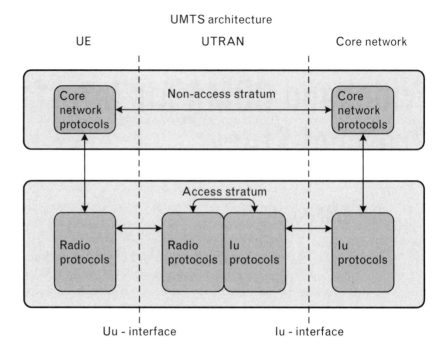

layers will be similar to the future GSM MM and CM layers. The reader should understand that some changes have to be made to the legacy GSM CN protocols to meet the future GSM requirements. The upgrades to the current GSM CN will allow support for both the GSM and the UMTS RANs; GSM must be transformed into one of the UMTS radio modes. Present-day GSM operators would not accept any other solution; they want access to 3G. Therefore, it is interesting to notice that future GSM enhancements are being specified in GSM/EDGE Radio Access Network (GERAN) working groups that are part of the 3GPP organization.

Because the CN protocols already exist in GSM and are hardly new developments for 3G as such, they are not discussed thoroughly here. If necessary, they can be easily studied from numerous GSM and GPRS books (e.g., [1–5]). A short overview of each of the tasks is, however, offered here for continuity.

The lower layers (from the AS) are, as the reader can imagine, quite different from GSM. The radio access technology (RAT) used in the UTRAN is CDMA, but in the GSM it is TDMA. From this difference it follows that the protocols used are also very different. The packet-based GPRS protocol stack is also quite different from the UTRAN because of the difference in the RAT, even though they both are packet-based techniques. The widely stated claim about GPRS being a steppingstone to 3G is actually true, but only from the network and marketing points of view. GPRS actually provides a halfway step to a UMTS solution for the network infrastructure because the GPRS CN components can, in many cases, be reused in a 3G network. On the mobile station side, however, the truth is quite different, as a GSM/GPRS mobile and a WCDMA mobile don't have much in common in their protocol stacks, at least not in the AS parts of them. GPRS is also an important marketing test, allowing the wireless industry to see how subscribers accept new nonvoice content and applications. Think of UMTS as a GPRS network with an advanced and highly adaptive radio interface.

7.2 Control Plane

Radio interface protocols can be divided into two categories: horizontal layers and vertical planes (Figure 7.2).

There are three protocol layers in the AS: physical layer (L1), data-link layer (L2), and network layer (L3). The data-link layer can be further divided into several sublayers: medium access control (MAC), radio link control (RLC), broadcast/multicast control (BMC), and packet data convergence protocol (PDCP). The network layer also includes several sublayers, but among these only the radio resource control (RRC) belongs to the AS. The other sublayers within the network layer are part of the NAS (CN) protocols; these appear above the dotted line in Figure 7.2.

There are also two vertical planes; the control (C) and user (U) planes. The MAC and RLC layers exist in both the C and U-planes. The RRC is

FIGURE 7.2
Protocol tasks in the UTRAN air interface.

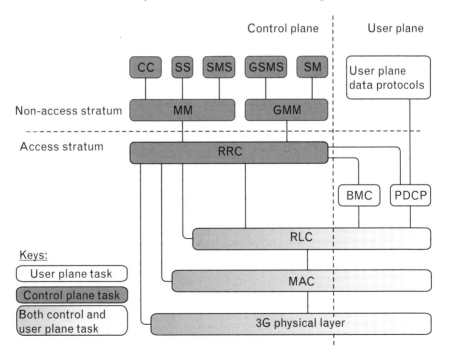

found only in the C-plane (i.e., RRC = Radio Resource Control), and the BMC and PDCP are found only in the U-plane. The C-plane carries control data, information that is needed by the protocol tasks to run the system. The U-plane, on the other hand, carries data that is generated by the user, or by a user application. The U-plane data is typically digitally coded voice, but increasingly also other forms of data.

All of these aspects are explained in subsequent chapters: first the C-plane (Figure 7.3) and then the U-plane protocols. The RRC gets the most thorough treatment in this book because it also manages the other protocol layers in the AS. Understanding the RRC is an essential prerequisite to understanding the air interface's inner workings. Each protocol layer performs strictly defined functions, possibly exchanging information with other layers via protocol interfaces.

In the following sections, try to notice the differences between the concepts of function and service. A function is something a protocol does for itself. This may require communication with its peer task and exploitation of the services provided to it by the layers below it. A service is something that is provided to higher protocol layers as a result of the functions of the protocol task itself. It is thus quite possible that the same process can be classified as both a function and a service.

FIGURE 7.3
WCDMA C-plane protocol stack.

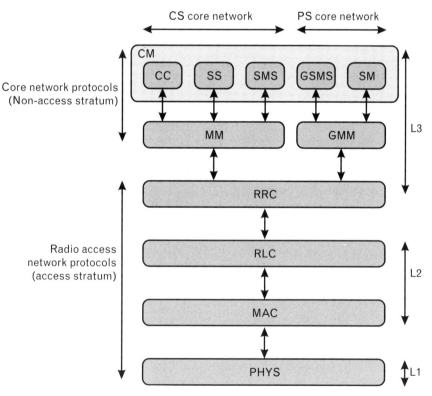

7.3 MAC

The UTRAN MAC is not the same protocol as the GPRS MAC, even though they both have similar names and handle similar tasks in similar ways. The UTRAN MAC can even contain different functionalities depending on whether it supports FDD, TDD, or both modes.

The MAC is not a symmetric protocol; the entities in the UE and in the UTRAN are different. A MAC task contains several different functional entities that are depicted in Figure 7.4. Note that this figure depicts the UE MAC. The UTRAN MAC is slightly different from the UE MAC and more complex.

- MAC-b handles the broadcast channel (BCH). The UTRAN has one MAC-b for each cell; the UE may have one or multiple MAC-b's, depending on the implementation. Several MAC-b's may be used for receiving neighbor cell BCHs. This entity is active in the downlink direction only. Note that in the UE this entity will be very simple.

- MAC-c/sh deals with common and shared channels except the HS-DSCH. It handles the paging channel (PCH), the forward access channel (FACH), the random access channel (RACH), and the downlink shared channels (DSCH). The uplink common packet channel (CPCH) in the FDD mode and the uplink shared channel (USCH) in the TDD mode are also handled by this entity. One MAC-c/sh exists in each UE and one exists in the UTRAN for each cell.

- MAC-d handles dedicated logical channels and the dedicated transport channels. The UE has one MAC-d, and the UTRAN has one MAC-d for each UE with assigned DCHs.

- MAC-hs handles the HSDPA functionality. The HS-DSCH is a high-speed downlink shared channel. The UE has one MAC-hs if it is HSDPA-capable; the UTRAN has one MAC-hs for each cell that supports HS-DSCH. Note that a UE does not have to support HS-DSCH and DSCH reception simultaneously. MAC-hs is a bit of a special case among other functional entities because it works with 2-ms subframes, whereas the other entities use 10-ms frames. This tight timing constraint also means that especially the HARQ function control cannot be handled via higher-layer protocols as usual, but must be handled directly from layer 1. In Figure 7.4 this is depicted as the associated downlink signaling. This data flow comes from an HS-SCCH physical channel. Correspondingly, the associated uplink signaling is mapped into an HS-DPCCH physical channel. From a

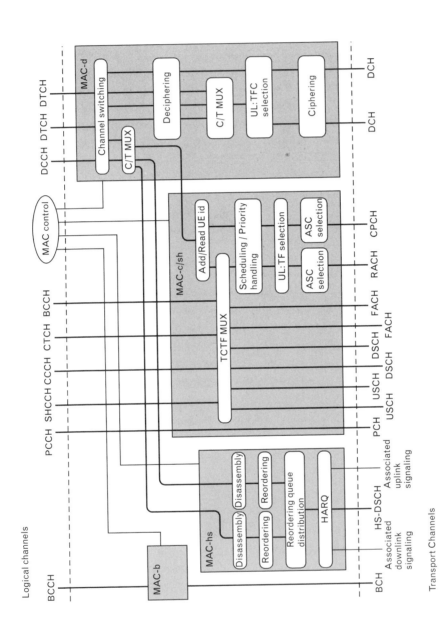

FIGURE 7.4 *MAC protocol layer functional entities.*

protocol stack architecture perspective, it would have been clearer to name these data flows as new transport channels.

The MAC operates on transport channels (see Section 3.2.2) between the MAC and layer 1. The logical channels are described between the MAC and RLC in Section 3.2.1. The internal configuration of the MAC is controlled by the RRC. The MAC is the lowest sublayer in layer 2; it has a thorough understanding of how to manipulate the physical layer on behalf of the layers above it.

7.3.1   MAC Services

The services MAC provides to the upper layers include the following:

- Data transfer;

- Reallocation of radio resources and MAC parameters;

- Reporting of measurements to RRC.

7.3.2   MAC Functions

MAC functions include the following:

- Mapping between logical channels and transport channels;

- Selection of the appropriate transport format for each transport channel depending on the instantaneous source rate;

- Priority handling between data flows of one UE;

- Priority handling between UEs by means of dynamic scheduling;

- Identification of UEs on common transport channels;

- Multiplexing/demultiplexing of higher-layer PDUs into/from transport blocks delivered to/from the physical layer on common transport channels;

- Multiplexing/demultiplexing of higher-layer PDUs into/from transport block sets delivered to/from the physical layer on dedicated transport channels;

- Traffic-volume monitoring;

- Transport-channel type switching;

- Ciphering for transparent RLC;

- Access service class selection for RACH and CPCH transmission;

• HARQ functionality for HS-DSCH transmission;

• In-sequence delivery and assembly/disassemby of higher layer PDUs on HS-DSCH.

Some of these functions are further explained in the following sections.

7.3.2.1   Priority Handling Between Data Flows of One UE

The priority of a data flow is used when the MAC layer chooses suitable transport format combinations (TFCs) for uplink data flows. Higher-priority data flows can be given higher bit rate combinations, and low-priority flows may have to use low bit rate combinations. A low bit rate can also mean a zero bit rate.

Note that there is not a single priority parameter attached to a data flow, but MAC has to derive it from at least two sources: the buffer occupancy parameter received from the RLC and the MAC logical channel priority received from the RRC.

At radio bearer setup/reconfiguration time, each logical channel involved is assigned a MAC logical channel priority (MLP) in the range 1, ..., 8 by the RRC. The details of the TFC selection algorithm are not defined in the MAC specification. Rather, the specification gives a list of constraints the algorithm implementation has to fulfill.

7.3.2.2   Identification of UEs on Common Transport Channels

If a UE is addressed on a common downlink channel or it uses the RACH, the UE is identified by the MAC layer. There is a UE identification field in the MAC PDU header for this purpose. If the message was addressed to this UE, it is routed further to the RLC, and from there, either to the RRC, the BMC, or the PDCP. Other messages are trashed.

7.3.2.3   Traffic-Volume Monitoring

The UTRAN-RRC layer performs dynamic radio access bearer (RAB) control. Think of the RRC as a kind of mediator between the network and the radio interface. The MAC layer is obliged to eventually react in some appropriate way to the RRC's needs. Based on the required traffic volume, the RRC can decrease or increase the allocated capacity. The task of monitoring the traffic volume is allotted to the MAC. The UE-MAC layer monitors the uplink transmit buffer, and the UTRAN-MAC layer does the same for the downlink buffer. If the queue in either entity goes out of range, the corresponding RRC is notified. The UE-RRC must further notify the UTRAN-RRC. It is the UTRAN-RRC that has to make decisions about radio resource allocations because only the UTRAN-RRC knows the total load situation of the whole system.

The monitoring of the traffic volume is controlled by the RRC. It may command the MAC to perform either periodic or event-triggered monitoring. In the case of periodic monitoring, the MAC sends a new report periodically after a timer has expired. In the case of event-triggered monitoring, the RRC gives a range of allowed buffer values, and once the transmission queue goes out of range, an alarm indication is sent back to the RRC (see Figure 7.5).

The purpose of the traffic-volume monitoring procedure is to allow for efficient radio resource usage. If the allocated resources are not sufficient for the generated traffic, the UTRAN may reconfigure itself and add resources. This may mean allocating a DCH instead of a shared channel or simply reducing the SF on a particular channel. Similarly, if the traffic-volume monitoring shows that the allocated resources are underutilized,

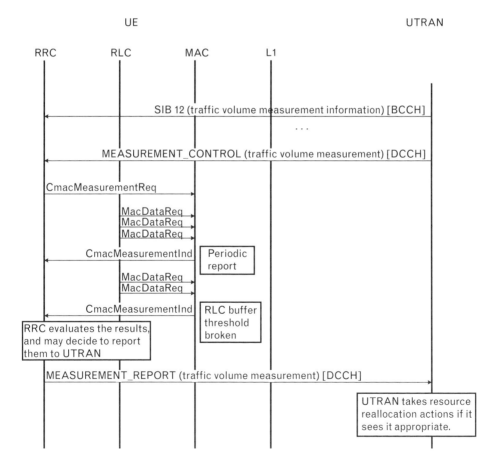

FIGURE 7.5 *Traffic-volume monitoring.*

the UTRAN may reconfigure the connection from a dedicated resource to a shared resource, or increase the current spreading factor.

Note that the transmission buffer to be monitored is actually in the RLC layer, but the buffer occupancy information is relayed down to the MAC layer with each MacDataReq signal. See Section 7 of [6], for a description of the traffic-volume-monitoring interlayer procedure.

7.3.2.4 Transport-Channel Type Switching

The MAC executes the switching between common and dedicated transport channels based on a switching decision made by the RRC. In the UE the dynamic-transport-channel-type switching function maps and multiplexes the DCHs (DCCH and DTCH) into logical channels. Note that in 3GPP jargon, the function *mapping between logical channels and transport channels* refers to a functionality in the MAC-c/sh, which has a more static nature, and *transport-channel type switching* refers to the more dynamic functionality in MAC-d.

7.3.2.5 Ciphering for Transparent RLC

If the RLC layer is in transparent mode (i.e., it is just a "pipe" between the PDCP and the MAC), then ciphering must be done in the MAC layer. Otherwise, it will be performed in the RLC layer. Ciphering prevents the unauthorized interception of data. The ciphering algorithm to be used is the same in the MAC and in the RLC; that is, the UE does not have to use more than one ciphering algorithm at a time. However, the ciphering algorithm may be changed according to commands from the UTRAN. See Section 7.5.2.16 for a further description of ciphering.

7.3.2.6 Access Service Class Selection for RACH Transmission

The MAC gets a set of access service classes (ASCs) from the RRC, and it chooses one of them to be used for the RACH transmission. These classes define the parameters used in a RACH procedure, including access slots and preamble signatures. The algorithm itself uses two variables: MAC logical channel priorities (MLP) and the maximum number of ASCs (NumASC). The MinMLP parameter is set as the highest logical channel priority assigned to the logical channel in question (note that a smaller MLP number means a higher priority; the range is from 1 to 8). The ASC number is obtained as follows:

If all the transport blocks in a transport block set have the same MLP, then select: $\text{ASC} = \min(\text{NumASC}, \text{MLP})$.

If the transport blocks in a transport block set have different MLPs, then select: ASC = min(NumASC, MinMLP).

The ASC enumeration is such that it corresponds to the order of priority (ASC 0 = the highest priority; ASC 7 = the lowest priority). ASC 0 would only be used for very important reasons, such as emergency calls.

7.3.2.7 Hybrid ARQ Functionality for HS-DSCH Transmission

Hybrid ARQ (HARQ) is an acknowledged retransmission scheme that is employed on the HS-DSCH channel. In the UE this functionality is relatively simple as the HARQ entity only has to check the correctness of the received packet and send either a positive or a negative acknowledgment back to the peer HARQ entity in UTRAN. However, the UTRAN HARQ has more complex duties. It has to take care of at least the following tasks:

- *Scheduling of data.* There are probably several active HSDPA UEs in the cell, and their data transmission processes may have different priorities. The scheduler has to select which data is sent first. Note that retransmitted data is probably of higher priority than data that is transmitted for the first time.

- *Buffering of data.* Because HS-DSCH employs acknowledged data transmission, the transmitting entity cannot discard the data as soon as it has been transmitted, but it must be buffered until a positive acknowledgment has been received

- *Retransmission functionality.* If a negative (or no) acknowledgment is received for a packet, the packet must be retransmitted. Because HARQ employs link adaptation, the retransmission may use a different modulation scheme, and a different redundancy version. This is to increase the probability of a successful transmission. Note that the UTRAN cannot select these quantities at will, but the result must comply with the allowed transport format combinations. In the case of HS-DSCH, the transport format selection is different from other channels because here the dynamic part of a transport format includes also the modulation scheme and the redundancy version.

Note that a failed packet will not be retransmitted forever in MAC-HARQ. If the data is important enough, it will also be protected by a higher layer (RLC) retransmission protocol, which takes care of the problem if it does not receive a positive acknowledgment in time.

7.3.2.8   In-Sequence Delivery and Assembly/Disassembly of Higher-Layer PDUs on HS-DSCH

Because of the HARQ retransmission protocol, it is possible that the UE receives the data packets via HS-DSCH in an order other than that in which they were originally transmitted. Thus, there has to be a reordering buffer in the UE. Assembly and disassembly functions are needed because the data packets in the HS-DSCH are probably a different size than in the RLC layer buffers. HS-DSCH is optimized for very high-speed data transfer; thus, the packets on this channel are typically very large.

7.3.3   TFC Selection

TFC selection is in fact a process that combines several functions from the earlier list. First some definitions:

- *Transport format* (TF) defines what kind of data and how much is sent on each transport channel in each transport time interval (TTI). TTI length is equal to the duration of a radio frame or a multiple of it.

- *Transport format combination* (TFC) is a set of TFs that are sent simultaneously (within the same TTI) on different active transport channels to or from the same UE. Indirectly, TFC gives the data rate used.

- *Transport format combination set* (TFCS) is a set of TFCs. The UE has to select one TFC from a set of allowed TFCs for data transmission in each TTI. The TFCS to be used is signaled to the UE via RRC signaling, but this set can be limited later by several different network procedures. As a result, only some TFCs from the original set are allowed TFCs at a given time.

The MAC layer has to choose a set of TFs, so that given the current channel conditions, the maximum amount of highest-priority data could be transmitted over the air interface. This is not a simple task. The MAC layer itself knows from the configuration data which transport formats and which combinations of transport formats are valid. However, not all such combinations are usable all the time. The current channel conditions could impose limitations on what TFCs can be used. Those combinations that could carry the highest amount of data also need the highest transmit power in the physical layer. This could be more than the maximum allowed transmit (TX) power the transmitter can use. In a CDMA system, more data basically means more power. And the more noise there is in the radio interface, the higher the transmitting power must be. Thus, it is quite conceivable that especially in a noisy environment, only some of the TFCs can be used. The UTRAN can signal a temporary TFC limitation to a UE via RRC layer signaling. But still, this does not remove the need to monitor the required TX power and limit the TFCs further if necessary.

On the other hand, the data to be transmitted is in the data buffers in the RLC layer. MAC has to get the buffer occupancy information from RLC, as well as the priority of the data in those buffers. It has to send as much data as possible, and at as high a priority as possible. And obviously, the MAC layer cannot send more data than there is in RLC buffers. Note that the data in RLC buffers is not a stream of bits, but a group of PDUs, and those cannot be divided or combined at will. Moreover, the MAC layer is not allowed to choose TFCs that require the RLC layer to add padding bits to its PDUs to make them match with the chosen TFC (i.e., to choose too-large TFCs). Further complexity is caused by the compressed mode as this will cause the transmitter to either send less data or use more power, in which case there is again the danger of exceeding the maximum allowed TX power. Furthermore, many applications employ variable rate adaptive codecs, and the MAC layer has to cooperate with them so that the produced bit rate exactly matches with some allowed TFC. And to make all this a bit more challenging, the reader must remember that the TFC selection must be made once every 10 ms, that is, the length of the radio frame. In fact, the selection frequency is equal to the length of the shortest configured TTI duration, so it could be 10, 20, 40, or 80 ms. But still, the algorithm must be based on the worst case, that is, 10 ms.

The TFC selection algorithm is not, and will not be, specified by the 3GPP. Only the TFC selection criteria are given in the MAC specification [7], and it can be implemented in a more or less efficient way. TF selection must be done on all DCHs, and also on RACH and CPCH channels.

The MAC protocol is specified in [7]. TFs were discussed in this book in Section 3.5.

7.4 RLC

One RLC task contains several different functional entities. For bearers using the transparent mode service or the unacknowledged mode (UM) service, there is one transmitting and one receiving entity for each bearer. For bearers using the acknowledged mode (AM) service, there is only one combined transmitting and receiving entity for each bearer. Different modes are used for different types of data. If the data is of an important nature, it needs lots of protection and AM service. On the other hand, some data is not suitable for AM service. For example, it is no good to use AM for voice. The AM retransmission protocol could guarantee that a voice packet does get through eventually, but a retransmitted voice packet cannot be used anymore because of the additional delay. A voice packet must be received in time without delays or it is worthless.

In general, the RLC layer is in charge of the actual data packet (containing either control or user data) transmission over the air interface. It makes sure that the data to be sent over the radio interface is packed into suitably sized packets. The RLC task maintains a retransmission buffer, performs ciphering, and routes the incoming data packets to the right destination task (RRC, BMC, PDCP, or voice codec).

The transparent mode is used for the BCCH, PCCH, SHCCH, DCCH, DTCH, and CCCH channels. For the CCCH and SHCCH, the transparent mode is used only in the uplink direction. Transparent mode means that very little processing is done to the data in the RLC. It contains transmission and receiver buffers and also, in some cases, segmentation and reassembly functions. Note that no RLC header is added to data units in the transparent mode (see Figure 7.6).

Despite this rather limited functionality, one instance of an RLC transparent entity is needed per direction and per bearer—one for the uplink and one for the downlink.

UM is used for the DCCH, DTCH, CTCH, and the downlink SHCCH and CCCH channels. The RLC adds a header to the PDU and ciphers/deciphers it. As in transparent mode, one instance is needed per direction and per bearer (see Figure 7.7).

AM can be used for the DCCH and DTCH channels. The SDUs are segmented or concatenated onto the PDUs of fixed length.

The multiplexer (MUX) chooses the PDUs and decides when they are delivered to the MAC. The MUX may, for example, send RLC control

FIGURE 7.6
Transparent entities in RLC.

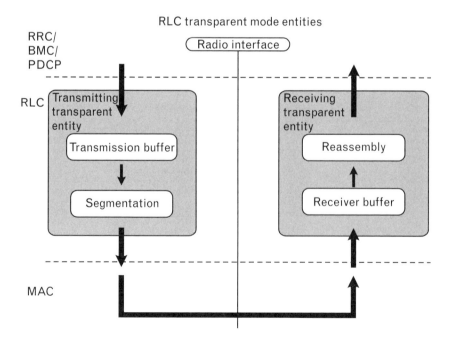

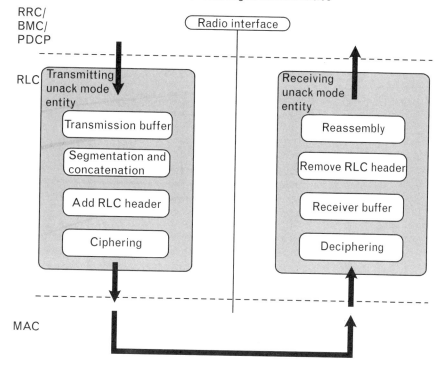

FIGURE 7.7
UM entities in RLC.

PDUs on one logical channel and data PDUs on another logical channel, or it may send everything via one logical channel.

If the data in AM mode does not fill the whole PDU, then padding is used to fill the rest of the PDU. This padding can be replaced with piggy-backed control information in order to increase the transmission efficiency.

There is only one AM entity per bearer in the UE that is common to both the uplink and the downlink (see Figure 7.8).

7.4.1 RLC Services

The following are services provided to upper layers:

Transparent Data Transfer Service

- Segmentation and reassembly;
- Transfer of user data;
- SDU discard.

Unacknowledged Data Transfer Service

- Segmentation and reassembly;

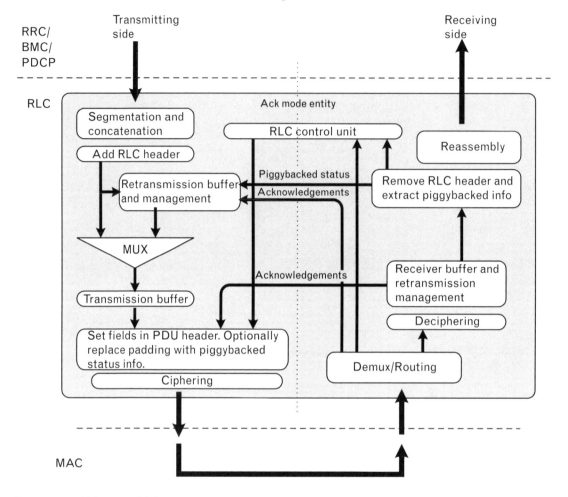

FIGURE 7.8   *AM entity in RLC.*

- Concatenation;
- Padding;
- Transfer of user data;
- Ciphering;
- Sequence number check;
- SDU discard.

Acknowledged Data Transfer Service

- Segmentation and reassembly;

- Concatenation;
- Padding;
- Transfer of user data;
- Error correction;
- In-sequence delivery of higher-layer PDUs;
- Duplicate detection;
- Flow control;
- Protocol error detection and recovery;
- Ciphering;
- SDU discard.

Maintenance of Quality of Service (QoS) as Defined by Upper Layers

Notification of Unrecoverable Errors

7.4.2 RLC Functions

The following functions are supported by the RLC:

- Segmentation and reassembly of higher-layer PDUs into/from smaller RLC payload units;
- Concatenation (RLC SDUs may be concatenated so that they will fill the RLC PUs);
- Padding;
- Transfer of user data;
- Error correction;
- In-sequence delivery of higher-layer PDUs;
- Duplicate detection;
- Flow control;
- Sequence number check (in unacknowledged data transfer mode);
- Protocol error-detection and recovery;
- Ciphering (in UM and AM modes);
- Suspend/resume function.

The RLC protocol is defined in [8].

7.5   RRC

We turn now to the most important subject of this chapter: the RRC. The RRC controls the configuration of the lower layers in the protocol stack, and it has control interfaces to each of the lower layers (PDCP, BMC, RLC, MAC, and layer 1). It is the conductor of the protocol stack orchestra.

7.5.1   RRC Services

The RRC provides the following services to the upper layers:

- *General control.* This is an information broadcast service. The information transferred is unacknowledged, and it is broadcast to all mobiles within a certain area.

- *Notification.* This includes paging and notification broadcast services. The paging service broadcasts paging information in a certain geographical area, but it is addressed to a specific UE or UEs. The notification broadcast service is defined to provide information broadcast to all UEs in a cell or cells. Note that the notification broadcast service seems to be quite similar to the general control service.

- *Dedicated control.* This service includes the establishment and release of a connection and the transfer of messages using this connection. These connections can be both point-to-point and group connections. Message transfers are acknowledged.

7.5.2   RRC Functions

The RRC functions include the following:

- Initial cell selection and cell reselection (includes preparatory measurements);
- Broadcast of information (system information blocks [SIBs]);
- Reception of paging messages;
- Establishment, maintenance, and release of RRC connection;
- Establishment, reconfiguration, and release of radio bearers;
- Assignment, reconfiguration, and release of radio resources for the RRC connection, which includes such things as the assignment of codes and CPCH channels;
- Handovers (HOs), which include the preparation and execution of HOs and intersystem HOs;

- Measurement control;

- Outer-loop power control;

- Security mode control (ciphering control, integrity protection, counter check);

- Routing of higher-layer PDUs (direct transfer);

- Control of requested QoS;

- Support for DRAC (fast allocation of radio resources on the uplink DCH);

- Contention resolution in the TDD mode;

- Timing advance in the TDD mode;

- Management of the CBS service (the service itself is implemented in BMC);

These functions are explained in the following sections. Some are quite similar to the corresponding GSM procedures, but there are also some new functions. When we compare the RRC to the GSM-RR functions, the most important changes include SHOs, intersystem HOs, and more flexible channel configuration management, which yields a more efficient usage of the available resources.

Some of these functions are depicted in the signaling flow diagrams in Chapter 11. The RRC protocol is specified in [9].

7.5.2.1   Initial Cell Selection

The initial cell selection, as well as other cell-evaluation procedures, is quite different from the GSM cell-selection procedures. Cell-selection procedures are discussed in [10], which offers a rather cryptic presentation. The purpose of the initial cell-selection procedure is to find a cell, not necessarily the best cell, but a usable cell, for the UE to camp on after power-on.

In the UTRAN, the number of carrier frequencies is quite small. One operator typically operates only on two or three frequency carriers. In the first phase of UMTS in Europe, the frequency allocation for UMTS-FDD is 2×60 MHz (uplink/downlink), which means that there can be, at most, only 12 carrier frequencies of 5-MHz bandwidth each. These carriers are then divided between up to six operators. Each carrier will only support one operator. This obviously forces the operators to coordinate their network-planning activities near national borders because the same frequency can be used by different operators in adjacent countries.

The specifications do not accurately dictate how the initial cell-selection procedure should be implemented; it is left for the UE manufacturers to decide. Most of the functionality, however, has to be in the physical layer, and the RRC layer has only a management role. The initial cell-selection procedure is performed on one carrier frequency at a time until a suitable cell is found. In principle the process includes the following:

1. Search for primary synchronization channels (P-SCHs);
2. Once such a channel is found, acquire time-slot synchronization from it;
3. Acquire frame synchronization from the corresponding S-SCH;
4. Acquire the primary scrambling code from the corresponding CPICH;
5. Decode system information from the cell to check whether it is a suitable cell for camping (i.e., it contains the right PLMN code and access to it is allowed).

The synchronization process in the physical layer is explained in Section 3.1.9. Here the issue is considered from the whole UE point of view.

All P-SCHs have the same fixed primary synchronization code. The search procedure should yield a set of P-SCHs in the area. Because the P-SCH is only transmitted during the first 256 chips of each time slot, the beginning of its transmission also indicates the start of a time slot in the corresponding cell.

In the second phase of the process, the received signal is correlated with all possible secondary synchronization code (S-SCH) words on the S-SCH. There are 16 different SSCs, and these can be combined into 64 different code words, each with a length of 15 SSCs. Once the right code word is found, this gives the UE the frame synchronization and the code group identity, which indicates eight possible primary scrambling codes for the control channels.

The third phase of the procedure consists of finding the right primary scrambling code for this cell. Each candidate cell's primary scrambling code (there are eight of them as shown in the second phase) is applied, in turn, to the common pilot channel (CPICH) of that cell. Because the CPICH carries a predefined bit/symbol sequence, the UE knows when it has found the correct primary scrambling code. The resolved primary scrambling code can then be used to detect the CCPCH, which carries the BCH, which contains the system information the UE is seeking. There are various ways to optimize this procedure to make it quicker.

Note that phase five actually contains another major procedure, PLMN (i.e., the operator) selection. PLMN is identified by a PLMN code, a number that is transmitted on the BCCH channel of that network. A UE tries to find its home PLMN, the operator it has a contract with. In principle, a UE should first scan through all UTRAN frequencies until a good

PLMN is found, and then start an initial cell-selection process on that frequency. Note that one frequency can only be used by one operator (except in areas near country borders). However, while looking for the right PLMN code, the UE has already obtained all the necessary information for camping on a suitable cell, and no new scanning procedure is necessary once the correct PLMN is found. The situation is different if the UE is roaming abroad, and the home PLMN is not found. In that case RRC has to report all available PLMNs to NAS and wait for its selection decision, which can be either automatic or manual (user selection). This is time consuming, and many readers may have noticed this phenomenon when arriving at an airport in a new country and switching their GSM phones on. It may take a very long time before the phone registers to a network, especially if the phone is a multimode model with several frequency bands to scan. PLMN selection process is probably best described in [11].

The initial cell-selection process is repeated as many times as necessary until the first suitable cell is found for camping. Once the UE has managed to camp on a cell, it decodes the system information from it, including the neighbor cell list. This information can be used to help the UE find the best cell to camp onto. Note that the initial cell-selection procedure only found a cell to camp on (the first possible cell). It is possible that this cell will not be the best possible cell. For example, there could have been other frequencies including better cells for this particular UE that had not yet been scanned.

The neighbor cell list immediately tells the UE which frequencies and neighbor cells should be checked while the best possible cell is being searched for. The list includes additional information that can be used to optimize the cell-synchronization procedure, information such as the primary scrambling codes and timing information (optional, relative to the serving cell). With this information it should be possible to quickly descramble the CPICH from a neighbor cell.

From the CPICH it is possible to calculate the received chip energy-to-noise ratio (Rx Ec/No) for this cell. This measurement is acquired for each neighbor cell in the list. Based on this information, the UE can determine whether there are better cells available. From a possible candidate cell, the UE must decode the system information to check that it is not barred for access.

If the neighbor cell list contains cells from another RAT—for example, GSM cells—and the serving cell quality level is worse than the Ssearch parameter, then the GSM cells must be taken into consideration in the cell-reselection procedure.

The initial cell-selection procedure described here is to be used in case there is no information on the current environment stored in the UE. However, normally the UE starts the cell selection with a stored information cell-selection procedure. The UE may have stored the necessary information of the cell it was previously camped on, such as frequency and scrambling code. The UE may first try to synchronize into that cell, and if it fails, it may trigger the initial cell selection.

7.5.2.2   Cell Reselection

The cell-reselection procedure, or as the 3GPP calls it, the cell reselection evaluation process, is performed in idle mode to keep the UE camped on a best cell. If the UE moves or the network conditions change, it may be necessary for the UE to change the cell it is camped on. This procedure checks that the UE is still camped on the best cell, or at least on a cell that is good enough for the UE's needs.

In normal idle mode, the UE has to monitor paging information and system information and perform cell measurements. The cell-reselection procedure will be triggered if the measurements indicate that a better cell has been found, or if the system information of the current cell indicates that new cell access restrictions are applied to the cell in question, such as cell barred. [2]

System information block 3 (SIB3) is an important message here because it tells the UE the quality parameter to measure, and also all the parameters for the cell-reselection evaluation algorithm.

The neighbor cells to be measured are given in the neighbor cell list (SIB11). The results of these measurements are evaluated periodically.

System information (SIB3) may also contain various optional threshold parameters that define when to perform various measurements (S_x is the measured quality parameter of the serving cell):

- If $S_{intrasearch}$ is given and $S_x \leq S_{intrasearch}$, then the UE must perform intrafrequency measurements.

- If $S_{intersearch}$ is given and $S_x \leq S_{intersearch}$, then the UE must perform interfrequency measurements.

- If $S_{searchRAT\ n}$ is given and $S_x \leq S_{searchRAT\ n}$, then the UE must perform inter-RAT measurements

If a threshold parameter is not given, then the UE must always perform the corresponding measurements.

Based on these measurements the UE periodically evaluates the best-cell status. If it seems that there is a better cell available, it will trigger a cell-reselection procedure.

---

2   An operator can bar a cell, for example, during maintenance or testing. Operators may have special test mobiles that can ignore the cell bar-flag and these can be used to test the base station functionality while access is refused for other mobiles.

7.5.2.3   Cell Reselection and Random Access

There is a potential problem with cell-reselection and network-access attempts. In the idle mode, the UE cannot monitor its environment continuously. That would quickly wear the battery down. Instead, the monitoring process is periodic, and the periodicity is set by the network. The longer the period, the bigger is the danger that the UE may not be camped on the best possible cell. Normally, this would not be a problem as the timers used here are pretty short and the next measurement process will fix the problem by triggering a reselection.

However, problems may be caused if the UE decides to launch an access attempt while it is not attached to the best cell. An access attempt includes the transmission of RACH bursts to a serving Node B. A WCDMA system is sensitive to interference, and sending RACH bursts to the "wrong" Node B could introduce severe and unnecessary interference to any Node B close to the UE. The UE ramps up the RACH transmission power until it receives a response from the base station, and if this RACH burst is addressed to a Node B far away from the UE, the received signal power in the nearby Node Bs will be unacceptably high. Notice that the RACH signal is interference for all other Node Bs except for the one to which it is addressed. A possible scenario could involve a user who walks around a corner and immediately starts a call. The mobile may still be camped on the old cell, although there might be a much better one available on the new street. If the RACHing process is done on the old cell, the new cell could be easily blocked because of it.

The very first 3GPP specifications tried to solve this problem with a procedure called the *immediate cell evaluation*. It was a procedure that was to be performed just prior to a random-access procedure. The purpose for this procedure was to make sure that the UE is camped on the best possible cell before it starts to send RACH bursts to the network. The problem with the immediate cell evaluation procedure was the delay it caused. The sending of RACH bursts should be started quickly, especially if the call setup procedure was triggered by a paging message. Immediate cell evaluation took some time, especially if as a result of the evaluation a cell reselection was required. The UE would then have to decode at least part of the SIBs in the new cell before it could start to send RACH bursts there because it needs to know the random-access parameters to be applied in the new cell.

Because of these problems, immediate cell evaluation was removed from the specifications. Now, if the random-access procedure fails, the UE is required to trigger a cell-reselection procedure immediately. However, this solution does not remove the problem described earlier. We can only hope that the presented scenario is rare enough.

7.5.2.4   Broadcast of System Information

Broadcast information (or system information, as it is also known) is information about the system and the serving cell that is sent by the network in a point-to-multipoint manner; the information is broadcast to all UEs. It is typically information that is common to all mobiles in a cell; thus, it can be sent using a broadcast service.

Broadcast information consists of messages called system information blocks (SIBs). A SIB contains system information elements of the same nature. There are 18 different blocks, named SIBs 1 through 18. In addition to the SIBs, there is a master information block (MIB) and up to two scheduling blocks (SBs).

Broadcast information is sent via the BCCH logical channel, which is mapped into the BCH in the idle mode and in the CELL_FACH, CELL_PCH, and URA_PCH in connected mode substates. However, in the CELL_FACH substate, the UE can also receive SIB 10 via the FACH.

SIBs are sent according to a certain schedule. The blocks that are more important than others are sent more often, and the blocks of lesser importance are sent less often. The schedule is not fixed, but it can be adjusted by the UTRAN according to the current loading situation. This provides a great deal of flexibility for air interface management.

A mobile station must find out the schedule of various SIBs so that it can wake up and receive only those blocks it needs and skip reception of the others. This is possible because the blocks are arranged as a tree. This tree always starts from a MIB, which must be received and decoded first. The location of a master block is easy to determine because in the FDD mode, it has a predefined repetition rate (8), and a position (0) within the repetition cycle. This means that once a mobile knows the current frame number (it is sent in every block), it can compute the cell system frame number (SFN) mod 8, and find out the position of this block within the 8-block rotation. In the TDD mode the MIB repetition cycle can be 8, 16, or 32 frames. The value that the UTRAN is using is not signaled; the UE must determine it by trial and error.

The MIB indicates the identity and the schedule of a number of other SIBs. These SIBs contain the actual system information the UE is looking for. The MIB may optionally also contain reference and scheduling information for one or two SBs, which give references and scheduling information for additional SIBs. Scheduling information for a SIB may be included only in the MIB or one of the SBs. This is depicted in Figure 7.9. The mobile must maintain this tree in its memory, so that it can decode only those blocks it needs and skip the rest. Note that this arrangement saves power and also provides the UTRAN the possibility to add new types of SIBs to the protocol if such are needed later. Based on the experiences with GSM, new system information messages are not just possible, but likely.

FIGURE 7.9
SIB scheduling tree.

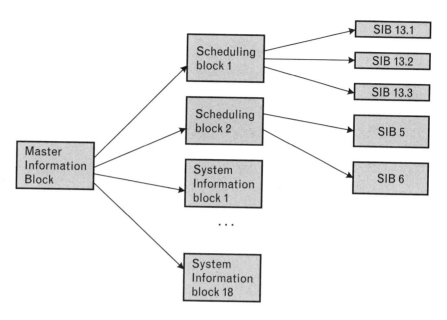

They can be added in later phases of the system as new services and functions are needed. If a mobile finds schedules of blocks it does not recognize, it simply ignores them. Other mobiles with updated protocol software can, however, use these. If a mobile notices that the schedule in its memory does not match the schedule used by the UTRAN, it must delete the stored schedule and start building the scheduling tree again beginning from the MIB.

The network may indicate that some information in a SIB has changed by setting the update flag (value tag) in a higher block; that is, in the same block that contains the schedule for this block. Once this tag changes, the mobile knows that it should recover the corresponding system information again.

Because of the tree structure of the scheduling information, the update flag scheme is always reflected to the value tag of the MIB; that is, if any SIB changes, then the MIB also changes. To keep the mobile from decoding the MIB continuously just to find out whether any value tag has changed, the value tag of the master block itself is sent on the paging channel. This information is sent in the BCCH modification information element within the paging type 1 message. The mobile has to monitor the paging channel in any case, or it couldn't receive incoming calls. If any SIB with a value tag is changed, this will be noticed from the changed value of the MIB value tag sent via the paging channel. The listening (camped) mobiles must then receive the MIB itself and examine which value tags have been changed, and further decode those blocks.

The values of some information blocks change too frequently for this value tag scheme to be practical. For these blocks without a value tag, a timer is used instead. Every time this timer expires, the corresponding SIB is decoded and the timer is started again.

Some modifications to SIBs are of greater importance than others. For example, if channels are being reconfigured, and mobile stations don't know about this at once, several malfunctions may take place. Therefore, a scheme has been devised that makes it possible for the mobiles to decode the new information immediately after the modification. In such a case, the BCCH modification information in the paging channel contains both the time of the change and the new value of the master block value tag after the change has occurred. The receiving mobile must start a timer using the given value as a time-out value, and once the timer expires, it has to decode the MIB, as well as all the changed SIBs.

Because the paging channel is only monitored when a mobile is in its idle CELL_PCH and URA_PCH states, it is also necessary to transmit the MIB value tag on the FACH, so that all mobiles in the CELL_FACH state can receive this information. This information is added to a system information change indication message on the FACH.

There are altogether 18 different SIBs plus the MIB, and they are briefly explained in the following paragraphs. For a more thorough description, consult the RRC specification [9]. Most probably, new SIBs will be added with the new specification releases.

MIB

- Contains PLMN identity;

- Includes references to other SIBs.

SB

- Contains scheduling information of number of SIBs.

SIB 1

- Contains NAS system information;

- Includes UE timers and counters to be used in idle mode and in connected mode.

SIB 2

- Contains a list of URA identities.

SIB 3

• Contains parameters for cell selection and reselection.

SIB 4

• Contains parameters for cell selection and reselection;

• To be used in connected mode only.

SIB 5

• Contains parameters for the configuration of the common physical channels (PhyCHs) in the cell.

SIB 6

• Contains parameters for the configuration of the common and shared PhyCHs in the cell;

• To be used in connected mode only.

SIB 7

• Contains the fast-changing parameters UL interference and dynamic persistence level;

• Changes so often, its decoding is controlled by a timer.

SIB 8

• Contains static CPCH information to be used in the cell;

• Used in FDD mode only;

• To be used in connected mode only.

SIB 9

• Contains CPCH information to be used in the cell;

• Used in FDD mode only;

• To be used in connected mode only;

• Changes so often, its decoding is controlled by a timer.

SIB 10

- Contains information to be used by UEs having their DCH controlled by a DRAC procedure;
- Used in FDD mode only;
- To be used in CELL_DCH state only;
- Changes so often, its decoding is controlled by a timer.

SIB 11

- Contains measurement control information to be used in the cell.

SIB 12

- Contains measurement control information to be used in the cell;
- To be used in connected mode only.

SIB 13

- Contains ANSI-41 system information;
- Includes four associated SIBs 13.1–13.4;
- Contains references (schedules) of the subblocks;
- To be used only when the CN of the system is ANSI-41.

SIB 14

- Contains parameters for common and dedicated physical channel (DPCH) uplink outer-loop power control information;
- Used in TDD mode only.
- Changes so often, its decoding is controlled by a timer.

SIB 15

- Contains assistance information for UE positioning methods;
- Allows the UE-based positioning methods to perform positioning without dedicated signaling;
- Allows the UE-assisted positioning methods to use reduced signaling;
- Includes five associated SIBs 15.1–15.5.

SIB 16

- Contains predefined channel configurations to be used during handover to UTRAN;

- Includes radio bearer, transport channel, and physical channel parameters to be stored by UE in idle and connected mode;

- There may be several different occurences of SIB 16 in each cell, but the UE is not required to read all of them before initiating RRC connection establishment.

SIB 17

- Contains fast-changing parameters for the configuration of the shared physical channels;

- Information becomes invalid after time specified by the repetition period (SIB REP) for this SIB;

- To be used in connected mode only;

- Used in TDD mode only.

SIB 18

- Contains the PLMN identities of the neighboring cells;

- To be used in shared access networks to help with the cell reselection process (see Chapter 12).

7.5.2.5 Paging

Paging is a procedure that is used by the UTRAN to tell a mobile that there is an incoming call waiting. Establishing the radio connection in the UTRAN is always initiated by the UE; thus, this procedure is needed to inform the UE that an establishment should, in fact, be attempted.

The paging information in the idle mode is carried by paging type 1 messages. One message may contain several paging records, each containing a paging request for a different mobile. It is also possible that a paging message does not contain any paging records at all, but only a BCCH modification information element, which contains the value tag for the MIB. Once a mobile receives a modified value tag, it knows that the MIB must be read from the BCCH.

In principle, the mobile must monitor the PCH continuously to make sure that it does not miss any paging messages and therefore lose incoming calls. However, a continuous reception scheme would soon wear down the battery; thus, a mechanism called discontinuous reception (DRX) is employed.

The DRX scheme is based on the fact that each UE (actually each USIM card) has a unique IMSI, and from that IMSI and the IE "CN domain specific cycle length coefficient" (received in SIB 1), it is possible to compute the paging occasions for this UE. These are frame numbers, and the network makes sure it will deliver paging messages to certain UEs only during the said frame numbers; the mobile knows when it is safe to fall asleep, confident that the network will hold to its paging schedule. Further enhancement (and substantial power savings) is achieved by introducing a paging indication channel (PICH). The UE actually listens only for this PICH periodically, and when a positive indication appears, then the UE knows to listen for the actual PCH as it may only now contain a message addressed to this mobile. There may also be several PCH/PICH pairs in one cell. This is indicated in SIB 5. The UE selects the one to be used based on its IMSI.

Several mobiles may listen for the same paging occasion. Thus, the UE (i.e., the UE's RRC layer) must check whether any of the paging identities of the received paging records matches its own identity. If a match occurs, the paging indication is forwarded to the MM function, which triggers a call-establishment procedure.

Note that there is a trade-off between mobile standby times and call setup times. If the discontinuous reception procedure uses a long DRX cycle (i.e., the UTRAN can send paging messages relatively seldom), then UEs do not have to listen for the PICH so often, which saves power. This results in longer UE standby times. However, the drawback is longer call setup time with mobile-terminated calls. These parameters can be set by the operator, and they can also be changed dynamically because they are continuously sent via the BCCH.

The paging description discussed so far has concentrated on idle-mode paging. It is also possible that the UTRAN pages the UE in the connected mode. In the CELL_PCH and URA_PCH states, the paging request triggers a UE state change. The UTRAN uses this procedure when it has some additional downlink data to be sent to the UE. The DRX cycle length to be used may be different from the idle mode in the CELL_PCH and URA_PCH states. The UE must use the shortest cycle length of any CN domain it is connected to, or the UTRAN DRX cycle length, whichever is shorter.

The actions the UE takes once a paging message is received in the RRC depend on the RRC state. In idle mode the UE shall react thus:

- If the IE "paging originator" is the CN, compare the included identities of type "CN UE identity" with all of its allocated CN UE identities.

- For each match, forward the identity and paging cause to the upper-layer entity indicated by the IE "CN domain identity."
- If the IE "paging originator" is the UTRAN, ignore that paging record.

In connected mode the UE shall behave thus:

- If the IE "paging originator" is the UTRAN, compare the included identities of type "UTRAN originator" with its allocated U RNTI.
  - For each match, the UE will enter CELL FACH state and perform a cell-update procedure with cause "paging response."
- If the IE "paging originator" is the CN, ignore that paging record.

So, if the paging originator is the CN, a possible paging response is initiated by the MM (RRC connection request, establishment cause = paging response). If the paging originator is the UTRAN, then a possible response is initiated by the RRC (cell update, cause = paging response).

Dedicated Paging
Dedicated paging is a paging message (paging type 2) that is sent in connected-mode states CELL DCH and CELL FACH. It is used, for example, to establish a signaling connection. Note that there is already an existing signaling connection in these states, but the dedicated paging procedure can be used in cases in which a CN other than the current serving CN wants to originate a dialogue with a UE. The RRC receives and decodes the message and forwards the paging identity and the cause to the NAS entity indicated by the CN domain identity. Note that for the access stratum, this is just another signaling message on a dedicated connection.

7.5.2.6 RRC Connection Establishment

The UMTS separates the concepts of a radio connection from a radio bearer (RB). The UE requests an RRC connection, but the network requests an RB setup. Experienced readers may recall that the bearer capability was attached to the radio channel in GSM. Once a resource was allocated it was not possible to change the bearer capabilities. The bearer capability can be changed dynamically during the radio connection in the UMTS provided that the network has the necessary resources. This is a very important enhancement.

An analogy to the relationship between a radio connection and an RB might be a train system, where an RRC connection is the track and a bearer connection is the railway carriages on the track. The track makes it possible to transfer goods, but the carriages define the kinds of goods delivered and in what quantity they can be delivered.

The RRC connection establishment procedure is quite simple in the RRC level (Figure 7.10), but note that a rather complex RACHing procedure (see Section 11.5) must take place in the lower layers before the RRC connection request message is received in a Node B.

Note that the various forms of this procedure are similar regardless of who initiated the connection: mobile or network. A network-originated connection is triggered by a paging procedure. Only the establishment cause field in the RRC connection request message indicates the reason for the connection establishment.

The RRC must store the value of "initial UE identity" encapsulated in the RRC connection request message, and check that it receives the same value back in the RRC connection setup message. It is possible that several UEs have sent RACH messages at the same time; thus, the response message may not be addressed to this particular UE. This procedure is also known as contention resolution, and it is required because the channels used for RRC connection request and setup messages are sent on common channels. If the received initial UE identity is the same as the sent value, then the UE can continue with the connection setup. Otherwise, the connection establishment must be started again.

The RRC connection setup message is a large one containing significant information to be used in the L1/L2 configuration. This includes transport format sets and transport channel information. The final message (RRC connection setup complete) is sent via the new DCH.

The network can also reject the connection setup and respond with an RRC connection reject instead. Note that an RRC connection is not the same thing as a dedicated connection. An RRC connection may also be mapped onto a common or shared channel.

FIGURE 7.10
*RRC connection
establishment.*

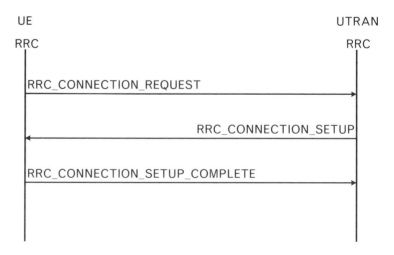

7.5.2.7   RRC Connection Release

Whereas an RRC connection is always initiated by the UE, its release is always initiated from the network side. There are two versions of this procedure depending on whether a DPCH exists or not. The RRC layer signaling is similar in both cases; the network sends an RRC connection release message, and the UE responds with an RRC connection release complete (Figure 7.11).

If a dedicated physical connection exists, then it is used for the transport of these messages. Both messages are sent using unacknowledged mode in the lower layers. The UE may send its response message several times (the number is defined by variable V308), and after the last transmission, it releases all its radio connections. Similarly, the UTRAN releases all UE-dedicated resources after receiving the complete message. In fact, this is done once a corresponding timer expires, even if it doesn't receive the connection release complete message.

If there is not a dedicated connection, then the downlink RRC connection release message must be sent via the FACH, and the uplink RRC connection release complete message via the RACH. Only one occurrence of both messages is sent, and the acknowledged mode is used. This means that a data acknowledgment message is sent as an acknowledgment in the RLC protocol level to the RRC connection release complete message via the downlink FACH.

7.5.2.8   RRC Connection Reestablishment

If a UE loses the radio connection suddenly, it may attempt a connection reestablishment. This requires a quick cell reselection and sending a request for reestablishment message to the UTRAN.

Once the UE detects that it has lost the connection, it starts timers T301 and T314/T315. The reestablishment attempt must succeed while these timers are still running. On their expiration a possible ongoing reestablishment is abandoned and the UE returns to its idle mode.

FIGURE 7.11
RRC connection release.

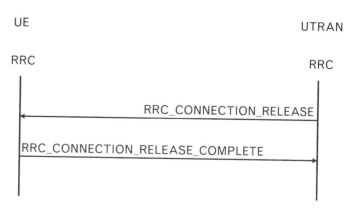

The reestablishment is requested by the NAS (MM) in the UE. The UE must find a suitable cell for reestablishment and do so quickly. The old serving cell will not be accepted, as it is still regarded as unusable.

In the new chosen candidate cell, an RRC connection reestablishment request is sent on the uplink RACH. If the UTRAN can reconnect the old connection, it responds with an RRC connection reestablishment message on the FACH. The UE configures its L1 according to the information obtained from this message, gains synchronization, and then responds with an RRC connection reestablishment complete message on the new DCCH (see Figure 7.12).

Note that a connection reestablishment may require a considerable number of tasks to be carried out by the network; the call may have to be rerouted via new switches and base stations. The probability of these being successful diminishes quickly with time, so a reestablishment procedure must be performed as quickly as possible.

7.5.2.9   Radio Bearer Establishment

As previously explained in the discussion of RRC connection establishment, radio connections and RBs are two separate concepts in UMTS.

A radio connection is a static concept. It is established once, and it exists until it is released. There is only one radio connection per (typical) terminal. On the other hand, the RB defines what kind of properties this radio connection has. There may be several RBs on one radio connection, each having different capabilities for data transfer. RBs are also dynamic; they can be reconfigured as necessary. This is not to say that radio connections cannot be configured. They will be reconfigured in many ways. The most common radio-connection–reconfiguration procedure is probably the HO procedure.

FIGURE 7.12
*RRC connection
reestablishment.*

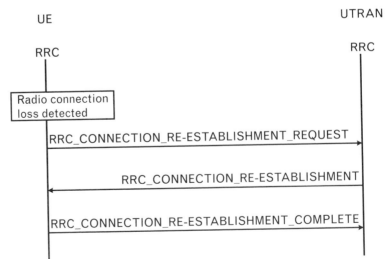

Indeed, it is also possible to have an RB without a dedicated radio connection. Circuit-switched bearers or bearers using real-time services typically need dedicated radio channels to meet their delay requirements. Packet-switched bearers or bearers using non-real-time services, however, often do not need a permanent association to a dedicated radio resource; they can use shared or common channels.

RB establishment is always initiated by the UTRAN. This is because an RB uses a certain amount of radio resources from the network and these resources are quite limited. Only the network knows what kind of resources it can grant to a UE.

The UTRAN RRC gets a request for a new bearer from the higher layers in the NAS. Down at the RRC level, the signaling is simple: the UTRAN sends a radio bearer setup message, and the UE responds with a radio bearer setup complete (Figure 7.13). However, the interlayer signaling to lower layers can be quite different depending on the requested QoS parameters and whether there already exists a suitable physical channel. These variations are discussed in Section 11.2.

7.5.2.10   Radio Bearer Reconfiguration

This procedure is used to reconfigure an RB if its QoS parameters have been changed or traffic-volume measurements indicate that more or fewer resources are required. Also, this procedure can either be synchronized or unsynchronized, depending on whether the old and new RB setups are compatible. The synchronized bearer-reconfiguration procedure includes an activation time parameter, which is used to ensure that the change takes place at the same time in both the UE and the UTRAN. At the end of the procedure, the old configuration, if such exists, must be released from Node B(s) (see Figure 7.14).

Note that the coexistence of the old and the new configuration in unsynchronized change does not mean that a UE has to maintain two

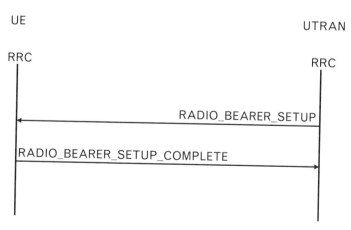

FIGURE 7.13
Radio bearer establishment.

FIGURE 7.14
*Radio bearer
reconfiguration.*

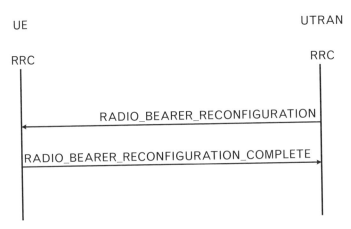

configurations at the same time. It means that a UE can use one configuration (the new one) and the network another (the old one) temporarily during the reconfiguration procedure, and they can still continue to communicate. A synchronized change is used when the old and the new configurations are incompatible; thus, the changes have to take place exactly at the same time.

7.5.2.11   Radio Bearer Release

This procedure releases one or more RB(s) (Figure 7.15). Again this procedure has two variations, synchronized and unsynchronized. The unsynchronized procedure is simpler because the old and new configurations can coexist and therefore the change does not have to take place simultaneously in the UE and in the UTRAN. The synchronized procedure, on the other hand, must use a time parameter to ensure synchronization.

The RB release procedure may include a physical channel modification or deactivation, depending on the requirements of the new situation (i.e., what kind of bearers still exist after this release).

FIGURE 7.15
Radio bearer release.

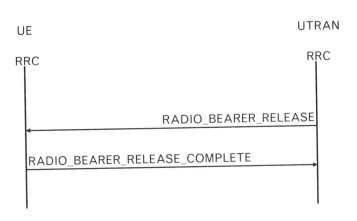

7.5.2.12   Management of the RRC Connection

This service includes assignment, reconfiguration, and release of radio resources for the RRC connection, for example, assignment of codes and CPCH channels.

The RRC layer handles the assignment of the radio resources (codes and CPCH channels) needed for the RRC connection, including the needs of both the control and the user planes. The RRC layer may reconfigure radio resources during an established RRC connection. This function includes coordination of the radio resource allocation between multiple RBs related to the same RRC connection. The RRC controls the radio resources in the uplink and downlink, such that the UE and the UTRAN can communicate using unbalanced radio resources (asymmetric uplink and downlink). The RRC signals the UE to indicate resource allocations for the purposes of an HO to GSM or other radio systems.

7.5.2.13   HOs

Being a mobile device, a UE may move during a connection. It is the responsibility of the RRC layer to maintain the connection if a UE moves from one cell area to another. In the case of a circuit-switched connection, this procedure is called an HO. There is no radio connection in packet-switched data transfer; thus, there is no need to maintain it. In this case a UE makes a normal cell reselection, and the data packets are then rerouted via the new cell.

An HO decision is done in the UTRAN RRC, and it is based on, among other things, the measurements done by the UE.

An SHO is a procedure in which a UE maintains its connection with the UTRAN via two or more Node Bs simultaneously. An SHO is a CDMA-specific procedure, made possible by the fact that all base stations use the same frequency; thus, it is relatively easy for the UE to receive several of them at the same time. The SHO procedure is a common condition in the life of a UE. A UE can be in an SHO state for most of the time it is in a call. A user cannot notice any kind of change in the voice quality while the UE enters or leaves the SHO state, except that the voice quality may actually be better in an SHO state than in a normal state just before entering the SHO state.

The base stations to which a UE is connected are said to belong to the UE's active set. SHOs are managed with active set update messages sent by the network (Figure 7.16). Note that the UE should not update its active set by itself, but only according to these messages. A UE in an SHO always consumes more network resources than a UE with a normal single connection to the network. Therefore, it must be the network side that decides which UEs need the additional gain from an SHO, and which UEs can do without.

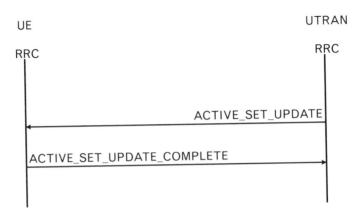

FIGURE 7.16
SHO management.

A hard handover (HHO) corresponds to the normal GSM HO procedure, which is always hard. The term *hard handover* indicates that the old connection is first released before the new one is set up. This may result in an audible break in a speech connection. Within UTRAN an HHO will be performed when the radio frequency channel changes. The UMTS also contains dedicated procedures for intersystem HOs, which are a form of HHO. In practice, these HOs will occur only between UMTS and GSM/GPRS networks at the first phase of UMTS.

Note that there is no special HHO procedure in the UTRAN air interface. An HHO can result from other procedures that reconfigure the air interface. If such a procedure changes the radio frequency of the connection, then an HHO takes place. Procedures that may trigger an HHO include a physical channel reconfiguration, a radio bearer establishment, a radio bearer reconfiguration, a radio bearer release, and a transport channel reconfiguration.

The relocation procedure is a UTRAN internal HO procedure. It reroutes the connection in the UTRAN, but does not affect the radio connection in the air interface, even as there are some implications for higher layers in the UE (e.g., the PDCP may have to support a lossless relocation).

HO processes are further discussed in Section 2.5. Many HO procedures are depicted in Chapter 11.

7.5.2.14   Measurement Control

The measurements performed by the UE are controlled by the RRC layer, which decides what to measure, when to measure, and how to report the results, including both the UMTS air interface and other radio systems. The RRC layer also constructs reports of the measurement results from the UE to the network.

In the idle mode, the measurements are usually made to support cell-reselection procedures, which means they are internal to the UE. It is also possible that some of the measurement results must be reported to the network when the UE is in the connected mode.

The list of possible measurements the UTRAN may ask the UE to perform is a long one. The UE measurements can be grouped into seven categories or types:

1. Intrafrequency measurements;

2. Interfrequency measurements;

3. Intersystem (or Inter-RAT) measurements;

4. UE positioning measurements;

5. Traffic-volume measurements;

6. Quality measurements;

7. UE internal measurements.

This same grouping is also used in the measurement control message that is sent to the UE to ask it to perform measurements. Each control message can set up, modify, or release a measurement procedure. There may be several parallel measurement procedures running at the same time, so the control message also contains an identity number, which indicates the measurement process this message applies to. Each control message may contain control information for only one measurement type at a time.

The exact rules for how to process various measurement requests and how to perform the actual measurements in each state are given in Section 8.4 of [9].

The results of the measurements can be sent back to UTRAN in measurement report messages. This can happen either periodically, after some triggering event, or perhaps using a combination of these: the first report is sent after the triggering event, and the following messages are sent periodically, while the same triggering condition still applies.

Typically, the actual radio measurements will be performed in layer 1 according to instructions from the RRC layer. However, traffic volume measurements will be performed in the MAC layer. They are based on the RLC buffer occupancy information, which is regularly reported to the MAC layer.

The measurement model for air interface measurements is given in Figure 7.17.

A. *Measurements performed in L1.* These are filtered in L1 so that not all measurement samples will be sent to the RRC. The specification does not give exact filtering rules, but merely the performance objectives and the reporting rate at point B. L1 typically has to collect

Figure 7.17
Measurement model.
(Source: [12]).

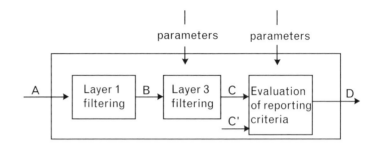

several samples from the same measurement process, average them, and send the results to the RRC.

B. *Measurement reports sent to the RRC.* L3 must further filter these reports based on the rules received from the UTRAN in a measurement control message.

C. *A measurement after processing in the RRC filter.* The specification states that the rate of reporting in C is the same as in B. These reports are then evaluated to find out whether they have to be sent further.

C'. *Reporting thresholds.*

D. *Measurement report sent over the air or the Iub interface.*

There is a long list of events that may trigger a measurement report. Normally, a UE does not have to report all these events. They form a pool of events from which the UTRAN can choose the events to be reported. In some cases the event triggers the first measurement report, which is then followed by periodic reporting. The list of possible events is given next.

Intrafrequency reporting events for the FDD mode:

1A. A primary CPICH enters the reporting range.

1B. A primary CPICH leaves the reporting range.

1C. A nonactive primary CPICH becomes better than an active primary CPICH.

1D. Change of best cell.

1E. A primary CPICH becomes better than an absolute threshold.

1F. A primary CPICH becomes worse than an absolute threshold.

Additionally for the TDD mode:

1G. Change of best cell.

1H. Time slot interference on signal code power (ISCP) below a certain threshold.

1I. Time slot ISCP above a certain threshold.

Interfrequency reporting events (for both FDD and TDD modes):

2A. Change of best frequency.

2B. The estimated quality of the currently used frequency is below a certain threshold, and the estimated quality of a nonused frequency is above a certain threshold.

2C. The estimated quality of a nonused frequency is above a certain threshold.

2D. The estimated quality of the currently used frequency is below a certain threshold.

2E. The estimated quality of a nonused frequency is below a certain threshold.

2F. The estimated quality of the currently used frequency is above a certain threshold.

Intersystem reporting events (for both the FDD and TDD modes):

3A. The estimated quality of the currently used UTRAN frequency is below a certain threshold and the estimated quality of the other system is above a certain threshold.

3B. The estimated quality of the other system is below a certain threshold.

3C. The estimated quality of the other system is above a certain threshold.

3D. Change of best cell in other system.

Traffic-volume reporting events:

4A. Transport channel traffic volume exceeds an absolute threshold.

4B. Transport channel traffic volume becomes smaller than an absolute threshold.

Quality reporting events:

5A. A predefined number of bad CRCs is exceeded.

UE internal measurement reporting events:

6A. The UE Tx power becomes larger than an absolute threshold.

6B. The UE Tx power becomes less than an absolute threshold.

6C. The UE Tx power reaches its minimum value.

6D. The UE Tx power reaches its maximum value.

6E. The UE RSSI reaches the UE's dynamic receiver range.

6F. The UE Rx–Tx time difference for an RL included in the active set becomes larger than an absolute threshold (for 1.28 Mcps: The time difference indicated by T_{ADV} becomes larger than an absolute threshold).

6G. The UE Rx–Tx time difference for an RL included in the active set becomes less than an absolute threshold.

UE positioning reporting events:

7A. The UE position changes more than an absolute threshold.

7B. SFN–SFN measurement changes more than an absolute threshold.

7C. GPS time and SFN time have drifted apart more than an absolute threshold.

7.5.2.15 Outer-Loop Power Control

The RRC layer controls the setting of the target of the closed-loop power control mechanism.

The closed loop is the power control method that is used during the UTRAN connection after the initial setup phase includes two subprocesses: the inner and the outer closed-loop power control. The inner-loop power control is an L1 internal procedure, and the outer-loop control also includes the RRC layer.

The inner-loop control uses the SIR_{target} value to adjust the transmission power levels in the air interface. L1 measures the received SIR and compares it to SIR_{target}. If the value is worse, a power increase request is sent to the base station; otherwise, a lower transmission power is requested. CDMA systems are always looking for ways to lower transmitter power. The problem here is that the SIR_{target} value cannot be kept constant. Different connections will have varying QoS targets, mobile terminals will have different speeds, and SHOs will also have an effect on the QoS (i.e., even if the set SIR_{target} is fulfilled, the result may still have too many errors for an application that requires high QoS). Therefore, an outer-loop power control is needed in the UE. This monitors the quality of the received signal and adjusts the SIR_{target} value accordingly; that is, if the received quality is too low, SIR_{target} is increased, and vice versa. A similar outer-loop power control scheme is used both in the UE and in the RNC. In the UE it adjusts the downlink quality, while in the RNC it adjusts the uplink quality.

Once an RB has been established, the UTRAN sends a radio bearer setup message with IE "added or reconfigured DL TrCH information." This IE contains a block error rate (BLER) value for each coded composite transport channel (CCTrCH) used. This is the quality target the UE attempts to maintain. If the received BLER is lower than the target BLER, then SIR_{target} is increased. If the quality exceeds the minimum required, SIR_{target} is reduced. The quality target control loop is run so that the quality requirement is met for each transport channel. The target BLER value can be dynamically changed by the UTRAN via a radio bearer reconfiguration message.

For the CPCH the quality target is set as the BER of the DL DPCCH as signaled by the UTRAN. This value is signaled to UEs in SIB 8 broadcasts. Similarly, the UE runs a quality target control loop such that the quality requirement is met for each CPCH transport channel that has been assigned a DL DPCCH BER target.

The UE sets the SIR_{target} when the physical channel has been set up or reconfigured. It will not increase the SIR_{target} value before the power control has converged on the current value. The UE may estimate whether the power control has converged on the current value by comparing the averaged measured SIR to the SIR_{target} value.

If the UE has received a DL outer-loop control message from the UTRAN indicating that the SIR_{target} value should not be increased above the current value, it should record the current value as the maximum allowed value for the power control function. Once the RRC receives a new DL outer-loop control message from UTRAN indicating that the restriction is removed, it is then free to continue using the standard power control algorithm.

7.5.2.16   Security Mode Control

The security mode control procedure is used by the UTRAN to control either the ciphering or the integrity protection processes. In the case of ciphering control, this procedure can either start the ciphering, or change the ciphering key. This section also discusses the counter check procedure, as it is closely related to the UE security (see Figure 7.18).

Ciphering Control
This is what is known in GSM as *ciphering key control*. It can trigger the start of ciphering or command the change of the cipher key.

This procedure is always initiated by the UTRAN. It sends a security mode command message to the UE containing the new ciphering key and the activation time. The activation time will be given as an RLC send sequence number (for AM and UM bearers) or as a CFN (for TM bearers). Once the activation time elapses, the UE starts to use the new configuration and sends a security mode complete message back to the UTRAN. Note that this message is still sent using the old ciphering configuration.

FIGURE 7.18

Security mode control.

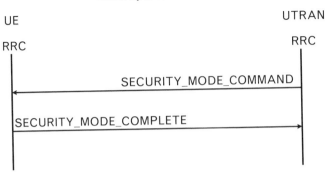

Several different ciphering algorithms will be defined by the 3GPP, but the UE only has to support one algorithm at a time (i.e., the same algorithm will be used both in the MAC and in RLC). However, the UTRAN may change the active algorithm at any time during the call. Although the algorithm is the same in the MAC and in the RLC, the used ciphering key sequence numbers will differ. There are three versions of the ciphering key sequence number depicted in Figure 7.19. In 3GPP jargon, the ciphering key sequence number is called *COUNT-C*.

The RLC layer uses the RLC sequence number as a part of the COUNT-C, but in transparent mode this cannot be used because in this case there are no sequence numbers in the RLC, so an eight-bit ciphering frame number from the MAC is used instead.

The actual ciphering process is depicted in Figure 7.20. Note that the deciphering process is similar, except that the input stream is ciphertext, and the output is plaintext.

See [13] for further information on ciphering.

Integrity Protection

Integrity protection is a scheme that guards the signaling traffic in the air interface against unauthorized attacks. An intruder could try to modify the message sequences (e.g., by means of a man-in-the-middle attack). Integrity

FIGURE 7.19

Ciphering key sequence numbers (COUNT-C).

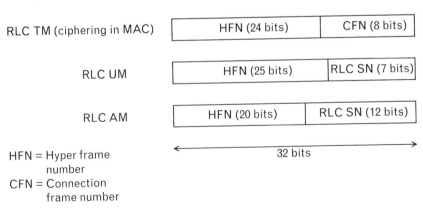

FIGURE 7.20
Ciphering process.

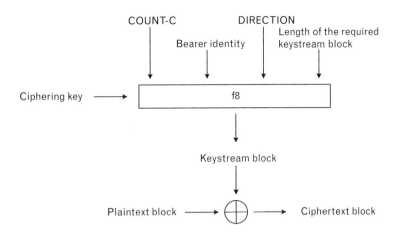

protection ensures that the signaling procedures are not tampered with (or at least makes it very difficult to break the security).

The integrity-protection process is started (and restarted) by the security mode procedure. The same procedure is also used for ciphering control.

To start or reconfigure the integrity protection, the UTRAN sends a security mode command message on the downlink DCCH in AM RLC using the present integrity-protection configuration.

Integrity protection is performed on all RRC messages except:

- HANDOVER TO UTRAN COMPLETE;

- PAGING TYPE 1;

- PUSCH CAPACITY REQUEST;

- PHYSICAL SHARED CHANNEL ALLOCATION;

- RRC CONNECTION REQUEST;

- RRC CONNECTION SETUP;

- RRC CONNECTION SETUP COMPLETE;

- RRC CONNECTION REJECT;

- RRC CONNECTION RELEASE (on CCCH only);

- SYSTEM INFORMATION;

- SYSTEM INFORMATION CHANGE INDICATION;

- TRANSPORT FORMAT COMBINATION CONTROL.

For the CCCH and for each signaling RB, two integrity-protection hyperframe numbers are used (both 28 bits):

1. Uplink HFN;

2. Downlink HFN.

And two message sequence numbers are used (both 4 bits):

1. Uplink RRC message sequence number;
2. Downlink RRC message sequence number.

By combining these numbers, we get two 32-bit integrity sequence numbers, COUNT-I, one for uplink and one for downlink, for each signaling radio bearer (RB 0–4). Once a UE receives a downlink signaling message, it calculates a message authentication code (MAC) based on the stored COUNT-I information and the received message. The algorithm for MAC calculation is given in Section 8.5.10.3 of [9] and in [13]. The calculated MAC must match with the received MAC, otherwise the message has been tampered with and must be discarded.

In the uplink direction, the UE calculates a MAC value and attaches it to the uplink signaling message. The UTRAN has to perform the integrity check in the same way as the UE.

Integrity protection is described in Section 8.5.10 of [9].

Counter Check

Counter check is yet another security procedure in the air interface. It is used to check that the amount of data sent in the air interface is similar in both the UE and in the UTRAN.

The UE must maintain a ciphering sequence number (COUNT-C) for each radio bearer. The UTRAN can query this number from time to time to ensure that there are no intruders taking part in the communication. It sends a counter check message to the UE and includes COUNT-C values for each RB. The UE must compare the received COUNT-C values with the actual COUNT-C values used, and include the number of mismatches in the response message (counter check response) (see Figure 7.21).

Once the UTRAN receives this message, it checks the number of mismatches to find out whether there is anything suspicious going on in the air interface. Note that counter check is used in the user plane, and integrity protection in the control plane. Thus, both procedures are needed, and they complement each other. Indeed, notice that counter check signal exchange is protected by the integrity protection.

The counter check procedure is defined in Section 8.1.15 of [9].

7.5.2.17 Direct Transfer

NAS messages (higher-layer messages) are relayed over the air interface within direct transfer messages. There are separate messages for the uplink and downlink directions, the uplink direct transfer and the downlink direct

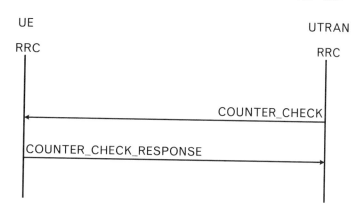

FIGURE 7.21
Counter check.

transfer, respectively. However, if there is not yet an existing signaling connection, then the initial direct transfer message must be used.

The initial direct transfer procedure is typically used when the upper layers request an initialization of a new session. This request also includes the initial NAS message. If there is no RRC connection, the RRC must establish it using the normal RRC connection-establishment procedure. After that the direct transfer message is transferred using AM. A successful signaling-connection establishment is also confirmed to UE-NAS.

In the RLC layer the direct transfer message should be transmitted using either radio bearer 2 (RB2) or RB3. Low-priority messages, such as GSM-SAPI3, should use RB3, and high-priority messages, such as GSM-SAPI0, should use RB2. Note that this priority is indicated by the NAS to the RRC.

Downlink direct transfer messages are routed to higher layers based on the value of the CN domain identity element in the message (CS domain/PS domain). This routing decision is done in the RRC.

Direct transfer is discussed in [9] in Sections 8.1.8, 8.1.9, and 8.1.10.

7.5.2.18 Control of Requested QoS

This function ensures that the QoS requested for the RBs can be met, which includes the allocation of a sufficient number of radio resources for an application. Different applications have different demands for the data transmission services they require. The QoS parameters may include the required bandwidth, the allowed transmission delay, and the allowed data error tolerance. QoS classes in UTRAN are discussed in Section 13.6.5.

The UTRAN air interface is a very flexible one, allowing for the dynamic allocation of system resources. The UTRAN allocates a minimum number of resources for each UE so that the negotiated QoS can be maintained. This chapter has in several earlier sections explained some of the individual procedures related to the QoS and the control of the system's resources, but the overall picture was not explained; it was not shown how the individual procedures are related to each other. This is the purpose of this section.

The individual procedures presented here include physical channel reconfiguration, transport channel reconfiguration, radio bearer reconfiguration, and traffic-volume measurements.

Increased Data

We begin by considering what happens when the amount of data transmitted increases. The UTRAN controls this procedure. The initial resource allocation (radio bearer setup) was discussed earlier. Once the data transmission is under way, the UE's MAC layer monitors the amount of unsent data in the RLC transmission buffer. If the buffer fills over a predefined threshold, then a measurement report is sent to the UTRAN. This message acts as a request message for additional resources.

The UTRAN may allocate additional resources to the UE depending on the availability of the resources and the negotiated QoS of the UE. There are several ways it can do this.

Small amounts of data are typically, but not necessarily, transferred over common transport channels. Even small amounts of data could be sent via dedicated channels if the data has strict delay requirements, for one cannot easily guarantee the delays in common channels. If the RLC's buffer fills up while the common channels are used, then the UTRAN may assign a dedicated channel to the UE (i.e., the case of RACH/FACH to DCH/DCH). This will be done using the PhyCH reconfiguration procedure; see Section 11.2.5.

If the UE already has a DCH assigned to its application, then the increase in data transmit capacity must be handled by reconfiguring the existing channels. For this purpose there are three possible procedures that could be employed. First among these is the physical channel reconfiguration procedure, which is called on when there are already suitable TFs and TFCs defined for the UE, and these can be used in the new configuration. Second, if the TFs and TFCs must be reconfigured, then the transport channel reconfiguration procedure (Section 11.2.4) must be used. Third, and the most powerful procedure of these is the radio bearer reconfiguration procedure, which can change QoS parameters, change the multiplexing of logical channels in the MAC layer, and reconfigure the transport channels and the physical channels. Note that the UTRAN decides which of the three procedures to use. The UE simply has to follow the orders given to it via the RRC signaling link.

Here we list the procedures in the order of the amount of change they make:

1. Radio bearer reconfiguration;
2. Transport channel reconfiguration;

3. Physical channel reconfiguration.

The radio bearer reconfiguration procedure can be used to change more parameters than the physical channel reconfiguration. A higher-order procedure may also include all the configuration parameters of a lower-order procedure.

It is also possible that new RBs can be allocated or the old RBs can be released, but these actions are not due to the traffic measurements in the RLC, but because a new application/service needs one. Then, of course, a release would mean that the UE no longer needs an RB.

The issues explained above also apply to downlink data, except that the transmission buffer to be monitored is now in the UTRAN.

Decreased Data

If the amount of transmitted data decreases, then the underutilized resources should be deallocated to make them available for other users. The triggering event is, as was the case with increasing data, the transmission buffer in the RLC. If the buffer contents remain less than a lower threshold for a certain time, then the traffic-volume-monitoring process sends a measurement report message to the UTRAN.

It is up to UTRAN to decide what kind of actions (if any) will be taken. As in the increased data case, there are three procedures to choose from: physical channel reconfiguration, transport channel reconfiguration, and radio bearer reconfiguration. If the amount of transmitted data is low enough, the UTRAN may command the UE to release its DCHs and use common channels instead (i.e., the case from the DCH/DCH combination to the RACH/FACH combination). This change can be included in all three reconfiguration procedures mentioned above.

Other Observations

Note that the UTRAN must consider the RLC buffer's content in both the UE and the UTRAN when it decides which procedure to use. For example, even if there is very little uplink traffic (i.e., the buffer in the UE-RLC is empty), it cannot move the UE from a DCH/DCH to a RACH/FACH if there is plenty of downlink traffic. The UE cannot have a DCH in only one direction (i.e., combinations of RACH/DCH and DCH/FACH are not allowed). DCHs must use fast power control, and it is impossible to implement an efficient power control loop using common channel resources in the other direction.

The UTRAN can also "fine-tune" the UE's data transfer resources with the transport format combination control procedure, which merely modifies the allowed uplink transport format combinations within the transport format combination set. Also note that even large changes in the amount of transmitted data do not necessarily trigger any of the reconfiguration procedures described earlier. The reader must remember that the transmitting entity has a set of TFCs it can choose the suitable one for the current situation. If there is lots of data in RLC buffers, a high-capacity TFC could be used, for example. On the other hand, if common channels are used in the physical layer, the configured TFCS most probably cannot include high-capacity TFCs, as those would be difficult to send over the air interface.

7.5.2.19   Support for DRAC

Many studies have shown that within a typical WCDMA cell, the capacity is uplink limited; that is, if the traffic is increased in a cell, it is the uplink that gets overloaded first. However, this conclusion may not be true in a cell with multimedia users, which will generate much more downlink than uplink traffic; a cell with significant asymmetric traffic.

In the uplink, the spreading codes are not orthogonal, but pseudorandom; thus, the user signals appear as interference to each other. To ease the situation in the uplink, the 3GPP has defined a scheme called dynamic resource allocation control (DRAC). It is a very fast method to spread the load in the uplink DCH while avoiding peaks in the interference level. The UTRAN may assign uplink DCHs with DRAC information elements indicating that the UE must use DRAC in that uplink DCH.

The UTRAN transmits the DRAC parameters regularly via the SIB 10 broadcast message. As this message must be received by the UE in the connected mode, it is mapped into a FACH (and further to an S-CCPCH) channel. These parameters indicate the allowed subset of TFCS according to the given maximum bit rate:

$$\sum_{DCHi\_Controlled\_by\_DRAC} TBSsize_i \,/\, TTI_i < MaximumBitRate \qquad (7.1)$$

After the first SIB 10 has been received, the UE starts the following process:

1. At the start of the next TTI, the UE will randomly select p, $0 \leq p \leq 1$.

2. If $p <$ Transmission_Probability parameter, then the UE will transmit on the DCH controlled by DRAC during $T_{validity}$ frames using the last stored allowed subset of TFCS, and then returns to step 1. Otherwise the UE will stop transmission on the DCH during T_{retry} frames and then return to step 1.

Transmission time validity ($T_{validity}$) and time duration before retry (T_{retry}) are indicated to the UE at the establishment of a DCH controlled by this procedure, which may be changed through radio bearer or transport channel reconfiguration. The UE will always use the latest received DRAC static parameters.

Most probably this scheme is used with bearers that do not have very strict real-time requirements. It is an ideal method to be used with relatively high data rate services with lax delay requirements. The UTRAN can use DRAC bearers to "fill" the free capacity in the uplink. It can measure the uplink data rates and interference, and once there seems to be unused capacity, it can immediately fill that using DRAC managed bearers. The DRAC scheme can modify the bearer data rate even in every frame if necessary. The allowed values for the $T_{validity}$ parameter are 1 to 256 frames.

Note that DRAC support requires simultaneous reception of the SCCPCH and the DPCH channels. DRAC is only applicable when the UE is in the CELL_DCH state. DRAC is also only applicable for FDD systems.

See also Section 14.8 in [9], Chapter 8 in [14], and Section 6.2.5 in [6]. However, I do not believe that DRAC will be the first feature to be implemented in 3GPP networks because most probably the uplink will not be the bottleneck that needs enhancing.

7.5.2.20   Contention Resolution

The RRC must store the value of initial UE identity encapsulated in the RRC connection request message and check that it receives the same value back in the RRC connection setup. It is possible that several UEs have sent on the RACH at the same time; thus, the response message might not be addressed to this UE. This procedure is known as *contention resolution*. If the initial UE identity is the same as the sent value, the UE can continue with the connection setup. Otherwise, the connection establishment must be started again.

7.5.2.21   Timing Advance

This functionality is only supported in the TDD mode. It is used to avoid large variations in signal arrival times at the Node B. Each radio frame in the TDD mode is divided into 15 time slots, which are further allocated either to the uplink or to the downlink. The transmissions from UEs in the cell must arrive at the Node B during their respective time slots. This means that UEs further away from the Node B have to transmit earlier than the nearby UEs so that both transmissions arrive at the Node B at the expected times.

The UTRAN may adjust the UE's transmission timing with timing advance. The initial value for timing advance will be acquired by the UTRAN from the timing of the PRACH transmission. The required timing advance will be given as a 6-bit number (the max value is 63) being the multiple of 4 chips, which is nearest to the required timing advance.

The RRC controls the timing-advance mechanisms. The UTRAN can send the IE UL timing advance with various configuration messages or within the uplink physical channel control message. The UTRAN will continuously measure the timing of transmissions from the UE and send the necessary timing-advance value as an adjustment. With the receipt of this IE, the UE adjusts the timing of its transmissions accordingly in steps of ± 4 chips.

The timing-advance mechanism is also needed in the TDD mode's HO procedure. If a TDD-to-TDD HO takes place, the UE transmision in the new cell is adjusted by the relative timing difference, Δt, between the new and the old cell:

$$TA_{new} = TA_{old} + 2\Delta t \qquad (7.2)$$

If UL synchronization is used (its support is optional for the UE), the timing advance is subchip granular and with high accuracy in order to enable synchronous CDMA in the UL. The functionality is otherwise similar to the nonsynchronized case, except that the units in the adjustment command represent multiples of $\frac{1}{4}$ chips.

7.5.2.22   Support for Cell Broadcast Service

Cell broadcast messages are text messages that are broadcast to everybody in a cell. These messages can be received by all mobiles capable of receiving cell broadcast service (CBS) and that are either in idle, CELL_PCH, or URA_PCH states. The user can choose which types of messages will be displayed and which will be discarded based on the message class type.

CBS is described in Section 7.11. Most of the CBS functionality will be put into the BMC task, but the RRC task must handle some configuration and allocation functions for the CBS. This service is a broadcast service; that is, only the downlink direction is used for message delivery. Thus, the supporting functionality in the RRC is different in the UE and the UTRAN.

Initial Configuration for CBS
This function performs the initial configuration of the BMC sublayer. The configuration is delivered to the UE via the broadcast system information; that is, it is received by the UE's RRC. This information is then used to configure the BMC and the L1 so that they can handle received CBS messages.

Allocation of Radio Resources for CBS
This functionality belongs to the UTRAN-RRC only. It allocates radio resources for CBS based on traffic volume requirements indicated by the BMC. The more queued CBS messages in the BMC buffers waiting to be

sent, the more resources should be allocated for the CBS. The radio resource allocation set by the RRC (i.e., the schedule for mapping of the CTCH onto the FACH/S-CCPCH), is indicated to the BMC to enable the generation of schedule messages. The resource allocation for CBS is broadcast as system information.

Configuration for CBS Discontinuous Reception

CBS messages can only be received while there is no active communication between the UE and the UTRAN; that is, in the idle, CELL_PCH, and URA_PCH states. In these states it is important that the UE saves as much power as possible to increase its standby time.

Therefore, the UTRAN only sends CBS messages during predefined times. The UE knows this schedule, and it only has to be prepared to receive CBS messages during those times. This function configures the lower layers of the UE when it will listen to the resources allocated for CBS based on scheduling information received from the BMC.

7.5.2.23 Capability Information

There will likely and hopefully be a wide variety of different types of UEs; multimedia applications imply a wide variety of terminals and appliances. They will have different capabilities, and the network must know the capabilities of a given UE before it can decide what kind of services and resources it can offer to any particular UE.

This information is typically sent to the network if the capabilities of a UE change. This may occur, for example, when a handheld UE is connected to a car kit; thus, its power class changes. The network may also require the UE to send along its capability information during the RRC connection setup procedure. The request can be sent in the RRC connection setup message and the response (the capability information) is included in the RRC connection setup complete message.

A typical message flow in this procedure consists of a UE capability information sent by the UE and a UE capability information confirm sent back to the UE from the network. The procedure can also be initiated by the network by sending a UE capability enquiry. This message triggers the UE to send its UE capability information (see Figure 7.22).

See also Sections 8.1.6 and 8.1.7 in [9] and Section 6.7.1 in [6].

7.6 RRC Protocol States

In GSM as in many other 2G systems, the radio resource protocol states were generally divided into two groups: the idle and the connected states. In

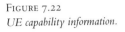
FIGURE 7.22
UE capability information.

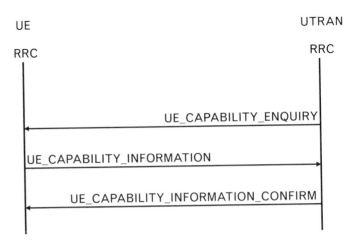

the idle state no dedicated radio resources existed between the UE and the base station. We should observe, however, that the idle state is a rather poor name, as the mobile station is far from being "idle." There are several idle-mode tasks it must handle, tasks such as neighbor cell monitoring, cell reselection, paging channel reception, and broadcast data reception. In the connected state, however, a duplex radio connection is in place. The boundary between the idle and the connected mode is pretty clear; it is the existence of a dedicated radio resource. But in the new UTRAN system, this division is blurred.

The idle state in UMTS is similar to GSM, as well as to those we find in other 2G systems: There is no uplink connection whatsoever. The UE has to monitor its radio environment regularly and, when necessary, perform a cell-reselection task. The reception of the broadcast system information and paging messages belong to the UE's idle-mode tasks.

The connected state is different from the corresponding state in circuit-switched 2G systems, but it has similarities with the packet-switched GPRS system. The connected mode is divided into four states (see Figure 7.23). In the connected state there exists a logical RRC level connection between the UE and the UTRAN, but not necessarily a dedicated physical connection.

1. CELL_DCH is a state in which a dedicated connection exists in both directions. This state is entered while an RRC connection is established with dedicated channels, and it is abandoned when the connection is released. This state is comparable to dedicated mode in the 2G circuit-switched networks.

2. CELL_FACH is a state in which there are no dedicated connections, but data can still be transferred. This data transmission is done via common channels. This feature is very useful if the amount of data transferred is small or it is bursty. The use of a common channel

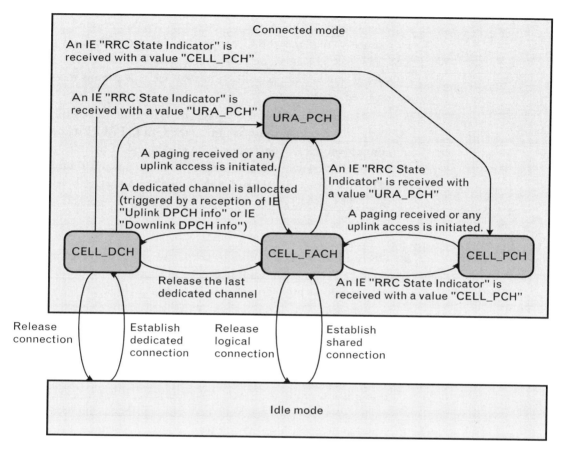

FIGURE 7.23   *RRC protocol states.*

preserves the radio resources in the cell. In the uplink direction, small data packets and control signals can be sent on a RACH or on a CPCH. In the downlink direction, the FACH can be used. In the TDD mode, the USCH and the DSCH can be used. The amount of data transmitted is monitored, and if necessary, dedicated resources can be allocated followed by a state change to the CELL_DCH state.

Because the CELL_FACH state requires the mobile to monitor the FACH channel, it consumes power, which is a scarce resource in handheld equipment. Therefore, if there is no data-transmission activity for a certain time, the RRC moves from the CELL_FACH state to the CELL_PCH state.

3. CELL_PCH state is much like the idle mode because only the PICH is monitored regularly. The broadcast data (i.e., the system

information and cell broadcast messages) are also received. The difference is that the RRC connection still exists logically in the CELL_ PCH state. The RRC moves back to the CELL_FACH state if any uplink access is initiated, or if a paging message is received.

Note that in order for the RRC to move from the CELL_PCH to the idle mode, it must first go to the CELL_FACH state so that connection release messages can be exchanged. If the UE makes a cell reselection while in the CELL_PCH state, it must inform the UTRAN about this. This also requires a temporary cell change to the CELL_FACH state. No uplink activity is possible in the CELL_PCH state itself.

4. URA_PCH is quite similar to the CELL_PCH state, except that every cell change does not trigger a cell-update procedure. In this state an update procedure is only initiated if a UTRAN registration area changes, which is not done with every cell reselection. A state change to this state is requested by the UTRAN if it sees that the activity level of the UE is very low. The purpose of this state is to reduce the signaling activity because of cell updates. The drawback of this arrangement is that if the UTRAN wants to initiate data transmission while the RRC is in this state, it has to expand the paging area from one cell to several cells, possibly to the whole registration area because the location of the UE is not known with great accuracy.

Note that the UTRAN registration area (URA) is a different concept from that of the CN GPRS routing area. The various location concepts in 3G (both in the core network and in the UTRAN) are further discussed in Section 7.7.

We should notice that the CELL_PCH state is actually a subset of the URA_PCH state. As discussed in Section 7.7, it is possible to define overlapping URAs to be used in the URA_PCH state. Thus, the UTRAN operator could define that each cell is a separate URA in addition to other larger URAs. Then the operator could assign small one-cell URAs for slow-moving mobiles, and larger URAs for mobiles with greater mobility. The small URAs could nicely perform the task of the CELL_PCH state. However, it has been decided to keep these states separate.

Generally, the state changes between these states are controlled by the UTRAN, but not always. For example, if a UE that is in the CELL_PCH or URA_PCH state wants to initiate a mobile-originated call, it moves to the CELL_FACH state before initiating the RACH procedure.

The higher layers in NAS, or applications, do not need to know about these states, as they are internal to RRC. The NAS only needs to known if the AS is in connected or idle mode. If the AS is connected, the NAS can always send data through it. If the internal RRC state happens to be, for instance, CELL_PCH, then RRC will move to CELL_FACH state and send the data via common channels or set up a dedicated connection.

7.7 Location Management in UTRAN

The cell-reselection procedure may also trigger the cell-update procedure. This procedure is used to update the UTRAN registers about the location of the UE. Note that in 3G, the mobility concept is handled separately in both the UTRAN and the CN. This is because the UTRAN and the CN are two separate logical entities, and at least in principle, the CN does not know what kind of access network is connected to it and vice versa. They cannot make use of the location concepts of each other because they do not know about them.

The CN level of mobility is handled with MM tasks. There are two location concepts at the CN level. Location areas are used by the circuit-switched network and routing areas by the packet-switched network. The location area of a UE is stored in the MSC/VLR, where it is used to route the paging messages to the right area. The routing area information of a UE is stored in the SGSN, where it is used for packet-switched paging. If a UE crosses a location area/routing area border, it initiates a location area/routing area update toward the CN. Optionally, the network may also demand that these registrations be done periodically. A successful registration will be acknowledged by the network, which may at the same time issue a new temporary mobile subscriber identity (TMSI) to the UE. If the CN has to page the UE, it will usually do so by referring to its recently assigned alias: the TMSI.

At the UTRAN level, the location concepts are the registration area and the cell area, which are independent of the CN location concepts. The UTRAN location concepts are only valid and used when the UE is in the RRC connected mode. They are also only visible within the UTRAN (see Figure 7.24).

As stated, the UTRAN-level location concepts are only maintained and used while the UE is in the RRC connected mode. For those only accustomed to circuit-switched networks, this may sound a bit strange. Why do we need location management if there is already a connection? In the WCDMA packet-based radio interface, the concept of "connected" is a bit different from that of the circuit-switched world. Here the UE may be in a connected state, even though it does not have a dedicated channel. Common transport channels may be used for data transfer if the data is sporadic or low in volume. As there is no dedicated channel, the network does not

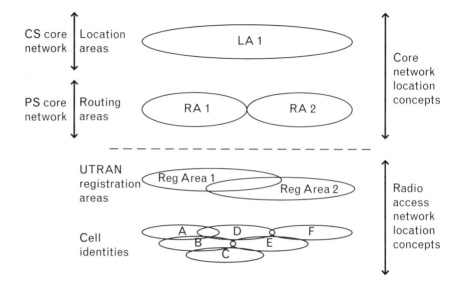

FIGURE 7.24
Location concepts.

directly know about the UE's movements. Therefore, the UE must inform the UTRAN if its location changes while it is in the RRC connected mode without a DCH. If the UE is in the CELL_FACH or CELL_PCH substates of the connected mode, it must inform the UTRAN of every cell change. This procedure is called the cell update (see Figure 7.25). If the UE has higher mobility, then the UTRAN may order the UE to the URA_PCH substate, in which the UE initiates a location registration only when it moves to a cell that belongs to another URA. This reduces the signaling overhead caused by the cell-area updates. The drawback is of course that a paging message may have to be sent to the whole URA, that is, to several cells. Both the cell-area and URA-update procedures can also be done periodically if the UTRAN so orders. Periodic updates are good in that they will remove the "ghost" users from the register of active users. Normally, when a UE is being switched-off, it will inform the UTRAN about this so that UTRAN can remove the UE from the register of active users, and, for

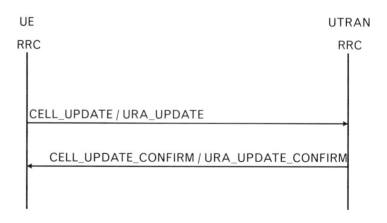

FIGURE 7.25
Cell/URA update.

example, in case of an incoming call, no paging is attempted. However, if the UE is outside the network coverage (for example, inside a thick-walled buiding) when it is switched off, then it cannot send the detach indication, and the UTRAN still assumes it to be active. Periodic updates remove this problem: If a UE does not send its scheduled location update message, it can be assumed to be switched-off.

The URAs can be overlapping or even hierarchical. The same cell may belong to several different URAs, and the UEs in that cell may have been registered to different URAs. SIB 2 contains a list of URA identities indicating which URAs this cell belongs to. This arrangement is done to further reduce the amount of location update signaling because now the UEs moving back and forth in the boundary area of two URAs do not have to update their URA location information if the boundary cells do belong to both URAs.

For example, in Figure 7.26 (left) users A, B, and C constantly cross the URA boundary, triggering URA update procedures. However, if URA areas are overlapping, such as in Figure 7.26 (right), then no update procedures are needed, if user A is assigned URA ID = 44 (or 45), user B is given URA ID = 45, and user C is given URA ID = 46.

As mentioned earlier, in addition to overlapping URA areas, there can also be hierarchical URA structures. One cell can have up to eight different URA identities.

The UTRAN location-registration procedures may also include the reallocation of the temporary identity of the UE. This identity can be included in the cell/URA update confirm message, and it is called the radio

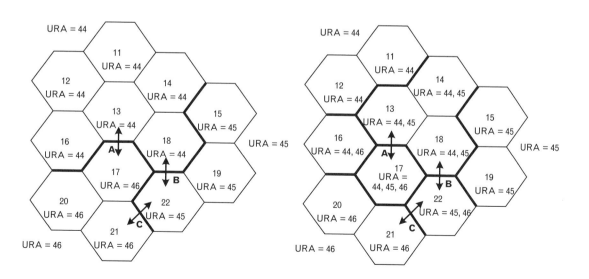

FIGURE 7.26   *Overlapping URA areas.*

network temporary identifier (RNTI). The RNTI is used to address the UE on common transport channels. If dedicated channels are used, then there is no need for an RNTI. The paging messages received in the RRC connected mode normally refer to the UE by using its RNTI. There are two variations of the RNTI: (1) cell RNTI (C-RNTI) and (2) UTRAN RNTI (U-RNTI).

The C-RNTI identifies a UE within a cell, so it can only be used in paging messages when the UE's location is known (i.e., it must be in the CELL_PCH state). This also implies that a new C-RNTI will be allocated to a UE every time it moves to a new cell and conducts a cell-update procedure. The U-RNTI is a UTRAN-wide identity that is used by the UTRAN for paging if it knows that the UE is in the URA-PCH substate (i.e., its location is only known at the URA level). Note that U-RNTI is not a URA-specific identity, but it identifies a UE within the whole UTRAN. Therefore, it can be optionally used for CN-originated paging [15].

The location concepts in the CN and in the UTRAN are not connected in any way. The operator can freely define them independently. Of course, this means that a routing area and a registration area could be defined to be the same, but this is not the intention of the specifications.

7.8   Core Network Protocols in the Air Interface

Only a short overview will be given of the CN protocols in the air interface. This is because these protocols already exist in the GSM/ GPRS system and they can be studied in other sources. A classic GSM reference book is [2]. Other useful GSM publications include [3–5].

7.8.1   Circuit-Switched Core Network

7.8.1.1   Mobility Management

As the name of this task states, one of the main functions of the MM task is location management. But the MM task has also been assigned network-registration and security functions. Typically, the MM procedures can be divided into three groups:

1.   MM common procedures;

2.   MM specific procedures;

3.   MM connection-management procedures.

The MM common procedures can always be initiated while an RRC connection exists. The procedures belonging to this type fall into two classes determined by what entity initiates them:

Initiated by the Network:

- TMSI reallocation procedure;
- Authentication procedure;
- Identification procedure;
- MM information procedure;
- Abort procedure.

Initiated by the Mobile Station:

- IMSI detach procedure.

An MM-specific procedure can only be initiated if no other MM-specific procedure is running or no MM connection exists. The procedures belonging to this type include:

- Normal location updating procedure;
- Periodic updating procedure;
- IMSI attach procedure.

The MM connection-management procedures are used to establish, maintain, and release an MM connection between the mobile station and the network over which an entity of the upper CM layer can exchange information with its peer. An MM connection establishment can only be performed if no MM-specific procedure is running. More than one MM connection may be active at the same time.

In the following a short description of each procedure from the previous list is given.

The TMSI reallocation procedure provides identity confidentiality, that is, protects a user against being identified and located by an intruder. TMSI stands for temporary mobile subscriber identity. It can be used instead of globally unique IMSI to identify a user to the network. Usually the TMSI reallocation is performed at least at each change of a location area.

The authentication procedure permits the network to check whether the identity provided by the mobile station is acceptable or not. The network sends the UE two parameters, RAND and AUTN, and from these and its own secret parameters, it calculates a response parameter RES. The details of this procedure are explained in [13]. Authentication procedure also provides parameters enabling the mobile station to calculate new UMTS ciphering and integrity keys. The UMTS authentication procedure is always initiated and controlled by the network.

The identification procedure is used by the network to request a mobile station to provide specific identification parameters to the network. This parameter can be IMSI, IMEI, IMEISV, or TMSI.

MM information procedure can be used to provide the mobile station with subscriber specific information. In practice this information includes various time-zone and clock information.

The abort procedure may be invoked by the network to abort any on-going MM connection establishment or already established MM connection.

The IMSI detach procedure is performed by the UE if it is being deactivated or if the SIM is detached from the UE. The detach procedure is optional, and a flag (ATT) broadcast in SIB 1 message on the BCCH is used by the network to indicate whether the detach procedure is required. The procedure causes the mobile station to be indicated as inactive in the network.

The normal location updating procedure is used to update the location area (LA) information in the network registers. This procedure is triggered when in the idle mode the UE selects a cell that belongs to a different LA than the previous cell. Periodic updating procedure is similar to the normal location updating procedure, except that it will be triggered periodically by a timer. It is used to notify the network about the availability of the UE.

The IMSI attach procedure is used to indicate the UE as active in the network. Typically, this happens when the UE is switched-on. IMSI attach is an optional procedure used whenever IMSI detach is used. This is indicated by a SIB 1 message. IMSI attach employs the normal location update procedure signaling; only the location updating type field in the message indicates that this is an IMSI attach.

The MM protocol is defined in [16].

7.8.1.2   GPRS Mobility Management

The GPRS mobility management (GMM) sublayer provides services to the session management (SM) entity and to the SMS (SMS) support entity for message transfer.

Depending on how they can be initiated, two types of GMM procedures can be distinguished:

1.   GMM common procedures:

2.   GMM-specific procedures.

GMM common procedures are initiated by the network when a GMM context has been established:

- P-TMSI (re)allocation;

- GPRS authentication and ciphering;

- GPRS identification;

- GPRS information.

GMM-specific procedures can be initiated either by the network:

- GPRS detach.

Or they can be initiated by the UE:

- GPRS attach and combined GPRS attach;

- GPRS detach and combined GPRS detach;

- Normal routing-area updating and combined routing-area updating;

- Periodic routing-area updating;

- Service request.

These procedures are very similar to the corresponding MM procedures described earlier. The main difference is that whereas MM procedures are used between a UE and a circuit-switched core network, GMM procedures are used between a UE and a packet-switched core network.

The GMM protocol is defined in [16]. In practice this protocol is considered to be an extension of the MM protocol and can be implemented within the same protocol entity.

7.8.1.3 Call Control

The call control (CC) protocol is one of several protocols in the connection management (CM) sublayer. This protocol includes the control functions for the call establishment and release.

A CC entity must support the following elementary procedures:

- Call-establishment procedures;

- Call-clearing procedures;

- Call-information-phase procedures;

- Miscellaneous procedures.

A call can be either a mobile-originated call (MOC) or a mobile-terminated call (MTC); that is, it can be initiated by either the mobile or by the network. Optionally the UE can also support a network-initiated MOC. This functionality can be used with the completion of calls to busy subscriber (CCBS) supplementary service.

The call-clearing procedure can be initiated either by the UE or by the network. Note, however, that this means the logical CC-level connection clearing. The actual radio connection (RRC level) is always released by the UE. A radio connection and a CC connection are separate concepts. One can use the radio connection for many other things besides the circuit-switched call, such as SMS and for packet-data applications. Therefore, releasing a call connection does not necessarily mean that the radio connection should also be released. There may be other applications that still need the radio connection.

While the call is active, the CC can perform various procedures. The user-notification procedure informs the user about call-related events, such as user suspension or resume. Support of multimedia calls will be an important procedure especially in UMTS. The dual-tone multifrequency (DTMF) control procedure enables the user to send DTMF tones toward the network. Key presses in the UE containing digit values (0–9, A, B, C, D, ⋆, #) are signaled over the air interface to the MSC, which converts them into DTMF tones and sends them onward to the remote user. Typical applications of the DTMF include various automated information services (e.g., telephone banking: "Press 1 if you want to hear your bank account balance; Press 2 if you want to settle your bills; Press 3 if you want to talk to the operator," and so forth). The support of DTMF is described in [17]. The support for the in-call modification procedure is optional for the UE. This procedure means that the same connection can be used for different kinds of information transfer during the same call, but not at the same time. In practice, this procedure is used for alternating the call between speech and fax services or between speech and data.

Miscellaneous CC procedures include in-band tones and announcements, status inquiry, and call reestablishment. The in-band tones and announcements procedure is used when the network wants to make the mobile station attach the user connection (e.g., in order to provide in-band tones/announcement) before the UE has reached the "active" state of a call. In this case, the network may include a progress indicator (IE) indicating user attachment in a suitable CC message. The status-inquiry procedure can be used to inquire about the status of the peer entity CC. This is a useful procedure in error handling. The call-reestablishment procedure is mostly an RRC-layer matter, as it involves setting up a new radio connection in place of the lost one. Within the CC level, however, this procedure includes provisions for the UE to make a decision as to whether a reestablishment should be attempted. The network-side CC must also identify and resolve any call states or an auxiliary state mismatch between the network and the UE.

The CC protocol is defined in [16].

7.8.1.4   Supplementary Services

Supplementary services (SS) are value-added services that may or may not be provided by the network operator. The list of various GSM supplementary services is long and ever increasing. These include, for example, the advice of charge (AoC), call forwarding (CF), and call waiting (CW) supplementary services. Because these services belong to the NAS, they are applicable to both GSM and the UMTS. It is likely that later on there will also be UMTS-only supplementary services.

One generic protocol is defined for the control of SS at the radio interface. It is based on the use of the facility information element or the facility message. The exact functionality triggered by this information element or message depends on the information it contains.

SSs are discussed further in Section 13.4. The SS protocol is defined in [18] and in 3G TS 24.08x and the TS 24.09x series of specifications.

7.8.1.5   Short Message Service

The purpose of the SMS is to provide a means to transfer short text messages between a UE and a short message service center (SMSC). These messages are sent using the control signaling resources, and their maximum length can be only 160 characters.[3] SMS is a non-real-time service; a store-and-forward service in which messages can be stored on the SMSC and delivered when the destination UE is available.

The term *SMS-MO* refers to a mobile-originated SMS message; *SMS-MT* refers to a mobile-terminated SMS message.

Note that a UTRAN 3G network will also include an enhanced version of the SMS called the multimedia messaging service (MMS); see Section 12.4.

The SMS protocol is defined in [19].

7.8.2   Packet-Switched Core Network

7.8.2.1   Session Management

The main function of the SM protocol is to support packet data protocol (PDP) context handling of the user terminal. Note that there is no "connection" concept in a (IP) packet-switched system as we know it in a circuit-

---

3   Nowadays users can send longer than 160-character text messages with their mobiles, but technically those are divided into several SMS messages, each with a maximum of 160 characters and sent separately over the air interface.

switched system. However, the communicating entities do need to know about the characteristics of the data to be transferred. This task is performed by the PDP context-activation procedure. Other functions this task must perform include PDP deactivation and PDP modification.

The SM procedures for identified access can only be performed if a GMM context has already been established between the UE and the network. If no GMM context has been established, the MM sublayer must initiate the establishment of a GMM context by use of the GMM procedures. After GMM context establishment, the SM uses services offered by GMM. Ongoing SM procedures are suspended during GMM procedure execution.

The SM protocol is defined in [16].

7.8.2.2   GPRS Short Message Service Support

The GPRS Short Message Service (GSMS) protocol task handles the SMS service while the UE is attached to the PS CN; that is, to the GPRS system. In practice this protocol is an extension of the circuit-switched SMS protocol, and both will typically be implemented within one protocol task entity. See [19] for further information.

7.9   User Plane

The lower layers of the U-plane are exactly the same as those of the C-plane (MAC and RLC). PDCP and BMC, however, exist only in the U-plane. The control of all the AS U-plane tasks is handled by the RRC. The U-plane is responsible for the transfer of user data, such as voice or application data, whereas the C-plane handles the control signaling and the overall resource management (see Figure 7.27).

7.10   Packet Data Convergence Protocol

As the name implies, the PDCP task is a convergence layer between the actual data protocol in the NAS and the radio access protocols in layer 2 (Figure 7.28). The PDCP itself is an AS protocol. This protocol entity is only used in the U-plane. The required control signaling for the PDCP is handled by the RRC. The PDCP handles the same functionality in the UTRAN as the SNDCP task does in the GPRS system.

The network layer in the NAS can accommodate several different data protocols. These current (and future) protocols must be transferred transparently over the UTRAN. This is the task of the PDCP, which must

FIGURE 7.27
*WCDMA U-plane proto-
col stack.*

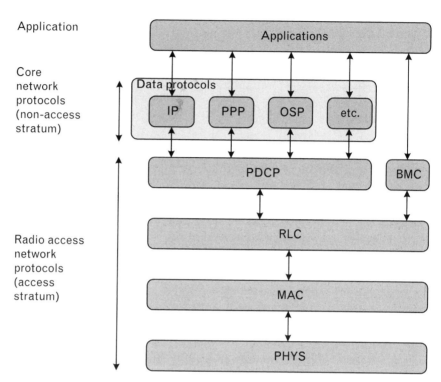

WCDMA air interface protocol stack
User plane

FIGURE 7.28
PDCP model.

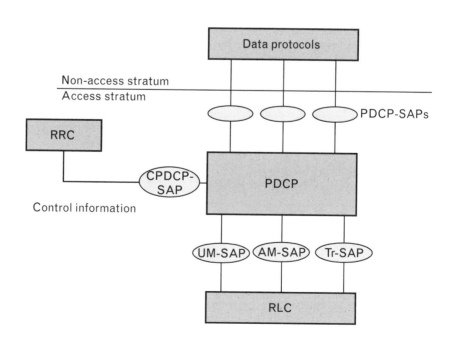

hide the particularities of each protocol from the UTRAN. The packets from all of these protocols will be conveyed over the UTRAN without any changes to the UTRAN protocols.

Therefore, the functions the PDCP shall perform include the following:

- Header compression and decompression of IP data streams;

- Transfer of user data;

- Maintenance of PDCP sequence numbering;

Header compression and decompression are performed by the PDCP to optimize the channel efficiency in the radio interface. The network data protocols are not especially designed for wireless environments; thus, they may have unnecessarily large header fields in their data packets. It is the task of the PDCP to compress these headers to more compact representations. Each data protocol has its own header format, so the PDCP must accommodate different compression algorithms. The particular algorithms and parameters are negotiated by the RRC protocol task, which indicates the result to the PDCP.

The transfer of user data includes forwarding the NAS data to the RLC layer and vice versa. Note that if acknowledged transfer mode is used in the RLC, then buffering of N-PDUs received from NAS is needed. They must be stored until the peer entity RLC acknowledges that they have been successfully sent.

The maintenance of PDCP sequence numbering is used in the SRNS relocation procedure only. If the SRNS changes in the UTRAN and AM is used in RLC, then an orderly continuation of data transfer requires that the PDCP sequence numbering information must be are exchanged between the UE and the UTRAN. This is called a "lossless SRNS relocation." The RRC may also command that a lossless SRNS relocation will not be supported.

The PDCP protocol is specified in [20].

7.11   Broadcast/Multicast Control

Broadcast/multicast control is a layer 2 sublayer that exists only in the U-plane. The necessary control information is received from the RRC, just as in the PDCP sublayer. This layer handles only downlink broadcast/multicast transmission (see Figure 7.29).

FIGURE 7.29
BMC protocol model.

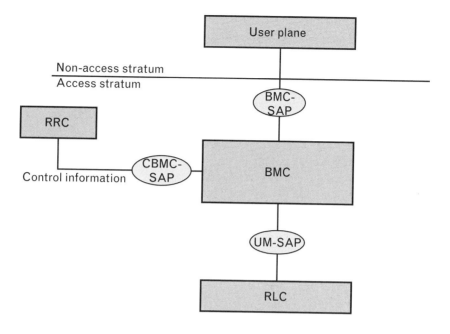

The BMC task implements the transfer of cell broadcast messages. Cell broadcast service (CBS) is not a UMTS-only service because it is also used in GSM, though not yet very widely. This service is specified for UMTS in [21]. Cell broadcast messages are SMS text messages (although they can be much longer than the normal SMS messages) that are broadcast to everybody in a cell (or in a set of cells). A CBS message consists of CBS pages. One page contains 82 octets, and if 7-bit characters are used, it is possible to send 93 characters in a page. A CBS message may contain up to 15 pages, which gives a maximum size of 1,395 characters for CBS messages. These messages can be received by all mobiles capable of receiving CBS. But the mobiles have to be in either the idle, the CELL_PCH, or the URA_PCH states. Cell broadcast messages are assigned a message class type, which can be used by the UE to filter and receive only those messages that are of interest to it. The categories to be subscribed to could include contents, such as news, traffic information, and weather forecasts. The user can, of course, reject all cell broadcast messages.

The functions of BMC are specified in [22]. These include the following:

- Storage of cell broadcast messages;

- Traffic-volume-monitoring and radio resource requests for CBS;

- Scheduling of BMC messages;

- Transmission of BMC messages to UEs;

- Delivery of cell broadcast messages to the upper layer (NAS).

The BMC entity in the RNC is responsible for storing the cell broadcast messages to be sent. They cannot usually be sent further right after they have been received from the core network. Their transmission to the UE must be scheduled, and also typically be repeated several times. Thus, the UTRAN-BMC needs some storage space.

The BMC in the RNC must also estimate the expected amount of traffic volume that is required for transmission of queued CB messages. This is indicated to the RRC so that it can allocate the necessary radio resources.

The CB messages are scheduled to enhance the performance of the UEs receiving them. A UE can listen for dedicated CB scheduling messages, and from those, extract the scheduling of the actual information bearing CB messages type by type. Thus, a UE does not have to receive all CB messages but only those that it knows belong to categories it has subscribed to. Outside the scheduled message sending times, the UE can enter DRX mode to save power. The BMC entity in the RNC is responsible for building the schedule and sending this information in schedule messages. The BMC in the UE must receive these messages and then inform the RRC so that it knows when to listen for the actual CBS messages. The configuration of layer 1 is done by the RRC, not by the BMC.

The actual transmission of the CB messages is done according to the defined schedule, and the BMC in the UE should forward to upper layers only those messages belonging to subscribed groups. The BMC also has to compare the message IDs and serial numbers of the received messages to the IDs and the numbers already received and stored. If they are identical, then the received message can be discarded.

We can see from this description that the BMC task in the UE is rather simple, but more involved in the RNC, which has many more functions to handle.

The broadcast/multicast protocol specification is in [22]. Broadcast services are further discussed in [21, 23].

7.12   Data Protocols

The PDCP layer connects to the standard data protocols in the NAS. Because these protocols are not specific to the 3G system, they are not discussed further here. The PDCP layer handles the header compression for these protocols because in some cases the size of the header would consume too large a share of the available transmission bandwidth.

The point-to-point protocol (PPP) is defined in [24, 25]. Both IPv4 and IPv6 will be supported by the PDCP. IPv4 is specified in RFC 791 and IPv6 in RFC 2460.

7.13   Dual-System Protocol Stack in UE

Figure 7.30 describes one possible implementation of the dual-system (GSM–3G) protocol stack in the UE.

As can be seen, the 3G UTRAN protocol stack is quite separate from the GSM–GPRS protocol stack in the AS level. Code reuse in the radio access protocols is not possible, but on the other hand, this kind of separation makes the implementation of new protocol tasks easier; the difficult dual-system issues do not have to be addressed in every line of new code. Once again, notice that the RLC/MAC in 3G and the RLC/MAC in GPRS are not the same protocols.

The NAS [i.e., core network protocols (MM and CM)] are, however, similar in both systems and they can be reused. Both the MM and CM layers from the GSM system will require some small modifications to accommodate the 3G radio access protocols below them. The GSM core network protocols will be upgraded to also support the UMTS features, at the same

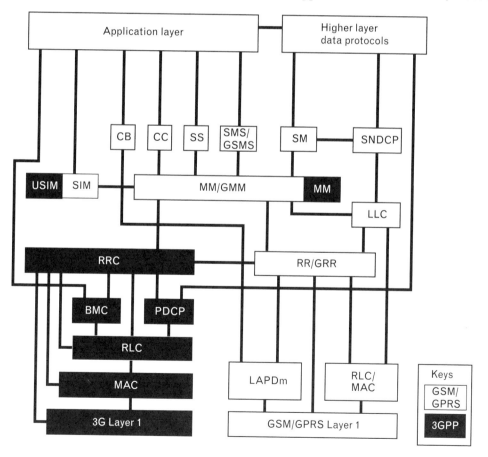

FIGURE 7.30   *Dual-system GSM/GPRS-3G (UTRAN) protocol stack.*

time retaining their GSM compatibility. The same core network can then support both the GSM and UMTS RANs. This means that these core network protocols must also be similar in the mobile station. The 3GPP will also handle the future specification work for the GSM core network protocols, so this makes the goal of common NAS protocols for both GSM and 3G an easier task.

References

[1]   Heine, G., *GPRS from A–Z*, Norwood, MA: Artech House, 2000.

[2]   Mouly, M., and M.-B. Pautet, *The GSM System for Mobile Communications*, published by the authors, 1992.

[3]   Walke, B., *Mobile Radio Networks*, New York: Wiley, 1999.

[4]   Mehrotra, A., *GSM System Engineering*, Norwood, MA: Artech House, 1997.

[5]   Redl, S., M. Weber, and M. Oliphant, *An Introduction to GSM*, Norwood, MA: Artech House, 1995.

[6]   3GPP TS 25.303, v 5.0.0, Interlayer Procedures in Connected Mode, 2002.

[7]   3GPP TS 25.321, v 5.0.0, MAC Protocol Specification, 2002.

[8]   3GPP TS 25.322, v 5.0.0, RLC Protocol Specification, 2002.

[9]   3GPP TS 25.331, v 5.0.0, RRC Protocol Specification, 2002.

[10] 3GPP TS 25.304, v 5.0.0, UE Procedures in Idle Mode and Procedures for Cell Reselection in Connected Mode, 2002.

[11] 3GPP TS 23.122, v 4.1.0, NAS Functions Related to Mobile Station (MS) In Idle Mode, 2001.

[12] 3GPP TS 25.302, v 5.0.0, Services Provided by the Physical Layer, 2002.

[13] 3GPP TS 33.102, v 4.3.0, 3G Security; Security Architecture, 2001.

[14] 3GPP TR 25.922, v 5.0.0, Radio Resource Management Strategies, 2002.

[15] 3GPP TS 25.301, v 5.0.0, Radio Interface Protocol Architecture, 2002.

[16] 3GPP TS 24.008, v 5.3.0, Mobile Radio Interface Layer 3 Specification; Core Network Protocols-Stage 3, 2002.

[17] 3GPP TS 23.014, v 4.0.0, Support of Dual Tone Multi-Frequency (DTMF) Signaling, 2001.

[18] 3GPP TS 24.010, v 4.2.0, Mobile Radio Interface Layer 3; Supplementary Services Specification; General Aspects, 2001.

[19] 3GPP TS 24.011, v 4.1.0, Point-to-Point (PP) Short Message Service (SMS) Support on Mobile Radio Interface, 2002.

[20] 3GPP TS 25.323, v 5.0.0, Packet Data Convergence Protocol (PDCP) Specification, 2002.

[21] 3GPP TS 23.041, v 4.2.0, Technical Realization of Cell Broadcast Service (CBS), 2001.

[22] 3GPP TS 25.324, v 5.0.0, Broadcast/Multicast Control BMC, 2002.

[23] 3GPP TR 25.925, v 3.4.0, Radio Interface for Broadcast/Multicast Services, 2001.

[24] IETF RFC 1661 "The Point-to-Point Protocol (PPP)," W. Simpson (ed.), July 1994.

[25] IETF RFC 1662 "PPP in HDLC-Like Framing," W. Simpson (ed.), July 1994.

CHAPTER 8

Network

8.1 General Discussion

A concise description of the 3GPP concept might be "a CDMA packet-based air interface combined with a GSM + GPRS core network." ITU's 3G concept, known as IMT-2000, included several other accepted technologies for 3G systems. However, the referenced WCDMA + GSM combination will be the most widely used. The simple reason for this is the fact that among the 2G mobile cellular networks, GSM is by far the most widely used technology. All of the significant 3G proposals for IMT-2000 are those that successfully protect the investments in their 2G legacy networks in a 3G world. The 3GPP work strives to protect the GSM investments and bring the markets into 3G. The present-day operators will not want to invest in new networks if they cannot recycle their existing GSM networks. Most big network manufacturers and operators are actively supporting this approach. So, for these operators, this 3G solution is a good deal because they can continue to use their upgraded GSM networks. In many cases the radio access network can be updated to conform to the 3G requirements. Mobile phone users, however, are not so lucky, as they will need new phones that are capable of accessing WCDMA base stations. Most probably, these phones will have to be dual-system GSM + WCDMA phones at first because the UTRAN radio access network coverage will be quite limited at service launch. The UTRAN may only provide coverage in urban hot-spots in the beginning, as the old GSM networks will be used to provide wide-area service.

The IMT-2000 network is divided into two logical concepts, the core network (CN) and the generic radio access network (GRAN). The noble idea behind this arrangement is that the GRAN will be capable of connecting, perhaps simultaneously, to several different CNs, such as GSM, B-ISDN + IN, or a packet-data network. The GRAN could be implemented, for example, as a GSM BSS, DECT, LAN, CATV, or Hiper-LAN2 network. 3GPP has also specified a new dedicated UMTS radio access network (RAN) called the UMTS Terrestrial RAN (UTRAN). An important requirement for the GRAN implementations is that they conform to the Iu interface specifications. Note, however, that the 3GPP

Release 99 specifications only contain provisions for the GSM–MAP (including GPRS) and the ANSI–41 core networks.

In GSM terms, the GRAN contains the base station subsystem, that is, the base transceiver station (BTS) and the base station controller (BSC). In the 3GPP specifications, the generic GRAN concept is translated into a concrete UTRAN network in which the BTS has the curious name Node B. The new name for the BSC is the radio network controller (RNC).

Between the GRAN and the core network we find the Iu interface, and between the GRAN and the UE we see the Uu interface (radio interface). See Figure 8.1.

8.2   Evolution from GSM

It must be noted that a GSM phase 2+ network provides a smooth transition path to UMTS, especially if the operator also operates a GPRS network. GSM networks have been updated little by little to include more and more features. Table 8.1 shows how well the future GSM 2.5G network will comply with the UMTS requirements.

As we can see from Table 8.1, a GSM network with all the add-ons very closely mimics a UMTS network. The only difference is the more flexible and capable UMTS air interface, which can handle different bearer types at the same time. Real-time services are confined to dedicated connections whereas non-real-time low-bandwidth services can quite easily use shared communication channels, which can more easily be changed dynamically.

UMTS can also achieve higher bit rates, but it must be noted that the differences between UMTS and 2.5 GSM are not so large from the user point of view. A GSM with all the 2.5G upgrades could achieve close to 200-Kbps user data rates. In theory (and in the marketing talk) GSM could

Figure 8.1
UMTS architecture.

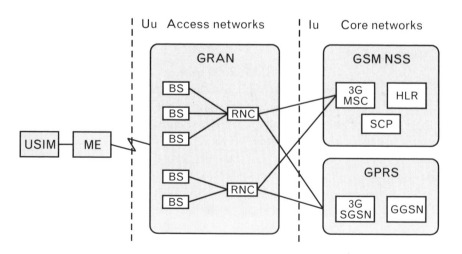

Table 8.1   GSM COMPLIANCE FOR UMTS TARGETS

UMTS TARGET	GSM COMPLIANCE
Small affordable hand portables	Yes
Deep penetration (50%)	Yes; already in some markets
Anywhere, anytime (indoor, office)	Yes (picocells, GSM office)
Anywhere, satellite mobile interworking	Yes
Hot spot capacity	Yes (cell hierarchies)
Wireline voice quality	Yes (EFR codec)
Global roaming	Yes (SIM, MAP)
IN services	Yes (CAMEL)
Multimedia, entertainment, nonvoice	Yes (TCP/IP transparency, GPRS, HSCSD)
Flexibility to mix different bearer types (non-real-time and real-time)	No
High bit rate services (200 Kbps)	No

From: [31].

approach 384-Kbps rates. If a UTRAN network wants to exceed this speed, it has to use very low spreading factors and allocate the resources of a base station mostly to one user. When a GSM base station offers close to 200-Kbps speeds to one user, it only uses one of its frequency carriers; there are probably several other carriers still available for other users. But a typical WCDMA base station has only one downlink (DL) frequency carrier, and if one user is provided with a DL connection of over 2 Mbps, then other users are left with very little capacity. In practice, the situation is not so bad because high-speed traffic is typically bursty and not continuous; thus, several users can have high data rates momentarily. There are also techniques to enhance the UTRAN's DL capacity further, like sectorization, smart antennas, and higher-order modulation schemes.

We can see that a GSM + GPRS combination provides a very good foundation for the UMTS core network building process. The biggest operator investment will clearly be building out the radio access network. Some of the latest GSM base stations are, however, said to be upgradeable to UTRAN standards. We are excluding the operator license fees here, as they are not technology related to the network's implementation.

Note that the easy accommodation of GSM for the UMTS requirements may also be a problem for new UMTS operators because the existing GSM (non-UMTS) operators can provide almost the same services without the extra UMTS investments. The competition will be hard for the new UTMS operators in the early phases, and it is especially hard for the new

UMTS operators, which do not have an existing 2G network. It is very expensive for them to build out wide-area-coverage 3G networks while they don't have any income from existing networks. Furthermore, in many countries the operating licenses are very expensive. The combined cost burden from the licensing fees, interest, and network construction can push a new green-field operator into a very unfavorable position. GSM operators can enhance their systems with EDGE and possibly WLAN hot-spot coverage, which are relatively inexpensive upgrades when compared to 3G investments. Telecommunications authorities may have to do some creative thinking to find ways to help these 3G companies. One way to ease the situation would be to force the current 2G network operators to lease their networks to new 3G operators so that they can provide wider coverage for their customers from the beginning. There are already successful examples of this concept in the GSM world. New GSM-1800 operators have been able to provide wider coverage by leasing GSM-900 capacity from existing operators in some countries. Note that these old operators are competitors for the new networks, so the telecommunication authority must be very careful in its decisions so that free competition is not obstructed more than absolutely necessary.

8.3   UMTS Network Structure

Figure 8.2 depicts the UMTS architecture at the very highest level. This chapter concentrates on both the core network (CN) and the UTRAN. Section 8.4 discusses the CN, and Section 8.5 handles the UTRAN. The interfaces between the UE and the UTRAN (Uu interface) and between the UTRAN and the CN (Iu) are open multivendor interfaces. Note that most of the first seven chapters of this book are dedicated to the Uu interface, with its WCDMA technology.

Figure 8.3 gives a much more detailed description of the UMTS architecture. The reader can see that the core network portion is the same as in the old GSM + GPRS core network combination. The same core network entities may serve both the UTRAN and GSM radio access networks. GSM's radio access network entities (the BSS) are included in the drawing to clarify the relationship of these two technologies. They are likely to linger in the networks to support traditional circuit-switched speech services.

FIGURE 8.2
*High-level UMTS
architecture.*

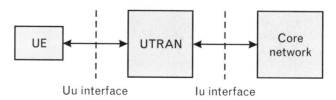

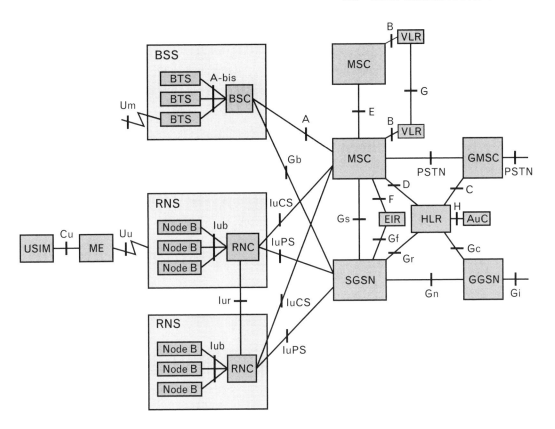

FIGURE 8.3   *UMTS network elements and interfaces.*

Note that this is the network as specified in Release 99. Release 5 will bring some changes to the core network, and these will be discussed in Section 8.9. The first live 3G networks are Release 99 compliant; thus, we will begin our discussion from this release.

The entities in this figure are briefly described in the following paragraphs. Because the core network entities are the same as in GSM/GPRS networks, these are not described in every detail, as there is already plenty of literature available for these networks. In Release 99 the core network is logically divided into two domains: circuit-switched (CS) and packet-switched (PS). The CS-domain handles circuit-switched connections, and the PS-domain handles the packet transfer. CS core network is built around MSCs, and PS core network around SGSNs. Note that the various core network registers are common for both domains, although VLR is typically employed by CS-domain only. In Release 5 there will be a new third domain, but this is discussed in Section 8.9.

Note that this list of network elements is not comprehensive. New services will require new network elements. For example, location services (LCS) need various mobile location centers. There are also group call registers, gateway location registers, and so on. The number of different network elements also depends on the implementation. Some infrastructure vendors may combine small elements into bigger physical units. For a full and up-to-date description of all the core network elements and interfaces, please refer to [5].

8.4 Core Network

8.4.1 Mobile Switching Center

The mobile switching center (MSC) is the centerpiece of the circuit-switched core network. The same MSC can be used to serve both the GSM-BSS and the UTRAN connections. A GSM-MSC must be upgraded to meet the 3G requirements, but the same MSC can be used to serve both access networks. In addition to the radio access networks, it has interfaces to the fixed PSTN network, other MSCs, the packet-switched network (SGSN), and various core network registers (HLR, EIR, AuC). Physically, the VLR is implemented in connection with the MSC, so the interface between them (the B interface) exists only logically.

Several BSSs can be connected to an MSC. The number and the size of MSCs also vary; a small operator may only have one small MSC, but once the number of subscribers increases, several large MSCs may be needed.

The functions of an MSC include the following [1]:

- Paging;

- Coordination of call setup from all MSs in the MSC's jurisdiction;

- Dynamic allocation of resources;

- Location registration;

- Interworking functions (IWFs) with other type of networks;

- Handover management (especially the complex inter-MSC handovers);

- Billing of subscribers (not the actual billing, but collecting the data for the billing center);

- Encryption parameter management;

- Signaling exchange between different interfaces;

- Frequency allocation management in the whole MSC area;

- Echo canceler operation and control.

The MSC terminates the MM and CM protocols of the air interface protocol stack, so the MSC has to manage these protocols, or delegate some responsibilities to other core network elements.

8.4.2 Visitor Location Register

The visitor location register (VLR) contains information about the mobile stations roaming in this MSC area. It is also possible that one VLR handles the visitor register of several MSC areas. Note that a VLR contains information from all active subscribers in its area, even from those to whom this network is their home network, so the name VLR is misleading as most entries in that register are not visitors, but users in their own home network. The VLR contains pretty much the same information as the home location register (HLR), the difference being that the information in the VLR is there temporarily, whereas the HLR is a site for permanent information storage. When a user makes a subscription, the subscriber's data is added to his home HLR. From there it is copied to the VLR the user is currently registered with. When a user registers with another network, the subscriber data is removed from the old VLR and copied to the new VLR. There are, however, some network optimization schemes, which may change this principle in the future. See the supercharger concept in Section 12.5. The VLR contains such data that the normal call setup procedures can be handled without consulting the HLR. This is important especially if the user is roaming abroad, and the signalling connection to the home network is expensive.

A VLR subscriber data entry includes the following information:

- International mobile subscriber identity (IMSI);
- Mobile station international ISDN number (MSISDN);
- Mobile station roaming number (MSRN);
- Temporary mobile station identity (TMSI), if applicable;
- Local mobile station identity (LMSI), if used;
- Location area where the mobile station has been registered;
- Identity of the SGSN where the MS has been registered, if applicable;
- Last known location and the initial location of the MS.

In addition, there can be lots of optional data, depending on what features the network supports [e.g., CAMEL or local service area (LSA)].

The VLR may also contain supplementary service parameters. The procedures the VLR has to perform include the following:

- Authentication procedures with the HLR and the AuC;

- Cipher key management and retrieval from the home HLR/AuC;

- Allocation of new TMSI numbers;

- Tracking of the state of all MSs in its area;

- Paging procedure support (retrieval of the TMSI and the current location area).

The organization of the subscriber data is described in [2].

8.4.3   Home Location Register

The HLR contains the permanent subscriber data register. Each subscriber information profile is stored in only one HLR. The HLR can be implemented in the same equipment as the MSC/VLR, but the usual arrangement is to have the MSC/VLR as one unit, and the HLR/AuC/EIR combination as another unit. One PLMN can have several HLRs.

The subscriber information is entered into the HLR when the user makes a subscription. There are two kinds of information in an HLR register entry, permanent and temporary. The permanent data never change, unless the subscription parameters are changed. An example of this is the user who adds some supplementary services to his/her subscription. The temporary data contain things like the current (VLR) address and ciphering information, which can change quite often, even from call to call. Temporary data are also sometimes conditional; that is, it is not always there. A subscriber data entry can be accessed by either IMSI or MSISDN.

The permanent data in the HLR include among others:

- International mobile subscriber number (IMSI), which identifies the subscriber (or actually his or her SIM card) unambiguously;

- MSISDN [the directory number of the MS (e.g., +44–1234–654321)];

- MS category information;

- Possible roaming restrictions;

- Closed user group (CUG) membership data;

- Supplementary services parameters;

- Authentication key;

- Network access mode (NAM), which determines whether the user can access the GPRS networks, non-GPRS networks, or both.

In addition, if GPRS is supported, PDP addresses are included. Again, there may be lots of other entries, depending on what features the network supports.

The temporary data include the following:

- Local mobile station identity (LMSI);

- Triplet vector; that is, three authentication and ciphering parameters: (1) random number (RAND), (2) signed response (SRES), and (3) ciphering key (Kc);

- Quintuplet vector; that is, five authentication and ciphering parameters: (1) random challenge (RAND), (2) expected response (XRES), (3) cipher key (CK), (4) integrity key (IK), and (5) authentication token (AUTN);

- MSC number;

- VLR number (the identity of the currently registered VLR).

In addition, if GPRS is supported, SGSN and GGSN numbers (SS7 addresses) are included.

Note that these lists are not exhaustive; the subscriber data registers can contain a lot of information (dozens of different entries). The subscriber data organization in the core network is specified in [2]. The tables in the end of that specification give a good picture of what information is stored and where.

The HLR also forwards the charging information to the billing center.

8.4.4 Equipment Identity Register

The equipment identity register (EIR) stores the international mobile equipment identities (IMEIs) used in the system. An EIR may contain three separate lists:

1. *White list:* The IMEIs of the equipment known to be in good order;

2. *Black list:* The IMEIs of any equipment reported to be stolen;

3. *Gray list:* The IMEIs of the equipment known to contain problems (such as faulty software) that are not fatal enough to justify barring them.

At a minimum an EIR must contain a white list. It is unfortunate that the black list and the checks against it are not mandatory, as stolen mobile phones can now be used in some networks that have a weaker security policy. And it is even more unfortunate that changing the IMEI code of a handset is not yet illegal in many countries.

Typically a PLMN has only one EIR, which then interconnects to all HLRs in the network. Note that EIR handles IMEI values, not IMSIs or any other identities. The IMEI is (or should be) a unique identity of a mobile handset assigned when it is manufactured.

8.4.5 Authentication Center

The authentication center (AuC) is associated with an HLR. The AuC stores the subscriber authentication key, Ki, and the corresponding IMSI. These are permanent data entered at subscription time. The Ki key is used to generate an authentication parameter triplet (Kc, SRES, RAND) during the authentication procedure. Parameter Kc is also used in encryption algorithms.

An AuC physically always exists with an HLR. The MAP interface between them (the H interface) has not been standardized.

8.4.6 Gateway MSC

The Gateway MSC (GMSC) is an MSC that is located between the PSTN and the other MSCs in the network. Its function is to route the incoming calls to the appropriate MSCs by first interrogating the appropriate HLR. If the operator allows the outside networks to access its HLRs, then a dedicated GMSC is not necessary as the other networks can route the calls to the right MSC by themselves.

In practice it is also possible that all MSCs are also GMSCs in a PLMN.

8.4.7 Serving GPRS Support Node

The serving GPRS support node (SGSN) is the central element in the packet-switched network. It contains two types of information:

- Subscription information:
  - IMSI;
  - Temporary identities;
  - PDP addresses.
- Location information:
  - The cell or the routing area where the MS is registered;
  - VLR number;
  - GGSN address of each GGSN for which an active PDP context exists.

The SGSN connects to the UTRAN via the IuPS interface and to the BSS via the Gb interface. It also has interfaces to many other network elements as seen in Figure 8.3.

8.4.8 Gateway GPRS Support Node

The gateway GPRS support node (GGSN) corresponds to the GMSC in the circuit-switched network. Whereas the GMSC only routes the incoming traffic, however, the GGSN must also route the outgoing traffic. It has to maintain the following data:

- Subscription information:
    - IMSI;
    - PDP addresses.
- Location information:
    - The SGSN address of the SGSN where the MS is registered.

The GGSN receives this information from the HLR and from the SGSN.

8.5 UMTS Terrestrial Radio Access Network

The UTRAN is the new radio access network designed especially for UMTS. Its boundaries are the Iu interface to the core network and the Uu interface (radio interface) to user equipment (UE).

The UTRAN is just one realization of the GRAN concept. The other possible implementations in the future may include, for example, the Broadband Radio Access Network (BRAN) and the UMTS Satellite Radio Access Network (USRAN).

The UTRAN consists of radio network controllers (RNCs) and Node Bs (base stations). Together, these entities form a radio network subsystem (RNS). See Figure 8.4.

The internal interfaces of the UTRAN include the Iub and Iur. The Iub connects a Node B to the RNC and the Iur is a link between two RNCs.

The Iub is intended to be an open interface, but it is situated in so delicate a position in the network infrastructure that it is also possible that it will, in practice, become a manufacturer proprietary interface. The corresponding interface in GSM (A-bis) is like that; one has to use compatible equipment from the same manufacturer on both sides of the A-bis interface. The

FIGURE 8.4
UTRAN components and interfaces.

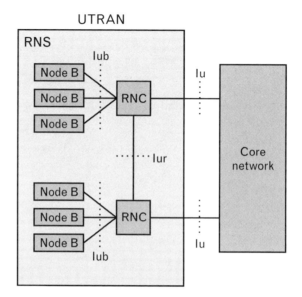

Iub interface has to manage difficult issues like power control; thus, manufacturers are tempted to use their own proprietary solutions here.

8.5.1 Radio Network Controller

The RNC controls one or more Node Bs. It may be connected via the Iu interface to an MSC (IuCS) or to an SGSN (IuPS). The interface between RNCs (Iur) is a logical interface, and a direct physical connection doesn't necessarily exist. An RNC is comparable to a BSC in GSM networks.

Functions that are performed by the RNC include the following:

- Iub transport resources management;

- Control of Node B logical operation and maintenance (O&M) resources;

- System information management and scheduling of system information;

- Traffic management of common channels;

- Macro diversity combining/splitting of data streams transferred over several Node Bs;

- Modifications to active sets; that is, soft handover;

- Allocation of DL channelization codes;

- Uplink outerloop power control;

- DL power control;

- Admission control;

- Reporting management;

- Traffic management of shared channels.

8.5.2   Node B

Node B is the UMTS equivalent of a base station transceiver. It may support one or more cells, although in general the specifications only talk about one cell per Node B. The term *Node B* is generally used as a logical concept. When physical entities are referred to, then the term *base station* is often used instead.

Functions that are performed by a Node B include the following:

- Node B logical O&M implementation;

- Mapping of Node B logical resources onto hardware resources;

- Transmitting of system information messages according to scheduling parameters given by the RNC;

- Macrodiversity combining/splitting of data streams internal to Node B;

- Uplink innerloop power control (in FDD mode);

- Reporting of uplink interference measurements and DL power information.

In addition, because Node B also contains the air interface physical layer, it has to perform the following functions related to it (these are further discussed in Chapter 3):

- Macrodiversity distribution/combining and soft handover execution;

- Error detection on transport channels and indication to higher layers;

- FEC encoding/decoding of transport channels;

- Multiplexing of transport channels and demultiplexing of CCTrCHs;

- Rate matching;

- Mapping of CCTrCHs on physical channels;

- Power weighting and combining of physical channels;

- Modulation and spreading/demodulation and despreading of physical channels;

- Frequency and time synchronization;

- Radio measurements and indication to higher layers;

• Innerloop power control;

• RF processing.

Network manufacturers are also offering solutions where the same physical base station equipment will offer both the GSM and the WCDMA transmitter/receiver capability (i.e., they are combined GSM-BTS and WCDMA-Node Bs).

8.6   GSM Radio Access Network

The GSM radio access network is also known as the base station subsystem (BSS). It consists of one BSC and one or more BTS, as in Figure 8.5. The BSC controls the functionality of a BTS over the A-bis interface. The A-bis interface is not a multivendor interface, but it contains solutions that are proprietary to each manufacturer. The functional split between the BSC and the BTS is such that the BTS should contain only the transmission equipment and related functions, and the managing equipment and everything else should be in the BSC. Generally it can be said that the intelligence in this system lies in the BSC. The BTS is purposely left quite dumb, as it is then cheaper to build. Note that the number of BTSs in a mobile network is much greater than the number of BSCs, so designing a "super BSC" and a "simple BTS" makes sense. A good presentation of base station subsystem architecture can be found in [1].

8.6.1   Base Station Controller

A BSC controls a group of BTSs connected to it via the A-bis interface. The number of BTSs under its control depends on the network configuration. The BSC functions include the following:

• Radio resource management for BTSs;

• Intercell handovers (for inter-BSC handovers, help is needed from the MSC);

• Frequency management (allocation of frequencies to BTSs);

• Management of frequency-hopping sequences;

• Time-delay measurements of uplink signals with respect to the BTS clock;

• Implementation of the O&M interface;

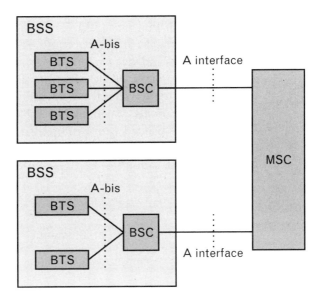

FIGURE 8.5
BSS subsystem.

• Traffic concentration to reduce the number of required lines to BTSs and an MSC;

• Power management.

8.6.2   Base Transceiver Station

The BTS consists of one or more transceivers (TRXs). Each TRX can support one carrier, that is, eight timeslots. Eight timeslots on a radio carrier constitute eight physical channels. Note that it is not possible to use all timeslots for traffic channels, as common control channels do require part of the capacity. Typically, a BTS serves one cell. There are also configurations in which several sectored cells are transmitted from the same BTS site. This can be regarded as one BTS with several sectored cells, or several BTSs each with a sectored cell. The radius of BTS cells can vary a great deal. The smallest BTS cells are indoor cells with a radius of just a few meters. At the other extreme, the maximum theoretical radius of a basic-GSM cell is just over 30 km. In practice this can be used only in open rural areas. There are also modified BTSs with a radius of 70 km. These have, however, a rather poor spectral efficiency; thus, they are only used in special circumstances when a large coverage is required, but the expected traffic density is very low. Ensuring that the layout of BTSs provides wide-enough coverage and simultaneously enough capacity in traffic hot spots is a major task for network operators. Network planning is discussed further in Chapter 9, although the discussion there is mostly about WCDMA networks.

The GSM specifications define that the transcoder/rate adapter unit (TRAU) is also part of the BTS. However, it is common practice that this unit is located at the MSC. The function of the transcoder is to convert the digitized speech [full rate (FR) or enhanced full rate (EFR) coded ~ 13 Kbps] from the GSM air interface into 64-Kbps PCM speech used in telephone networks and vice versa. It makes sense to locate this unit as close as possible in the middle of the network because this preserves the required transport capacity between the BTS and the MSC. One can transfer four times more GSM FR coded channels than PCM-coded channels. A 13-Kbps channel is padded with extra bits to make it a 16-Kbps channel, and four of these can be carried over a single 64-Kbps channel. Therefore, the most common location for TRAUs is at the MSC, although logically it is still part of the BSS.

The BTS functions include the following:

- Scheduling of broadcast and common control channels;
- Detection of random and handover access bursts sent by the mobile stations;
- Timing advance calculations;
- Uplink measurements;
- Channel coding (error protection) and encryption/decryption;
- LAPDm protocol (layer 2);
- Frequency hopping;
- Transcoding and rate adaptation (although this is usually handled by an MSC).

8.6.3 Small Base Transceiver Stations

The latest trend in GSM base station systems is the development of very small base stations. Traditionally, GSM has not been a very suitable system for low-tier environments, such as homes and indoor office systems. The GSM infrastructure has been relatively complex and expensive compared with simpler systems designed for the wireless office and similar applications, and it has, therefore, been out of reach for domestic use. The traffic density can be quite high in offices, and GSM has not been able to easily provide coverage for traffic hot spots. Also, network planning in GSM is a complex operation. One cannot just nail a GSM home base station to one's living room wall and expect that all of one's neighbors with similar systems (possibly using the same frequency) would remain friendly. However, these problems must be solved as, in the future, the GSM network needs to expand to just these kinds of areas. This issue is discussed here because even though the low-tier GSM BSS is not a 3G issue as such, it will be increasingly used in the future and it will be a competitor of 3G networks. Many WCDMA operators will have GSM licenses and, thus, will have access to this technology when planning indoor networks. Actually there are several alternatives for indoor technologies: the FDD and TDD modes of UTRAN, GSM, WLAN-cellular interworking systems, DECT, Bluetooth, and many other short-range RF systems. Silventoinen [9] presents two different indoor scenarios—the home base station and the office base station—analyzing their problems and proposing possible solutions.

The home base station (HBS) is a small GSM base station that is aimed at residential use. Currently, the customer looking for cheap call rates at home has to use either a fixed-line telephone or some cordless technology, like DECT or CT2. GSM mobile phone calls are more expensive, although the general trend is toward lower tariffs. Using one's GSM phone at home is, nevertheless, an attractive concept as the handsets take on more and more PDA features. An HBS must be affordable; therefore, its capabilities are less than those of a "real" base station. There are many technical problems with the HBS concept. The two major ones are frequency allocation and the HBS synchronization. The problem with frequency allocation results from the fact that HBSs are probably set up without any input from the local operators; therefore, there cannot be any centralized frequency planning. Because the HBS is an indoor system, the interference caused to the normal GSM network will probably be small. But the stronger outdoor GSM network could severely interfere with the functionality of an HBS. Automatic search receivers could solve the interference problem, but they are still quite costly for home users. Such a receiver would monitor the radio environment interference during the HBS setup process and choose the GSM frequency with the lowest noise level.

HBS synchronization is a problem because an HBS is most likely connected to a PSTN, which cannot provide a suitable time or frequency reference. This may cause a drift in both the time and frequency domains, which can increase system interference and reduce spectral efficiency. There are solutions to all of these problems, but the low-cost character of the HBS restricts the viable methods, so that the price tag of this system remains low.

The two main HBS scenarios presented are the cordless approach and the base station approach. In the cordless approach, the HBS serves only as an access point to the PSTN. Therefore, no GSM-specific services can be used when the mobile is connected to an HBS. This approach is easier to implement, but it is not an attractive technology choice for the GSM operators as they do not get any revenues from the calls made via cordless HBSs. ETSI promotes this approach under the name of the cordless telephone system (CTS); see [10] and [11]. In this system the HBS is called CTS Fixed Part (CTS-FP), and it would be a completely private system. However, if the HBS is tuned to use licensed frequency band, then there must be some sort of agreement with the licence owner (i.e., the GSM operator) because it has paid for the right the use this band and is probably quite unhappy to let others use it even if they are not interfering with the public GSM network.

The most probable solution to this problem would be a fixed monthly licence fee that is paid to the operator. A more elegant solution would be to specify licence exempt bands for GSM carrier frequencies. These could be used by low-power home systems for free.

In the base station approach, the HBS is connected to the normal GSM network via the PSTN. This would require specifying a new protocol for the HBS-GSM PSTN interface. Specification work is always slow; thus, this approach may never be implemented. However, the GSM operators would certainly like this alternative much more as the calls would go via their infrastructure and bills could be sent out for the service.

Office base stations, although they are also indoor systems, differ from home base stations in many ways. Whereas with the HBS the traffic density most probably will not be a problem, office base stations are usually built just to tackle the problem of traffic hot spots. Because of the high number of users, the cost of the base station equipment is not an important factor, as it is with HBSs. Silventoinen [9] presents three possible alternatives for office systems: single cell, multicell, and a hybrid system called the in-building base station system (IBS).

In a single-cell system there is only one cell for the entire office. This is a technically simple system to implement. No network planning is needed because there is only one cell. Thus, there are no intraoffice handovers. The drawback is that this system has a poor spectral efficiency. The capacity can be increased only by allocating more carriers to the system. Small GSM operators may not have enough bandwidth available, especially if the office is large and the traffic density high.

In multicell systems, the office is divided into several small cells. The capacity can be increased more easily in this kind of system just by increasing the number of cells. But this is technically a rather complex system, as it requires tedious network planning. Intraoffice handovers do happen frequently, and in relatively modest systems there cannot be any frequency reuse if the number of cells is lower than the minimum reuse factor. Setting up a multicell system is certainly a much costlier solution than implementing a single-cell system.

The hybrid IBS office system tries to combine the good properties of the previous two approaches. It contains a single logical cell, which is then divided into several radio subcells. These subcells can use much simpler transmitters than "real" BTSs. The low-cost transmitters are called RF-heads. TRXs are kept in a centralized hub that corresponds to the BTS. The problem with this approach is that the RR protocol task in the hub needs some modifications to cope with it. The old definition of the channel (i.e., frequency/timeslot pair) is no longer sufficient in the IBS because the channel now has a third dimension: place. The same frequency/timeslot can be reused in some other RF head. Thus, although the IBS approach is technically better than the two other approaches, it requires changes to the current specification; therefore, it is not readily available. Single-cell and multicell systems can be deployed immediately.

8.7 Interfaces

The interfaces in the UMTS system follow the GSM/GPRS naming convention, where applicable. The UTRAN contains some new interfaces and, thus, some new names.

From the specifications point of view, there are three kinds of interfaces in the UMTS/GSM network. The first category contains those interfaces that are truly open. This means that they are well-specified, and the specification is such that the equipment on different ends of the interface can be acquired from different manufacturers. In an old GSM network, only the A interface and the air interface are truly open interfaces.

The second category includes those interfaces that are specified at some level, but the interface is still proprietary. The equipment for such interfaces must come from the same manufacturer, as the implementation is specific to a manufacturer. The A-bis interface is a good example of such an interface. It is rather well-specified as a whole, but some issues are left open; thus, it is not an open interface. Sometimes an interface exists only logically if two devices are physically only one entity. Quite often the MSC and the VLR are combined; thus, the B interface doesn't physically exist.

The third category contains those interfaces for which there is no specification at all. The interface has only a name, and possibly a description of the tasks it should be able to handle. The H and I interfaces in the GSM core network belong to this category. Obviously, these interfaces are not open. They are either proprietary or they are not used at all in some cases.

The following sections contain short descriptions of these interfaces. For a detailed description see [3] and [4] for the GSM interfaces, and [5] and [6] for the UMTS interfaces.

8.7.1 A Interface

The A interface exists between the MSC and the BSC, which is logically the BSS. This interface is an open multivendor interface, which should mean that an operator can buy the MSC and the BSS equipment from different manufacturers and connect them together over the A interface. This interface is specified in the 08-series GSM specifications. Though the A interface is a pure GSM interface and not part of the UMTS concept, it can connect a BSS subsystem to a 3G-MSC, which makes it eligible for examination in this section.

The protocol stack for the A interface is depicted in Figure 8.6. This diagram shows the protocol stacks for the whole BSS subsystem, as well as the A interface as discussed here. The other protocols are pure GSM protocols, which can be studied in [1], [3], and [7].

8.7.2 Gb Interface

The Gb interface is a non–UMTS interface, which will often be present in the UMTS core network. The Gb interface connects the packet-switched core network to the GSM network. It is used when the GSM mobile station uses GPRS services. GPRS-capable GSM phones of the future will be able to use at least some of the UMTS packet-based services, especially once enhanced GPRS (EGPRS) is launched. EGPRS can expand the user-data speeds in the GSM air interface up to rates as high as 200 Kbps. This will certainly give EGPRS phones the ability to use many of the 3G services and applications if the operator so allows. Note that the combined RLC/MAC protocol used in the radio interface in GPRS is not the same protocol as the separate RLC and MAC protocols used in the UTRAN air interface. See Figure 8.7.

8.7.3 Iu Interface

This interface connects the core network and the UMTS Radio Access Network (URAN). A truly open, multivendor interface, it is the most important and central interface for the 3GPP concept. The Iu can have two different physical instances, Iu-CS and Iu-PS, and there will probably be

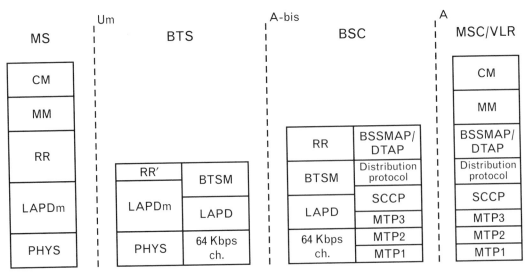

FIGURE 8.6   *GSM BSS protocols.*

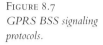

FIGURE 8.7
*GPRS BSS signaling
protocols.*

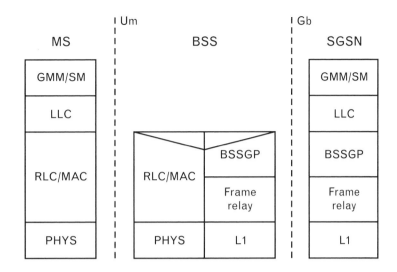

more in the future. The Iu-CS connects the radio access network to a circuit-switched core network, that is, to an MSC. The Iu-PS connects the access network to a packet-switched core network, which in practice means a connection to an SGSN.

The URAN can have several kinds of physical implementations. The first one that was implemented is the UTRAN, which uses the WCDMA air interface technology. Thus, the URAN is a generic concept and the UTRAN will be the first concrete implementation of it. Specification work is also under way for the Broadband Radio Access Network (BRAN), which connects a HiperLAN2 radio access network to a core network. The URAN concept and BRAN are depicted in Figure 8.8. AP stands for access point, and APC for access point controller.

BRAN is being specified by ETSI under the name of HiperLAN2, and it can support user-data rates of around 30 Mbps. The maximum physical rate is 54 Mbps. HiperLAN2 uses unlicensed radio spectrum in the 5-GHz radio band. It can provide a coverage range of 30 to 50m indoors and up to 150 to 200m outdoors. A typical application for HiperLAN2 includes laptops with wireless modems in office and campus environments.

However, the maximum data rates are so high that they provide lots of possibilities for totally new kinds of applications. HiperLAN2 is well explained in [12].

It will be possible to execute handovers between UTRAN and BRAN, provided that the user terminal is a dual-system UMTS-HiperLAN terminal.

The UMTS Satellite Radio Access Network (USRAN) connects a satellite network to the core network. This access network had not been specified as of 2002, and it will not be implemented in the near future. Several different satellite access networks have been proposed to the ITU for a 3G system, but only time will show which of those will survive to develop into

FIGURE 8.8
*BRAN and UMTS
interworking.*

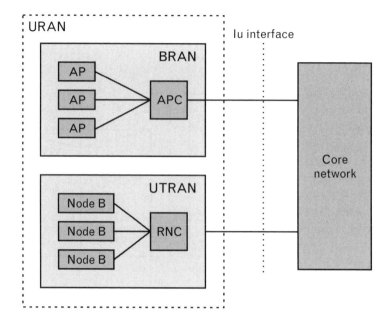

concrete systems, if any. Satellite cellular business during the last few years has been especially difficult, and we have seen some remarkable failures. Some uncertainty remains as to whether there will be enough customers for a commercially viable satellite cellular system.

The protocol model in the Iu interface is divided into two horizontal layers, the radio network layer and the transport network layer. This is depicted in Figure 8.9. The split is made to separate the transport technology (in the transport network layer) from the UTRAN-related issues (in the radio network layer).

This picture may look a bit confusing at first. A protocol stack diagram usually has two planes, control and user. The control plane transfers signaling information, and the user plane transfers application data. This is also the case in the Iu interface, but this requires some explanation.

In the vertical direction, the Iu protocol model is divided into three planes, the (radio network) control plane, the (radio network) user plane, and the transport network control plane. Both radio network layer planes, control and user, are conveyed via the transport network layer using the transport network user plane.

The signaling bearer in the transport network layer is always set up by O&M actions. The signaling protocol for the Access Link Control Application Protocol (ALCAP) may be the same type as the signaling protocol for the Application Protocol, or it may be different. Once the signaling bearers are in place, the Application Protocol in the radio network layer may ask for data bearers to be set up. This request is relayed to the ALCAP in the transport network layer. The ALCAP is responsible for the data bearer setup, and it has all the required information about the user plane technology. It is also

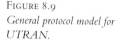

FIGURE 8.9
General protocol model for UTRAN.

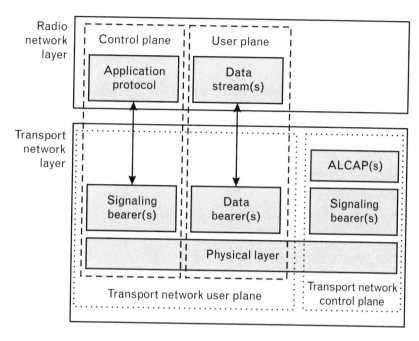

possible to use preconfigured data bearers, as is done in the Iu-PS interface, in which case no ALCAP is needed. Because the signaling bearer in the transport network control plane is only needed for the ALCAP, the entire transport network control plane is unnecessary in this case.

So what is the purpose of this rather complex protocol model? The complexity strives for the total separation of the control plane from the user plane. If the radio network layer control plane had set up the user plane data bearers by itself, it should have had its own knowledge of the underlying technology and its capabilities. The radio network layer control plane doesn't have to know anything about the transport technology. The bearer parameters it requires are not directly tied to any user plane technology, but they are general bearer parameters. Thus, the radio network layer and the transport network layer are logically independent of each other.

As indicated earlier, there are two different physical instances for the Iu interface: Iu-CS and Iu-PS. The corresponding protocol stacks are given in Figure 8.10 and in Figure 8.11 for the UTRA network. The various protocols in these stacks are too numerous to be discussed thoroughly, but a short description is given of all of them in the next section.

Both versions of this interface use the asynchronous transfer mode (ATM) transport technology. In the case of the CS domain control plane, there are SS7-based protocols on top of the ATM layers. In the CS domain user plane, only an ATM adaptation layer 2 (AAL2) task is needed to handle the transport of audio and video streams.

In the PS domain control plane, there are two alternative protocol stacks to use. The first one is the same as in CS domain, and the second one

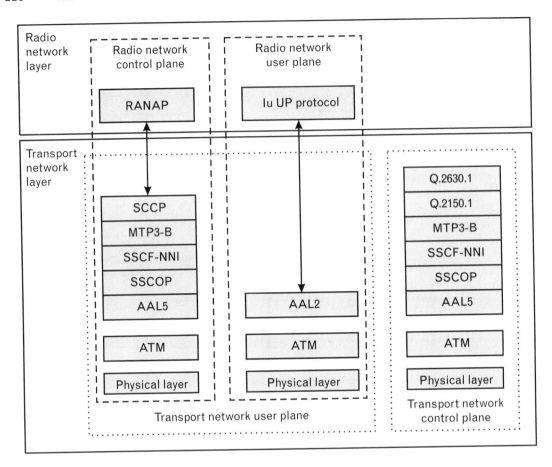

FIGURE 8.10   *Iu interface/CS domain.*

is more IP-oriented. This version can be used once the data transmission is based on the IP technology. The user plane in this domain is different from the one in the CS domain. The data packet forwarding is handled by the GPRS Tunnelling Protocol for user plane (GTP-U). The Iu interface is specified in the 25.41x series of the 3GPP specifications. A good starting point is [13].

8.7.4 Iub Interface

This interface is situated between the RNC and the Node B in the UTRAN. In GSM terms this corresponds to the A-bis interface between the BTS and the BSC. The Iub, like its A-bis counterpart, is hardly an open interface. The tasks Node B and RNC have to perform together are so complex that a proprietary solution is the most probable one.

FIGURE 8.11
Iu interface/PS domain.

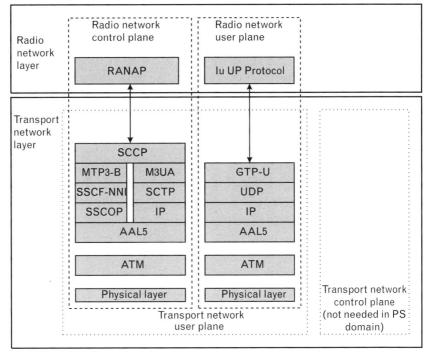

The protocol stack in this interface is based on the same principles as in the Iu interface; there are control and user planes and a transport network control plane, as well. The Iub separates the Node B from the RNC so that no internal details are visible over the interface as this could limit the future expandability of this technology.

The RNC manages Node B(s) over the Iub interface. The following list of functions to be performed over the Iub interface is presented in [8]:

- Management of Iub transport resources;

- Logical O&M functions of Node B;

- Implementation-specific O&M transport;

- System information management;

- Traffic management of common channels;

- Traffic management of dedicated channels;

- Traffic management of shared channels;

- Timing and synchronization management.

These issues are discussed in connection with the RNC and Node B presentations. See Sections 8.5.1 and 8.5.2. In Figure 8.12 FP stands for frame protocol.

8.7.5   Iur Interface

The Iur interface connects two radio network controllers. The applicable specification states that this interface should be open, but again only time will show whether this will really be the case. This interface can support the exchange of both signaling information and user data. All RNCs connected via the Iur must belong to the same PLMN. The protocol stack structure is based on the same principles as the Iu and Iub; that is, the radio network and

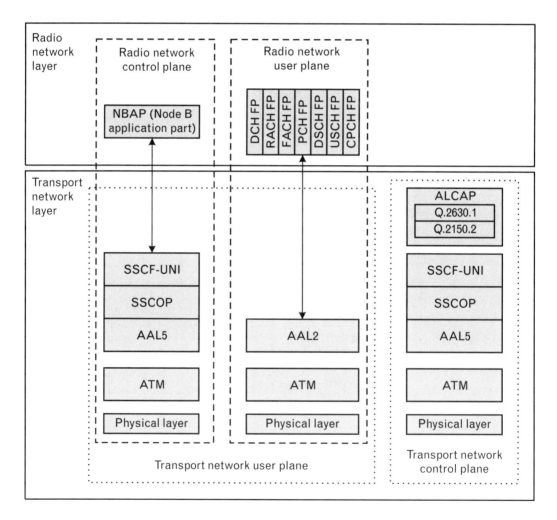

FIGURE 8.12   *Iub interface.*

the transport network are separated, so that one of these technologies can be changed without the other having to be changed.

The Iur interface exists to support macrodiversity. The reader may recall that several base stations can have an active connection with the same mobile station at the same time in a CDMA network. It is possible that these base stations are controlled by different RNCs. Without an Iur interface, this situation would have to be controlled via the Iu interface (i.e., via the MSC), which would be a very clumsy method indeed. Macrodiversity is a purely radio-access-technology-related phenomenon and the MSC should not be bothered with these kinds of issues. The Iur interface is needed so that the UTRAN can manage the problem of soft handovers by itself.

There is always only one RNC in control of a UE connection: This is the managing RNC. This managing RNC is called the serving RNC (SRNC). The connection to MSC is routed via the SRNC. Any other RNC involved in the connection is a slave RNC, which is called a drift RNC (DRNC). There may be more than one DRNC per UE connection. See Figure 8.13.

An associated concept is the controlling RNC (CRNC). Every Node B is controlled by only one RNC. This RNC has sole control of a group of Node Bs on a UE's behalf. Therefore, this RNC is the CRNC of the Node Bs in a connection. Depending on its role in a connection, a CRNC can also be either a SRNC or a DRNC.

FIGURE 8.13
Serving and drift RNCs.

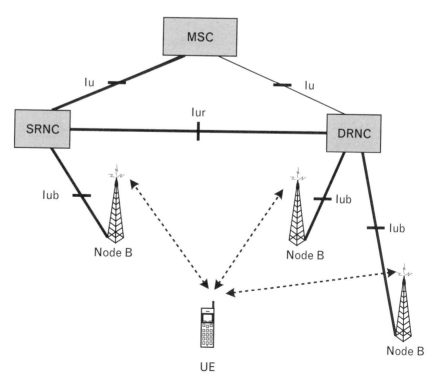

The DRNC handles the macrodiversity combining/splitting of data streams sent via its cells. This means that only one data stream for each UE is needed over the Iur interface. The SRNC can, however, explicitly request separate Iur interface connections, in which case the macrocombining is done in the SRNC. Those data streams that are communicated via DRNC(s) and the SRNC are combined, or split, by the SRNC.

The power-control issues (i.e., the uplink outer-loop power-control and the DL power-control commands) are managed by the SRNC, even for those data streams that are communicated via a DRNC.

The signaling information over the Iur interface is transferred using the Radio Network Subsystem Application Part (RNSAP) Protocol; see Figure 8.14. The Iur interface functionality is further discussed in the RNSAP paragraph in Section 8.8. See also [15].

8.7.6   MAP Interfaces

The interfaces between the core network entities are called the MAP interfaces as they generally use the Mobile Application Part (MAP) protocol as a signaling protocol. The "old" interfaces, which have been inherited from the GSM standard, are named with a single capital letter (MAP-A through MAP-M).

The introduction of GPRS into GSM networks brought a batch of new interfaces, which were named using a capital G and a small letter. For

FIGURE 8.14
Iur interface.

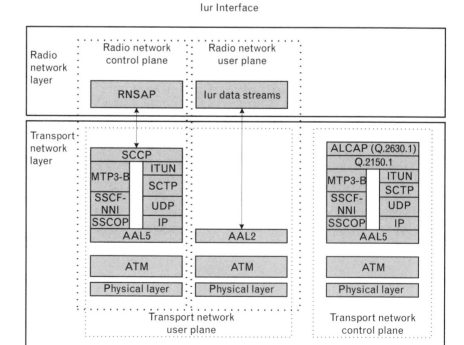

example, the interface between the SGSN and the HLR is named as Gr (r for roaming); see Figure 8.15. The meaning of the other G*x* interfaces could be described as follows:

- Gf = fraud interface;

- Gi = Internet interface;

- Gp = PLMN interface;

- Gc = context interface;

- Gn = node interface;

- Gb = base interface.

If location services (LCS) are used in a PLMN, then we will get still more interfaces. These interfaces are named using a capital L and a small letter. The LCS interfaces are described in Figure 8.16.

Most, but not all, of the interfaces in the core network use the MAP protocol stack for their signaling traffic. The generic MAP protocol stack is

FIGURE 8.15
MAP interfaces.

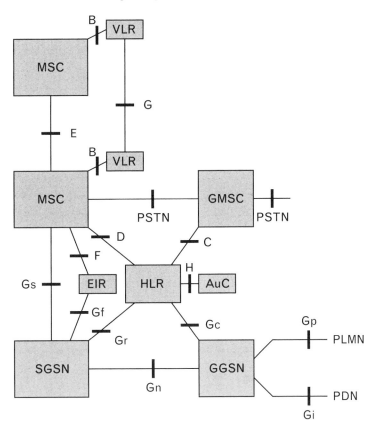

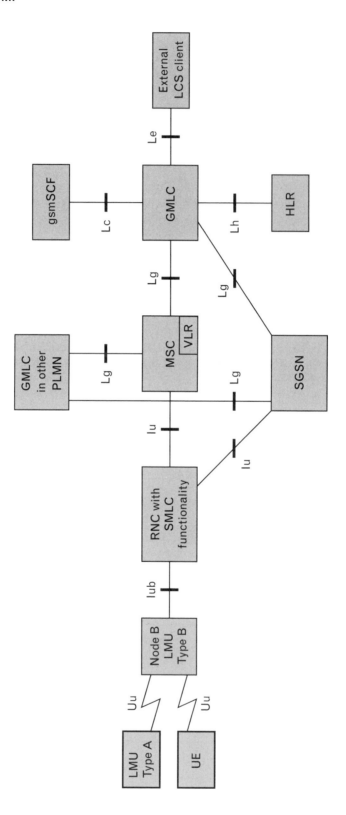

FIGURE 8.16 *LCS network elements and interfaces.*

described in Figure 8.17. For example, the topmost protocol layer is named MAP-D in the D interface.

Note that this presentation is not a comprehensive one. There are also plenty of other interfaces and their associated details in the core network. The MAP specification [6] itself is a true mammoth. The version 4.5.0, for example, contains well over 1,200 pages. But one has to remember that this specification contains descriptions for all MAP interfaces. A typical MAP protocol is quite simple; it doesn't contain too many messages, and there are only a few procedures. One should not be intimidated by the size of this specification; it is actually quite readable and helps elucidate the inner workings of a PLMN.

8.8   Network Protocols

The network protocols are briefly described below in alphabetical order. This section should be read in connection with the previous section, which describes the interfaces and the protocol stacks used in these interfaces. Because of the number of different protocols in an UMTS network, these descriptions are by necessity rather short. Each protocol description, however, includes references to further sources of information.

Figures 8.18 and 8.19 depict the protocol stacks for the control and user planes of the basic line of communications, that is, UE-Node B-RNC-MSC. Note that these pictures describe only one possible implementation of the protocol stacks. For example, different channel types often have their own protocol stack variations. Also note that the protocol stack in the Iu interface is different if the UTRAN connects to the PS domain (a connection to an SGSN).

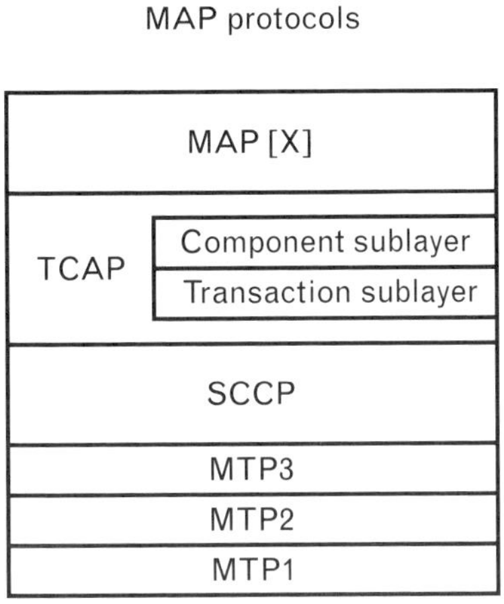

FIGURE 8.17
MAP protocols in core network interfaces.

MAP protocols

MAP [X]	
TCAP	Component sublayer
	Transaction sublayer
SCCP	
MTP3	
MTP2	
MTP1	

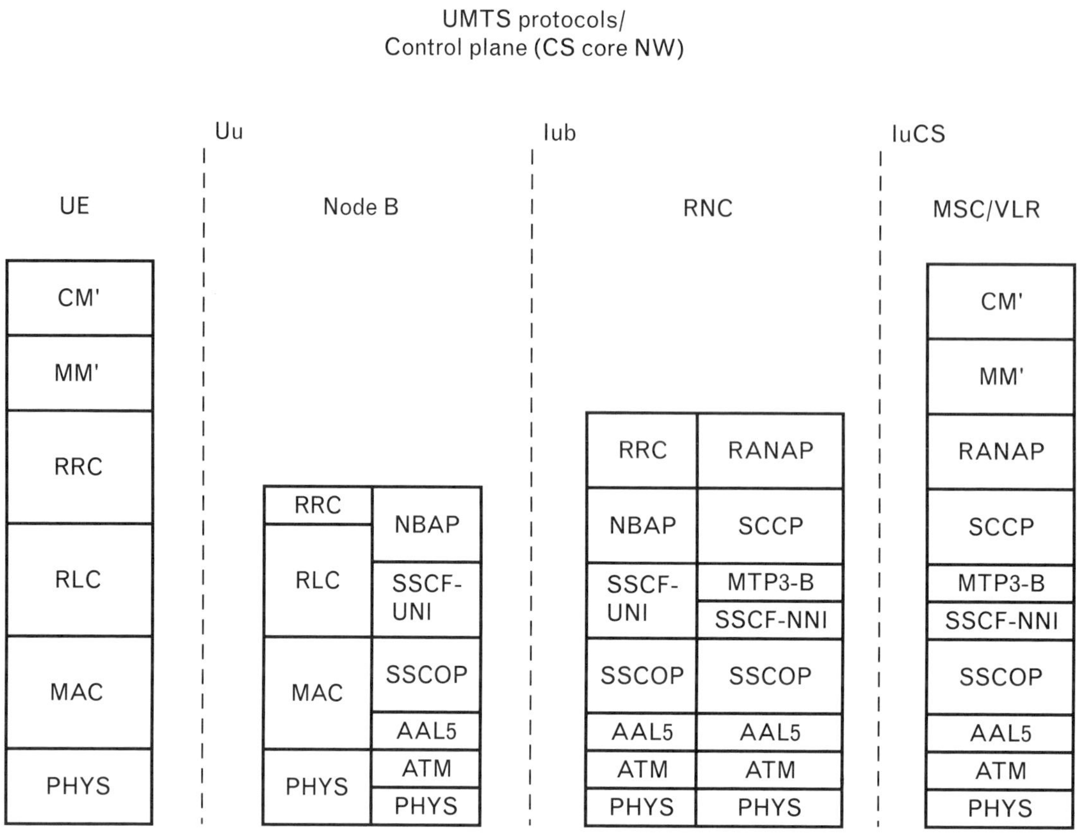

UMTS protocols/ Control plane (CS core NW)

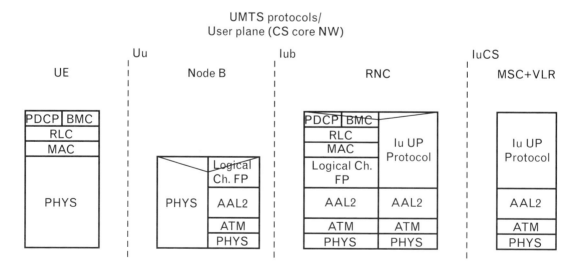

FIGURE 8.18   *Control plane, circuit-switched core network.*

UMTS protocols/ User plane (CS core NW)

FIGURE 8.19   *User plane, circuit-switched core network.*

8.8.1 Asynchronous Transfer Mode

The core network transport is based on ATM. ATM is a transmission proce-dure based on asynchronous time-division multiplexing using small, fixed-length data packets. These data packets have a length of only 53 bytes, of which 5 bytes are for the packet header and 48 bytes are reserved for the payload.

The fixed packet[1] length makes it possible to use very efficient and fast packet switches. The chosen packet length (53 bytes) was a compromise between the requirements of speech transfer and data transmission. Filling up long ATM packets with speech samples yields delays, which reduce the quality of real-time speech transmission. Thus, the shorter the packet, the better it suits speech transfer. However, pure non-real-time data transfer would be more efficient if longer packets were used. The length of 53 bytes was a suitable compromise as it allows (near)-real-time speech transmission, but doesn't hamper data transmission speeds too much. Also, the 5-byte header doesn't represent too much overhead if the payload is 48 bytes.

8.8.2 AAL2 and AAL5

Above the ATM layer we usually find an ATM adaptation layer (AAL). Its function is to process the data from higher layers for ATM transmission. This means segmenting the data into 48-byte chunks and reassembling the original data frames on the receiving side. There are five different AALs (0, 1, 2, 3/4, and 5). AAL0 means that no adaptation is needed. The other adap-tation layers have different properties based on three parameters:

- Real-time requirements;
- Constant or variable bit rate;
- Connection-oriented or connectionless data transfer.

The usage of ATM is promoted by the ATM Forum. There are numer-ous books written about the subject; see [16, 17] for a good example. The Iu interface uses two AALs: AAL2 and AAL5. AAL2 is designed for the trans-mission of real-time data streams with variable bit rates. AAL5 fulfils the same requirements except the real-time parameter.

8.8.3 Iu User Plane Protocol Layer

This protocol relays the user data from the UTRAN to the CN and vice versa. Each radio access bearer is associated with one Iu user plane (UP)

---

1 In ATM jargon, these packets are called *cells*. To prevent the obvious confusion, this naming convention is not used herein.

protocol task. This means that there will be several Iu UP protocol tasks allocated for one user if a user has several radio access bearers. These Iu UP tasks are established and released together with their associated radio access bearers.

The Iu UP protocol can operate in two modes:

1. Transparent mode;

2. Support mode.

The particular mode is decided by the CN when this protocol task is created. It cannot be modified later unless the associated radio access bearer is modified at the same time.

The transparent mode is, as the name indicates, transparent. In this mode the only function of this task is to transfer user data across the Iu interface. No special Iu UP frames will be generated for this transfer, but lower-layer PDUs can be used instead.

The CN creates a support mode Iu UP task if any other particular feature in addition to the ordinary user data transfer is needed. The following functions are possible in the support mode:

• Transfer of user data;

• Initialization;

• Rate control;

• Time alignment;

• Handling of error events;

• Frame-quality classification.

In support mode, a special Iu UP frame is created to relay the user data across the Iu interface. The In UP protocol is described in [27].

8.8.4   GPRS Tunnelling Protocol-User

GPRS Tunnelling Protocol-User is often referred to as GTP-U. GTP is the protocol between GPRS support nodes (GSNs) in the UMTS/GPRS backbone network. It includes both the GTP signaling (GTP-C) and data transfer (GTP-U) procedures.

GTP is defined for the Gn interface (i.e., the interface between GSNs within a PLMN), and for the Gp interface between GSNs in different PLMNs. Only GTP-U is defined for the Iu interface between the serving GPRS support node (SGSN) in the PS domain and the UTRAN.

On the Iu interface, the RANAP Protocol performs the control function for GTP-U. In the user plane, GTP uses a tunnelling mechanism (GTP-U) to provide a service for carrying user-data packets. GTP is defined in [30].

8.8.5   SS7 MTP3-User Adaptation Layer

SS7 MTP3-User Adaptation Layer is often referred to as M3UA. This protocol supports the transport of any SS7 MTP3-User signaling (in the UTRAN case this will be SCCP) over IP using the services of the Stream Control Transmission Protocol. The specification for M3UA was still under work at the time of this writing; the draft is available from [29].

8.8.6   MAP (MAP-A Through MAP-M)

MAP is actually a set of protocols used by the core network elements for their mutual communication. One could describe the core network as a set of database registers, and the MAP protocol as a database query language. The principles and tasks of the various MAP protocols are well discussed in [6].

8.8.7   Message Transfer Part

Message transfer part (MTP) provides message routing, discrimination and distribution, signaling link management, and load sharing. Its usage is defined in [32].

8.8.8   Node B Application Part

The Node B Application Part (NBAP) is used to manage the Node B by the RNC via the Iub interface. The NBAP can support several parallel transactions.

The NBAP protocol has the following functions:

- *Cell configuration management*. The RNC can manage the cell configuration information in a Node B.

- *Common transport channel management*. The RNC can manage the configuration of common transport channels in a Node B.

- *System information management*. The RNC manages the scheduling of system information to be broadcast in a cell.

- *Resource event management*. The Node B can inform the RNC about the status of Node B resources.

- *Configuration alignment.* The CRNC and the Node B can verify and enforce that both nodes have the same information on the configuration of the radio resources.

- *Measurements on common and dedicated resources.* The RNC can initiate measurements in the Node B. It is then Node B's task to report the results of these measurements back to the RNC.

- *Physical shared channel management* (only in TDD mode). The RNC manages the physical resources in the Node B belonging to shared channels (USCH/DSCH).

- *Radio link management.* The RNC manages radio links using dedicated resources in a Node B.

- *Radio link supervision.* The RNC reports failures and restorations of a radio link.

- *Compressed mode control* (only in FDD mode). The RNC controls the usage of compressed mode in a Node B.

- *DL power drifting correction* (only in FDD mode). The RNC has to adjust the DL power level of one or more radio links in order to avoid DL power drifting between the radio links.

- *Reporting general error situations.*

- *DL power timeslot correction* (only in TDD mode). The Node B can apply an individual offset to the transmission power in each timeslot according to the DL interference level at the UE.

There are two kinds of NBAP procedures:

1. NBAP dedicated procedures are those related to a specific UE in Node B.
2. NBAP common procedures are those that request initiation of a UE context for a specific UE in Node B, or those not related to a specific UE.

The NBAP protocol is defined in [18].

8.8.9 Physical Layer (Below ATM)

The ATM standard does not dictate any special physical medium to be used with it. The physical layer in the Iu interface consists of physical media dependent (PMD) and transmission convergence (TC) sublayers. There is a wide variety of different standards (17 altogether) in [14], which can be used to implement PMD.

8.8.10 Q.2150.1

This protocol task is a converter between the ALCAP and MTP3-B protocols.

8.8.11 Q.2630.1

A generic name for this protocol is Access Link Control Application Part (ALCAP). It will be used to establish user plane connections toward the CS domain. This is also known as the AAL2 signaling protocol.

8.8.12 Radio Access Network Application Part

The Radio Access Network Application Part (RANAP) is defined in [25]. This is a sizable protocol, comparable to the BSSAP protocol in the GSM system. This protocol deserves a longer discussion, as it is the glue between the core network and the UTRAN.

RANAP provides the signaling service between the UTRAN and the CN that is required to fulfil the RANAP functions described later. RANAP services are divided into three groups:

1. *General control services.* These are related to the whole Iu interface.
2. *Notification services.* These are related to specified UEs or all UEs in a specified area.
3. *Dedicated control services.* These are related to only one UE.

Signaling transport (i.e., the transport network layer in the Iu interface below RANAP) provides two different service modes for the RANAP:

1. *Connection-oriented data-transfer service.* This service is supported by a signaling connection between the RNC and the CN domain. It is possible to dynamically establish and release signaling connections, one for each active UE. The connection provides a sequenced delivery of RANAP messages. RANAP is notified if the signaling connection breaks.
2. *Connectionless data-transfer service.* RANAP is notified in case a RANAP message did not reach the intended peer RANAP entity.

The RANAP protocol has the following functions:

• *Relocating serving RNC.* This function moves the serving RNC functionality, as well as the related Iu resources (RABs and signaling connection), from one RNC to another.

- *Overall RAB management.* This function is responsible for setting up, modifying, and releasing RABs.

- *Queuing the setup of RAB.* This function allows requested RABs to be placed into a queue and to inform the peer entity about the queuing.

- *Requesting RAB release.* While the overall RAB management is a function of the CN, the UTRAN can request the release of RAB.

- *Release of all Iu connection resources.* This function releases all resources related to one UE from the corresponding Iu connection.

- *Requesting the release of all Iu resources.* While the Iu release is managed from the CN, the UTRAN can request the release of all Iu resources from the corresponding Iu connection.

- *SRNS context forwarding function.* This function transfers a Serving Radio Network Subsystem (SRNS) context from the RNC to the CN for an intersystem forward handover in case of packet forwarding.

- *Controlling overload in the Iu interface.* This function adjusts the load in the Iu interface.

- *Resetting the Iu.* This function resets an Iu interface.

- *Sending the UE common ID (permanent NAS UE identity) to the RNC.* This function makes the RNC aware of the UE's common ID.

- *Paging the user.* The CN sends a paging request to the UE.

- *Controlling the tracing of the UE activity.* This function sets the trace mode for a given UE.

- *Transport of NAS information between the UE and CN.* This function has two subclasses:

  1. *Transport of the initial NAS signaling message from the UE to the CN.* This function transparently transfers the NAS information. As a consequence, the Iu signaling connection is set up.

  2. *Transport of NAS signaling messages between the UE and CN.* This function transparently transfers the NAS signalling messages over the existing Iu signaling connection.

  - *Controlling the security mode in the UTRAN.* This function sends the security keys (ciphering and integrity protection) to the UTRAN and sets the operating mode for security functions.

  - *Controlling location reporting.* The CN sets the mode in which the UTRAN reports the location of the UE.

- *Location reporting*. This function transfers the actual location information from the RNC to the CN.

- *Data volume reporting*. This reports unsuccessfully transmitted DL data volume over the UTRAN for specific RABs.

- *Reporting general error situations*. This function reports general error situations for which function specific error messages have not been defined.

8.8.13   Radio Network Subsystem Application Part

The Radio Network Subsystem Application Part (RNSAP) specifies the radio network layer signaling procedures between two RNCs. The managing RNC in these procedures is called the serving RNC (SRNC) and the slave RNC is called the drift RNC (DRNC).

The RNSAP offers the following services:

- *RNSAP basic mobility procedures*. These procedures handle mobility within the UTRAN.

- *RNSAP DCH procedures*. These procedures handle DCHs, DSCHs, and USCHs between two RNSs. In general, only one RNSAP DCH procedure per UE can be active at any time.

- *RNSAP common transport channel procedures*. These procedures control common transport channel data streams over the Iur interface. This excludes DSCH and USCH streams because they are already handled by the previous service class.

- *RNSAP global procedures*. These are not related to a specific UE, but they involve two peer CRNCs.

The RNSAP protocol includes the following functions:

- *Radio link management*. The SRNC can manage radio links using dedicated resources in a DRNS.

- *Physical channel reconfiguration*. The DRNC reallocates the physical channel resources for a radio link.

- *Radio link supervision*. The DRNC reports failures and restorations of a radio link.

- *Compressed mode control* (only in FDD mode). The SRNC controls the compressed mode within a DRNS.

- *Measurement of dedicated resources.* The SRNC triggers measurement of dedicated resources in the DRNS. This function also allows the DRNC to report the results.

- *DL power-drifting correction* (only in FDD mode). The SRNC can adjust the DL power level of one or more radio links in order to avoid DL power drifting between the radio links.

- *CCCH signaling transfer.* The SRNC and DRNC can pass information between the UE and the SRNC on a CCCH controlled by the DRNS.

- *Paging.* The SRNC can page a UE via the DRNS.

- *Common transport channel resources management.* The SRNC can use common transport channel resources within the DRNS (excluding DSCH resources for FDD).

- *Relocation execution.* The SRNC can finalize a relocation procedure previously prepared via other interfaces. A relocation procedure merely means that the serving RNC has changed.

- *Reporting general error situations.*

- *DL power timeslot correction* (only in TDD mode). The DRNS can apply an individual offset to the transmission power in each timeslot according to the DL interference level at the UE.

The RNSAP is defined in [26].

8.8.14 Signaling ATM Adaptation Layer

Sometimes AAL, SSCOP, and SSCF are considered as one layer: the signaling ATM adaptation layer (SAAL).

8.8.15 Service-Specific Coordination Function

Service-Specific Coordination Function (SSCF) is also referred to as SSCF-NNI on the Iu interface, where NNI stands for network node interface. This protocol task maps the requirements of the higher layer to the requirements of the SSCOP. Its usage is defined in [33].

8.8.16 Service-Specific Connection-Oriented Protocol

Usage of the Service-Specific Connection-Oriented Protocol (SSCOP) is specified in [34]. The SSCOP defines mechanisms for the connection establishment, release, and reliable exchange of signaling information between signaling entities.

8.8.17   Signaling Connection Control Part

Signaling Connection Control Part (SCCP) is defined in [19-24]. This protocol shall comply with the ITU-T White Book. Here two SCCP message transfer service classes are used: class 0 and class 2. Class 0 provides a connectionless service and class 2 a connection-oriented service. Each mobile has its own signaling link while the connection-oriented service is used.

8.8.18   Stream Control Transmission Protocol

The Stream Control Transmission Protocol (SCTP) can transmit various signaling protocols over IP networks. The SCTP is defined in [28].

8.8.19   UDP/IP

User Datagram Protocol (UDP) is specified in RFC 768. Both IPv4 and IPv6 shall be supported. IPv4 is specified in RFC 791 and IPv6 in RFC 2460.

8.9   UMTS Network Evolution—Release 5

So far, the discussion in this chapter has been about Release 99. However, Release 5 brings considerable changes to the core network architecture. Whereas in Release 99 there are only two domains in the core network, CS and PS, in Release 5 a new IP Multimedia Subsystem (IMS) domain is introduced.

The IMS domain does not have its own new switch in the way the other domains have. Instead, IMS employs an enhanced PS domain, and uses its services to offer IMS multimedia services. Because of the complexity of the new architecture, it is depicted in two figures: Figure 8.20 shows the Release 5 architecture minus the new IMS domain. Figure 8.21 then shows only the new IMS domain-specific control elements. Note that the Home Subscriber Server (HSS) is the connecting element between PS and IMS domains, and it can be found in both figures. Again note that Figure 8.20 shows only the basic architecture. New elements and functions will have to be added if the network supports LCS, CAMEL, or CBS services.

Some observations from the IMS domain include the following:

- Control and data paths are separated.

- It implements an All-IP network.

- Voice and data can be handled in a similar way.

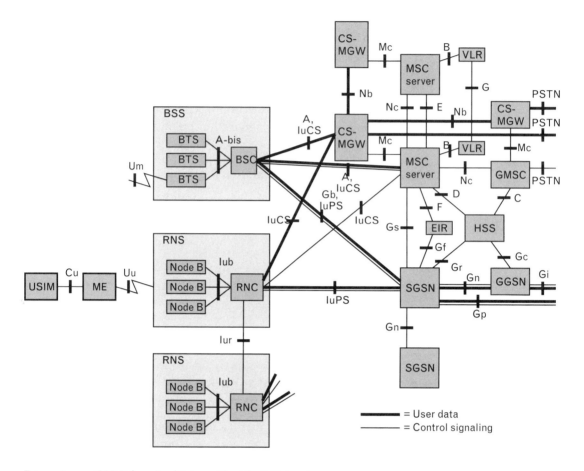

FIGURE 8.20　*GPP Release 5 architecture without the IMS domain parts.*

- There is no need for separate MSCs and SGSNs.

- HSS is a new super-HLR.

- Release 5 is backward compatible with earlier releases.

To make the high-speed data transfer more efficient, IMS has a new approach to the network design. Previously, data was transferred through several network elements on its way to its destination. In the new system, data typically bypasses the control logic in the core network. The old CS switch, MSC, has been divided into two logical entities, a media gateway (MGW), and an MSC server (MSC). The control logic is in MSC, and the actual switching matrix in MGW. These logical entities can be implemented in the same or separate physical units. The separation of control and data traffic enables the network to employ more efficient routers for the high-speed data, as the small-sized control messages are handled elsewhere.

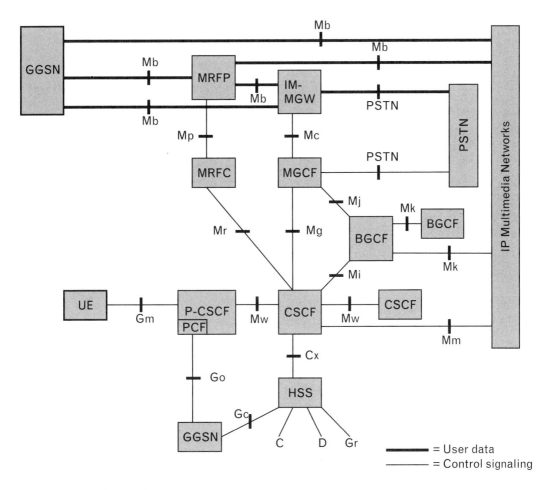

FIGURE 8.21 *IMS domain architecture.*

An *All-IP network* means that all traffic data, including voice, is transferred as IP packets. This opens the 3G mobile environment to the large IP applications industry. The problem of mobile networks has been to find revenue-generating applications, and IP Multimedia Domain will make this task easier. The new applications do not necessarily have to be developed for the mobile environment anymore, at least not because of the transport technique used. Another important argument for All-IP networks is that the technology makes the separation of PS and CS domains obsolete. All-IP networks make the transport technology uniform, and that should reduce network-deployment costs.

Voice can also be handled as packets in an All-IP network. This kind of service is called Voice over IP (VoIP), and it is discussed in Chapter 12. Note that VoIP as such is hardly an improvement for voice transfer. Circuit-switched systems were originally designed for voice transfer; they can do it quite efficiently, and provide high-quality results. However, All-IP networks bring lots of advantages; thus, voice too has to be transformed into a packet service.

The problem from the network point of view is that it has to be backwards-compatible with earlier releases. There will be lots of Release 99 devices in use when Release 5 is deployed, and those would become useless if there were no backwards compatibility. A UE can use IMS domain services if it has the capability; otherwise, the network has to provide Release 99/4 services. The problem with this is that the operator still has to keep the old CS and PS domain elements in its network. This is costly, and makes the network architecture a very complex one, as seen from Figures 8.20 and 8.21.

IMS domain traffic is all packet traffic, and it is transported via SGSN/GGSNs. The new IMS domain functions are typically accessed by the HSS. HSS combines the functions of HLR and AuC and holds the profiles of all subscribers.

The Call Session Control Function (CSCF) is the centrepiece in IMS. There are three kinds of CSCFs:

1. Serving CSCF (S-CSCF);

2. Proxy CSCF (P-CSCF);

3. Interrogating CSCF (I-CSCF).

The S-CSCF provides session-control services for a UE. This includes routing decisions and establishment, maintenance, and release of multimedia sessions. It also generates charging information for the billing system. The S-CSCF is located in the home network.

The P-CSCF is the first IMS entity that is contacted by the UE when it initiates a session in a visited network. The P-CSCF is located in the same network as the GGSN. The P-CSCF passes the session control to the S-CSCF that is located in the home network.

The I-CSCF is the contact point within an operator's network for all connections destined to a subscriber of that network, or a roaming subscriber currently located within that network's service area. It finds the correct S-CSCF within the network for the incoming Session Initiation Protocol (SIP) request.

The Breakout Gateway Control Function (BGCF) selects the network in which PSTN/CS domain interworking is to be performed. If the interworking is to occur in the same network in which the BGCF is located, then the BGCF selects an MGCF that will be responsible for the interworking with the PSTN/CS domain. If the interworking is in another network, the BGCF forwards this session signaling to another BGCF in the selected network.

The Media Gateway Control Function (MGCF) is an interworking management entity. It performs protocol conversions between ISUP (PSTN) and the IMS call control protocols. The MCFC also controls the conversions on the user plane, for example conversions between different voice coding schemes. It also selects the CSCF for incoming calls from legacy networks.

The Multimedia Resource Function Controller (MRFC) controls the media-stream resources in the MRFP. It interprets information coming from an application server and S-CSCF and controls MRFP accordingly. As a result, it also generates charging data records (CDR) for the billing system.

The Multimedia Resource Function Processor (MRFP) handles bearers on the Mb reference point. MRFPs provide resources to be controlled by the MRFC. It can process and manipulate various media streams.

IP Multimedia-Media Gateway Function (IM-MGW) terminates bearer channels from a CS network and media streams from a PS network. The IM-MGW may support media conversion, bearer control, and payload processing (e.g., codec, echo canceler, conference bridge). It interacts with the MGCF for resource control and owns and handles resources, such as echo cancelers.

Subscription Locator Function (SLF) is only needed if there are several HSS entities in a network. As a response to a query, it responds with an identity of the HSS that contains the profile of the given user.

Application Server (AS) offers value added IM services and resides either in the user's home network or in a third-party location. The third party could be a network or simply a standalone AS.

Note that in Figure 8.21 the elements present functions that have to be implemented in the IMS Domain. It does not mean that each of them will be implemented in its own physical equipment. Most probably there will only be a few new physical network elements, in which case many of the interfaces presented here will be internal logical interfaces. Note that at the time of this writing, IMS Domain specifications are not yet ready; thus, there is still no clear picture of how these features will be implemented in practice.

IMS Domain is introduced in [35] and [36]. A thorough presentation of IP technology in 3GPP can be found from [37].

REFERENCES

[1]   Mehrotra, A., *GSM System Engineering*, Norwood, MA: Artech House, 1997.

[2]   3GPP TS 23.008, v 3.6.0, "Organization of Subscriber Data," 2001.

[3]   Mouly, M., and M.-B. Pautet, *The GSM System for Mobile Communications*, Published by the authors, 1992.

[4]   GSM 09.02, version 7.5.1, "Mobile Application Part (MAP) Specification," 2000.

[5]   3GPP TS 23.002, v 3.5.0, "Network Architecture," 2002.

[6]   3GPP TS 29.002, v 4.5.0, "Mobile Application Part (MAP) Specification," 2001.

[7]   Walke, B., *Mobile Radio Networks*, New York: Wiley, 1999.

[8]   3GPP TS 25.430, v 3.7.0, "UTRAN Iub Interface: General Aspects and Principles," 2001.

[9]   Silventoinen, M., "Indoor Base Station Systems," in *GSM—Evolution Towards 3rd Generation Systems*, Z. Zvonar, P. Jung, and K. Kammerlander (eds.), Norwell, MA: Kluwer Academic Publishers, 1999, pp. 235–261.

[10]   GSM 42.056, version 4.0.0, "GSM Cordless Telephony System (CTS), Phase 1; Service Description; Stage 1," 2001.

[11]   GSM 03.56, version 7.1.0, "GSM Cordless Telephony System (CTS), Phase 1; CTS Architecture Description; Stage 2," 2000.

[12]   Johnsson, M., "HiperLAN/2—The Broadband Radio Transmission Technology Operating in the 5 GHz Frequency Band," HiperLAN2 Global Forum White Paper, 1999, , accessed January 25, 2001 at http://www.hyperlan2.com/web/pdf/white-paper.pdf.

[13]   3GPP TS 25.410, v 3.6.0, "UTRAN Iu Interface: General Aspects and Principles," 2001.

[14]   3GPP TS 25.411, v 3.5.0, "UTRAN Iu Interface Layer 1," 2001.

[15]   3GPP TS 25.420, v 3.5.0, "UTRAN Iur Interface: General Aspects and Principles," 2002.

[16]   Kyas, O., *ATM Networks*, 2d ed., London: International Thomson Computer Press, 1996.

[17]   Roberts, J., U. Mocci, and J. Virtamo, "*Broadband Network Teletraffic*," COST 242 report, Berlin: Springer-Verlag, 1996.

[18]   3GPP TS 25.433, v 3.9.0, "UTRAN Iub Interface NBAP Signalling," 2002.

[19]   ITU-T Recommendation Q.711, "Functional Description of the Signalling Connection Control Part," July 1996.

[20]   ITU-T Recommendation Q.712, "Definition and Function of Signalling Connection Control Part Messages," July 1996.

[21]   ITU-T Recommendation Q.713, "Signalling Connection Control Part Formats and Codes," July 1996.

[22]   ITU-T Recommendation Q.714, "Signalling Connection Control Part Procedures," July 1996.

[23]   ITU-T Recommendation Q.715, "Signalling Connection Control Part User Guide," July 1996.

[24]   ITU-T Recommendation Q.716. "Signalling Connection Control Performance," July 1996.

[25]   3GPP TS 25.413, v 3.9.0, "UTRAN Iu Interface RANAP Signalling," 2002.

[26]   3GPP TS 25.423, v 3.9.0, "UTRAN Iur Interface RNSAP Signalling," 2002.

[27]   3GPP TS 25.415, v 3.10.0, "UTRAN Iu Interface User Plane Protocols," 2002.

[28]   Stewart, R., et al., IETF RFC 2960 "Stream Control Transmission Protocol," October 2000.

[29]   Sidebottom, G., et al., "SS7 MTP3-User Adaptation Layer (M3UA)," IETF Internet Draft, February 2002.

[30]   3GPP TS 29.060, v 3.12.0, "General Packet Radio Service (GPRS); GPRS Tunnelling Protocol (GTP) Across the Gn and Gp Interface," 2002.

[31]   Ojanperä, T., "UMTS Data Services," in *GSM—Evolution Towards 3rd Generation Systems*, Z. Zvonar, P. Jung, and K. Kammerlander (eds.), Norwell, MA: Kluwer Academic Publishers, 1999, p. 338.

[32]   ITU-T Recommendation Q.2210, "Message Transfer Part Level 3 Functions and Messages Using the Services of ITU-T Recommendation Q.2140," July 1996.

[33]   ITU-T Recommendation Q.2140, "B-ISDN ATM Adaptation Layer—Service Specific Co-ordination Function for Signalling at the Network Node Interface (SSCF at NNI)," February 1995.

[34]   ITU-T Recommendation Q.2110, "B-ISDN ATM Adaptation Layer—Service Specific Connection Oriented Protocol (SSCOP)," July 1994.

[35]   3GPP TS 23.002, v 5.6.0, "Network Architecture," 2002.

[36]   3GPP TS 23.228, v 5.4.1, "IP Multimedia Subsystem (IMS); Stage 2," 2002.

[37]   Hameleers, H., Johansson, C., "IP Technology in WCDMA/GSM Core Networks," *Ericsson Review* Volume 79, 2002, athttp://www.ericsson.com/review.

CHAPTER 9

Network Planning

9.1   Importance of Network Planning

Network planning is a major task for operators. It is time consuming, labor-intensive, and expensive. Moreover, it is a never-ending process, which forces a new round of work with each step in the network's evolution and growth. Sometimes extra capacity is needed temporarily in a certain place, especially during telecommunications conferences, and network planning is needed to boost the local capacity. Changes in the network are also needed with changes in the environment: A large new building can change the multipath environment, and a new shopping center can demand new cell sites, and a new highway can create new hotspots.

The quality of the network-planning process has a direct influence on the operator's profits. Poor planning results in a configuration in which some places are awash in unused or underused capacity and some areas may suffer from blocked calls because of the lack of adequate capacity. The income flow will be smaller than it could be, some customers will be unhappy, and expensive equipment will possibly be bought unnecessarily.

9.2   Differences Between TDMA and CDMA

The network-planning processes in a 3G WCDMA network and in a GSM network are similar in many ways, but there are some fundamental differences. In both systems a lot of data needs to be collected and processed before a proper network plan can be produced. In GSM, a lot of work is done with frequency planning. In 3G WCDMA, however, there is no frequency planning because all base stations use the same frequency. Actually, there will typically be two to three frequencies per operator in the first phase of UMTS, but TDMA-type frequency planning is not possible with so few frequencies. In WCDMA the different frequencies are typically used for different levels of the network hierarchies; for example, one frequency for macrocells, another for microcells, and a third for picocells. So frequency planning is rather trivial in UMTS. On the other hand, a WCDMA network needs to combine and balance coverage and capacity planning to make the network functional. Also, the deployment of base stations must be done with special care to prevent them from interfering too much with one another.

The TDMA networks, like GSM, require frequency planning to make the network workable. Because the network is divided into many cells, the same frequency can be reused over and over again in different cells. This increases the capacity the network can provide. However, the same frequency cannot be used in adjacent cells, as this would increase the interference level, especially in the border areas of these cells. In the area where these cells overlap, communication would be almost impossible.

In GSM (and in other TDMA networks) the cells are grouped into clusters. The same frequency is used only once within a cluster. All clusters are identical (in principle), as can be seen from Figure 9.1. These clusters are then repeated over the landscape to provide the required coverage. Because of the cluster shape, only certain cluster sizes are practical, for example, 3, 4, 7, 12, 15, 21 (and so on) cells. If the cluster is small, the same frequency can be used more often and the network capacity is higher (in theory). But small clusters also increase the cochannel interference from other cells using the same frequency, which reduces the network's capacity. This suggests that some kind of optimum cluster size may be lurking in the complexity of the system. In this example, the clusters contain four cells (cell numbers 1, 2, 3, and 4). The cochannel interference is mainly caused by the six nearest cells using the same frequency. Cell 1 in the middle of Figure 9.1 will receive cochannel interference from the six closest other cell 1s. Note that the situation is rarely this bad. Of the six interfering cells, the particular frequency may not be used at the same time in all cells; some may support DTX, for example. Quite often it is possible to identify only one dominant interferer (DI) at a time.

The network operator must find the optimum cluster size for the network. Note that in the real world the case is much more complex than that presented in Figure 9.1. This particular example would only be true in a flat

FIGURE 9.1
Cell clusters and cochannel interference.

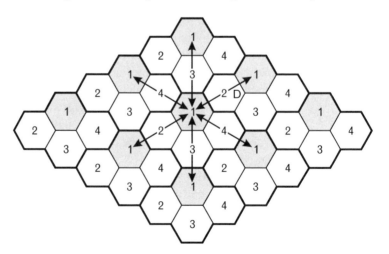

desert environment. And, of course, the cells are not hexagonal, but more or less round in a flat environment. In the more typical real world the radio signal may find lots of different obstacles in its way from the transmitter, which would make all kinds of complicated shapes for cell coverage. The local environment has to be considered, as well as the expected traffic density. The cluster size can be different in different areas. In open countryside, where the signal has fewer obstacles and the required capacity is smaller, the operator may use larger clusters. In cities, where the capacity demand is higher but where high buildings quite often block the signals, the operator may decide to use small clusters. Cells can also be directed such that the cell area forms a narrow but long beam. A cell can also be divided into several sectors with each sector forming its own subcell.

Hierarchical cell structures (Figure 9.2) bring yet another new dimension to the frequency planning discussion. The operator may use hierarchical cell structures in certain places where traffic may need some kind of partitioning among cells with overlapping coverage. For example, macrocells handle fast-moving mobile stations to reduce the number of handovers (HOs) made in underlying smaller cells optimized for pedestrian traffic. A macrocell's size is typically calculated in kilometers. Microcells are used in cities to increase capacity. The typical microcell users are city-center pedestrians or sometimes office workers inside buildings. The diameter of a microcell is a few hundred meters. Picocells are used in traffic hot spots, such as offices or shopping centers, where we find an enormous number of slow-moving or stationary users demanding service. The radius of such a cell can be only tens of meters, and we often find these in indoor locations. Picocells shouldn't be used in environments where there are fast-moving mobile stations, like cars, because the generated HO control signaling traffic would quickly exhaust the network's signal-processing capacity.

FIGURE 9.2
Hierarchical cells.

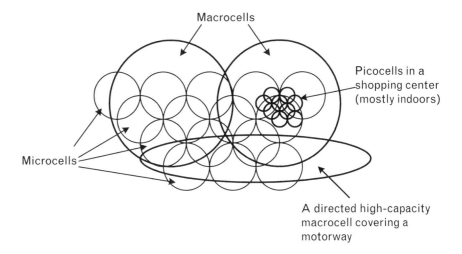

There is one ultimate hierarchical cell level that is quite often over-looked in these discussions, namely, the satellite networks. The IMT-2000 concept also includes a provision for deploying 3G satellite networks. These satellite cells would form the highest level in a cell hierarchy. The radio access technology wouldn't be the same in the terrestrial and satellite cells, so dual-system mobile stations would be needed. Moreover, the infrastructures in both networks must be planned so that they can support intersystem HOs.

In GSM, the cochannel interference is typically minimized by means of power control, discontinuous transmission (DTX), and frequency hopping. There is also an interesting idea about using the multiuser detection (MUD) scheme in GSM; see [1]. The suggested scheme would reduce the interference inflicted by the nearby cells that are using the same frequency.[1] Cell sectorization also reduces the cochannel interference, as can be seen intuitively from Figure 9.3. In this example some of the cells are divided into three sectors each. The interference level could be reduced even more with six sector cells, but of course increasing the number of sectors also increases the cost as more transmission equipment is needed.

In CDMA networks there is little or no frequency planning because all cells use the same frequency, or only a very few frequencies. A typical WCDMA operator may be given, for example, a 2×15-MHz frequency slice, which is enough for 3×5MHz WCDMA duplex frequency channels. The UMTS Forum recommends 2×15 MHz (FDD) + 5 MHz (TDD) operator licenses as a minimum allocation.

In GSM, the channel bandwidth is only 200 kHz, which provides dozens of frequency channels even for small operators. This makes network

FIGURE 9.3
Sectorized cells with three sectors per cell.

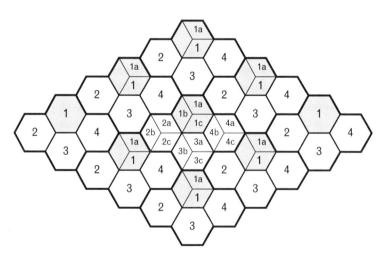

1   The algorithm detects the cochannel interferers through their different training sequences, and then deducts them from the overall received signal. This method would, however, require changes to the current GSM specification.

planning easier as a traffic increase can often be accommodated just by adding new TRX units (on new frequencies) to existing base stations. This may include a redistribution of existing frequencies by means of frequency planning, but this is still relatively easy and inexpensive.

In WCDMA, the cell sizes are not fixed, but depend on the required capacity (i.e., pulsating cells or breathing cells). So coverage and capacity parameters are dependent on each other. This means that both parameters have to be planned together. If new capacity is needed in a WCDMA network, it is most probable that it cannot be accommodated just by adding new channel elements to the existing base stations. A 3G operator will have only two to five frequency channels; thus, a 2G-like channel element addition is not a good solution to capacity problems. New base station sites will most likely be needed to ease the capacity shortage, and once a new base station is added to the network, its influence will reach even distant base stations. The parameters in the nearest base stations must be changed a lot, which triggers changes in the neighboring base stations. Hierarchical cell structures can help add new capacity without forcing a replanning of a large surrounding area. Hierarchical cell structures (HCS) in WCDMA are explained in Section 9.5.4.

9.3 Network Planning Terminology

This section explains some concepts and terms used in network planning. Even though some of us may never take part in actual network planning, it is still good to know what is going on when somebody talks about Erlangs or blocking probability.

- *Traffic intensity* is measured in Erlangs. One Erlang is equivalent to one call lasting one hour. Thus, the traffic intensity can be calculated from

  - [Number of calls (per hour) × average call duration (in seconds)]/ 3,600
  - If the result is smaller than 1 Erlang, then quite often the appropriate unit is the mErlang (= 0.001 Erlang).

- *Traffic density* measures the number of calls per square kilometer (Erlang/km^2). This is only usable for circuit-switched voice calls. For data services, the traffic density is better measured using Mbps/km^2.

- *Spectral efficiency* is defined as the traffic that can be handled within a certain bandwidth and area. This can be written as

  - Traffic intensity (Erlang)/(Bandwidth × Area) = bps/(MHz × km^2)

- *Outage* is the probability of a radio network not fulfilling a specified QoS target.

- *Cell loading* indicates the relative occupancy of the cell. This is given as a percentage of the maximum theoretical capacity.

- *Loading factor* defines the amount of interference loaded into the cell by surrounding cells. This is given as a ratio of the power received by a base station from other cells to the power it receives from mobiles in its own cell. Notice that all power received from outside the home cell is interference.

9.4   Network Planning Process

Network planning is not just frequency planning, but a much broader process. The network planning process includes things like traffic estimation, figuring the proper number of cells, the placement of base stations, and frequency planning. First, the amount of expected traffic is estimated, and then a radio network that can handle this traffic is designed. There are three phases in the design process. It starts with (1) the preparation phase, which sets the principles and collects data, followed by (2) the high-level network-planning phase (network dimensioning), and (3) the detailed radio-network planning phase.

9.4.1   Preparation Phase

The preparation phase sets the principles for the planning process. The first thing to be defined is the coverage the operator is aiming for. One operator may aim to have adequate coverage only in big towns and nothing in the countryside. Another operator may also try to cover the main roads in the rural areas. A third operator may aim for countrywide coverage as soon as possible. The chosen alternative depends on the available resources and the selected marketing strategy. Notice that the telecommunication authorities may quite often place certain requirements on the coverage, which may state, for example, that the network has to cover x percent of the population (or area) within y years from the network launch (or from the date the operating license was granted). But note that in WCDMA, the coverage is not given by simple footprints of cells. The operator must decide what kind of coverage it is aiming for. In a WCDMA cell, the available data rate depends on the interference level—the closer the UE is to the base station, the higher the data rates that can be provided (see Figure 9.4). Thus, an operator that is aiming to provide 384-Kbps coverage must use more base stations than an operator that is aiming for 64-Kbps coverage.

FIGURE 9.4
Different cell coverage for different data rates.

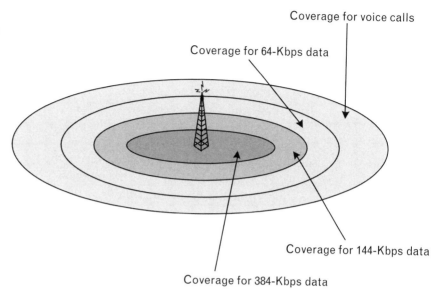

Coverage for voice calls

Coverage for 64-Kbps data

Coverage for 144-Kbps data

Coverage for 384-Kbps data

Other planning parameters set in this phase include the allowed blocking probability, migration aspects (if the operator already has an existing cellular network), the quality of service (QoS), and so on. If call blocking is allowed with a non-negligible probability, then less capacity needs to be allocated, and the network will be cheaper to implement. On the other hand, this will affect customer satisfaction. If an operator has an existing 2G cellular network, it may be best to decide to provide the wide-area coverage using this earlier network. The new network is first built in cities and towns where the demand for the new capacity is greatest and where the new investment provides a positive cash flow in the shortest amount of time because the user density is high. If base station sites are expensive to acquire, or their deployment is restricted for environmental reasons, then the operator may decide to use fewer cell sites with high-capacity base stations. If the operator has an existing GSM-1800 network, its base station sites are probably quite suitable for the initial UMTS deployment. The preparation phase defines what kind of network will eventually be built.

The other important task in the preparation phase is the data gathering. This is required for traffic estimation, which should be as accurate as possible. An operator must acquire population and vehicle traffic information from the planned coverage area. How many people live in an area? How many people work there? What is the vehicle traffic density on main roads during rush hours? Are there any special places that may require lots of capacity at certain times? These could include sports arenas, conference centers, and sites of public festivals. Then the operator must estimate the mobile-phone penetration and the amount of traffic generated by each user. Note that an average business user probably generates more traffic than an average residential user. The business calls will probably be longer, and many of those calls may include data traffic. The problem with WCDMA is that many applications in the new network will be new. Voice traffic patterns are easier to estimate because they will most probably follow 2G voice traffic patterns. But the traffic patterns generated by data applications are more difficult to estimate as there are very few precedents. GPRS networks have not found much success, most probably because the GPRS operators tried to sell the technology and not the services. I-mode is a packet-based data network, which has really become very popular in Japan, but again we should not make too many predictions about this for Europe and the United States. PC ownership in Japanese households is not so common as in Europe and the United States; thus, i-mode is the primary method for Internet access for many Japanese. Again, the success of data services will depend a lot on the pricing model employed (packet-based, time-based, flat fee, or combination of these?).

Once these demographic parameters have been calculated or estimated, the operator has a good idea of the expected traffic. Note that the calculations/estimations must be based on the peak traffic rates. And they should include some room for the future growth as there is no point in building a network that is only good for today. Next the operator has to establish a crude estimation of the number of cells needed. This depends on the capacity of the cells and the area to be covered. The preliminary deployment of cells on a map is done here.

9.4.2   Network Dimensioning

Network dimensioning is a process that aims to estimate the amount of equipment needed in a telecommunications network. In the case of a WCDMA network, this includes both the radio access network and the core network. This process includes calculating radio link budgets, capacity, and coverage, and then estimating the amount of infrastructure needed to satisfy these requirements. The output of the process should be an estimation of the required equipment and a crude placement plan for the base stations.

The cell-count estimation procedure starts with the calculation of radio link budgets. This task involves setting the maximum allowable loading of the system. In a WCDMA radio network, this is not as straightforward a task as in a TDMA system. In a GSM network, it is possible to use (at least in theory) all channel elements in the whole network at the same time. The maximum theoretical system capacity is therefore easy to calculate. In a WCDMA system, the capacity is typically not limited by the exact number of channel elements, but by the amount of interference in the air interface. There are more radio resources than the network can ever use. This excess of resources (channel elements) gives CDMA networks their characteristic soft capacity limits. The maximum system capacity is reached well before all the equipment capacity is exhausted. However, notice that it is possible to use all (or at least many) channel elements in one cell and achieve high capacity locally if it is acceptable that the capacity will be very low in the neighboring cells. This arrangement is possible because of the low intercell interference level generated by the neighboring base stations. It allows the high-capacity cell to generate high levels of interference without disturbing the system's balance. This is depicted in Figure 9.5. It is not possible to use all channel elements in the whole network simultaneously because that would result in a very high interference level and the QoS would be very poor for most users. Only the users very close to the base stations would likely have an acceptable QoS.

The theoretical maximum capacity in a WCDMA network is called the pole capacity. Calculating the pole capacity is difficult and requires making many assumptions. One attempt to calculate it is made in [2].

The network planner must decide the maximum allowable loading of the system. This is usually given as a fraction of the pole capacity and is thus called the load factor (LF). Values between 0.4 and 0.6 are usually used in network planning [3]. Higher parameter values are not recommended as a certain margin to the theoretical maximum is needed. The reasons for this margin include the following:

- Interference margin;

- Power control margin.

An interference margin is required to prevent pulsating (breathing) cells. In WCDMA, the loading of a cell affects its coverage. The higher the load, the smaller the cell size. This may result in a nasty phenomenon, where the size of the cell becomes smaller as more and more traffic is conveyed via the cell. The smaller size means that some users will lose their connections, which eventually means less interference in the cell. Decreasing interference

FIGURE 9.5
High-capacity cell surrounded by low-capacity cells.

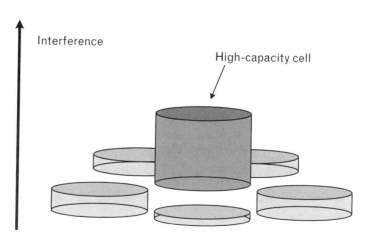

will increase the cell size, hence a pulsating cell. However, if the load factor is smaller than one unity, then there is some interference margin, which can be used to keep the cell size unchanged while the interference level changes.

A power control margin is required to give the mobile the possibility to perform fast power control (closed-loop power control) to counter the near-far problem and increase overall system capacity. The maximum cell size could be achieved if a mobile station transmits with full power, but then fast power control couldn't be used for this mobile. Fast power control is essential in UTRAN; it is required to keep interference levels as low as possible in a rapidly changing radio environment.

Other parameters to be specified at this stage include the data rates, mobile speeds, coverage requirements, terrain types, and asymmetry factors. These values can be based on empirical tests or assumptions.

The actual cell-loading algorithm is iterative. The aim is to find the largest cell size that can accommodate all the generated traffic for the parameters set earlier without exceeding the maximum allowable loading of the system. There are five phases in this algorithm:

1. Given all the parameters, calculate the radio link budget for a chosen traffic type at the provisional cell edge. The maximum system load should be used in this calculation.

2. Given the link budget, calculate the maximum cell range given the propagation model for the current terrain type.

3. Given the new computed cell area, calculate the number of users within the cell.

4. Given the number of users and their phone usage characteristics, calculate the actual cell loading.

5. (a) If the actual cell loading is greater than the maximum allowed loading, then there are obviously too many users in a cell. Reduce the cell radius and go to phase 4 (calculate the actual cell loading again). In this case, the system is said to be capacity limited.

   (b) If the actual cell loading is smaller than the maximum allowed loading, then the system is coverage limited. The cell could have accommodated more traffic. However, it is not possible just to increase the cell radius and calculate the new actual cell loading, as the link budget at the cell edge was calculated for the old cell radius. The correct way to solve this problem is to reduce the maximum system load value, and then rerun the algorithm starting from phase 1. Given the smaller maximum system load, the cell radius will increase, and furthermore the actual system load will increase.

   (c) If the actual system loading is equal to the maximum allowed system loading, then the optimum cell size has been found.

This algorithm is run until case 5(c) becomes true. This gives a cell size for this kind of scenario (terrain, user profiles, etc.). The algorithm must be rerun for all typical scenarios. Given the results, the network planner can then calculate the required number of base stations and also determine their approximate locations.

Also note that this algorithm must be run separately for both the uplink and the downlink, and the smaller cell size from those runs must be chosen as the optimum cell size. In a cell with symmetric traffic, it will typically be the uplink that determines the size of the cell. This is because WCDMA employs orthogonal spreading codes in the downlink channels, and those cause less interference than the nonorthogonal codes in the uplink. In this case, the system is said to be uplink limited. However, in a UMTS cell, the traffic can be very asymmetric, and there will possibly be much more downlink than uplink traffic. If the downlink load increases considerably, it will become the limiting factor for the cell sizes and not the uplink. There is also an additional factor in WCDMA dimensioning: the orthogonal codes. It is possible that the capacity in the downlink will be limited because of the lack of free codes. If all the orthogonal codes in a cell are already used, it is not possible to add users and traffic to that cell, even if the interference level would still be acceptable. It is, however, possible to start using additional scrambling codes in the downlink, each of them having their own orthogonal channelization code sets. These sets are, however, not fully orthogonal to each other, so interference will be increased.

If hierarchical cell structures (HCS) are used, then each hierarchical layer can be processed separately and a separate network dimensioning plan will be produced for each layer. This can be justified by noting that different layers will have different traffic profiles, and these should be processed separately in the dimensioning algorithm.

The mathematics behind this procedure, as well as a few examples, are given in [3] and [4]. ACTS STORMS (Software Tools for the Optimization of Resources in Mobile Systems) was a project running between 1995 and 1998 partially funded by the European Commission. The task of this project was to define, design, and develop a complete suite of network-planning software prototypes for UMTS. The deliverables of this project are worth reading for those interested in UMTS network planning. The final report of the project, including the list of other deliverables, is [5]. In addition, the European Union–funded research program COST 259 studied radio system aspects of personalised wireless systems, including network-planning methods and tools. The final report of this program is available as a book [16]. These documents have subtle differences in their handling of network planning, but the basic idea is the same.

As a summary, network dimensioning first attempts to estimate the amount of traffic to be generated in a WCDMA network, and then calculates the kind of equipment required, how much equipment is required, and where it is needed to handle all this traffic. With network dimensioning satisfactorily completed, we turn to the last step: detailed network planning.

9.4.3 Detailed Radio-Network Planning

The detailed network–planning phase includes the exact design of the radio network. Quite often it is not possible to obtain the optimum cell site. The owner of the site may not want to sell it, or it may be unusable (e.g., in the middle of a pond) or located in a restricted area. Environmental and health issues can also have an impact. Base station towers in an open country landscape may irritate some people. The radiation from base station transmitters is also a concern for some (with or without a good reason, most often without). All these issues have to be taken into consideration. The number of HOs has to be minimized as they create signaling traffic in the network. This can be done, for example, with large macrocells. Sectorization has to be considered and implemented where required.

Ojanperä [7] includes the following procedures in this phase:

- Detailed characterization of the radio environment;

- Control channel power planning;

- Soft handover (SHO) parameter planning;

- Interfrequency (HO) planning;

- Iterative network coverage analysis;

- Radio-network testing.

Note that in practice this phase is not done by hand, but by special computer tools. Both the network manufacturers and some independent software companies provide software packages that carry out this procedure for operators. You need a digital map, some population and traffic information, and a planning software package. The output is a cellular-radio-network plan.

9.5 Network Planning in WCDMA

In this section we will discuss some special issues in WCDMA network planning that have to be taken care of. The previous section presented the overall process; this one fills in the details. We will start with the matter of pilot pollution.

9.5.1 Pilot Pollution

Pilot pollution is a situation in which a mobile station receives several pilot signals with strong reception levels, but none of them is dominant enough that the mobile can track it. Remember that all these signals are on the same frequency and, thus, interfere with one another. Network planning strives to prevent this by ensuring that a dominant pilot signal usually exists for any mobile. The methods for this include directed cell beams, sectored cells, downward tilted antennas, and setting the pilot powers to different levels.

9.5.2 SHO Parameters

An SHO in a CDMA network is usually a preferable situation for a mobile station as it improves the quality of its connection. From the network point of view, the case is not necessarily positive. Of course, the quality of the individual connection improves, but on the other hand SHOs may increase the overall system interference level and, thus, also decrease the system capacity. An SHO also consumes more data transmission capacity in the network. An operator must find a suitable compromise between these extremes; an SHO must be provided to those mobiles that really need it, but not to others, to keep the level of system interference bearable. This can be accomplished by the correct setting of the SHO parameters. The 3GPP specification [8] gives the following SHO parameters:

- AS_Th: Threshold for macrodiversity (reporting range);
- AS_Th_Hyst: Hysteresis for the above threshold;
- AS_Rep_Hyst: Replacement hysteresis;
- ΔT: Time to trigger;
- AS_Max_Size: Maximum size of the active set.

These parameters control the SHO process, particularly the size of the active set, that is, the cells that participate in an SHO state with a certain mobile station. The SHO procedure is explained in Section 2.5.1.

9.5.3 HO Problems

Interfrequency HO is a difficult procedure for a mobile station as it has to perform preliminary measurements on other channels at the same time that it is receiving and transmitting on the current channels. There are two alternatives for accomplishing this procedure: (1) the use of two receivers in a mobile station, or (2) the use of compressed mode. As extra hardware is expensive, the most attractive method for achieving the interfrequency HO is the compressed mode. Compressed mode means that during some times-lots the data to be sent is squeezed, or compressed, and sent over a shorter period of time. This leaves some spare time, which can be used for measurements on other frequencies. This compression is achieved by temporarily using a lower spreading ratio. Compressed mode may also be necessary in the uplink if the measured frequency is close to it, as is the case with GSM-1800.

The problem with compressed mode for the network is the lower spreading ratio. A spreading code with a low spreading ratio directly uses more capacity from the corresponding cells. The more compressed slots, the more the capacity of these cells is wasted. Therefore, the network operator must find a compromise in which the interfrequency measurements and the compressed mode are used only as often as necessary to guarantee successful interfrequency HOs. Possible strategies could include assigning compressed slots less often to mobile stations with high reception levels in their serving cells, and more often to mobiles with weak reception levels. Note that intersystem HOs are one form of interfrequency HOs.

9.5.4 Hierarchical Cells

Hierarchical cell structures are by no means a WCDMA-specific issue. They are also used a lot in other network technologies, such as GSM. In WCDMA networks the hierarchical structures have some specific characteristics though.

The most straightforward way to implement a hierarchical cell structure in WCDMA is to allocate each hierarchy level on a different frequency. If an operator has been allocated a 15-MHz frequency area, this is enough for three frequency channels, each having a 5-MHz bandwidth. If the operator also has an unpaired TDD frequency slice, this can be used as one hierarchical level. One channel could be used for picocells, a second for microcells, and a third for macrocells. Another possibility is to use one frequency for macrocells and two frequencies for microcells. Picocells, if needed, could be provided on the TDD frequency.

This arrangement means that different hierarchies don't interfere (much) with each other. Adjacent channel interference (ACI) shouldn't be a problem, according to system level simulation results [7]. However, a later paper [9] claims that ACI may be a problem in cases where the adjacent carriers are employed by different hierarchy levels. A macrocell base station transmitter typically uses higher transmit power than a microcell transmitter, and a macrocell downlink may thus cause interference for UE receivers in overlapping microcells. The problem is even more serious if these carriers are allocated for different operators. An obvious drawback with the hierarchical cell scheme is that, especially for the frequency allocated for the macrocells, it is not used very efficiently. The frequency allocated for picocells probably handles several times as many calls as the macrocell frequency. Also, in many cases an operator may have been given less spectrum, for example, only 10 MHz, which is another limiting factor. Another factor that reduces the efficiency of hierarchical cells in different frequencies is that handovers between these frequencies are hard handovers requiring the use of the compressed mode.

If the different hierarchical levels use the same frequency channel, then some special problems may occur. Again, [7] argues that micro- and macro-cells can coexist on the same frequency if there is not much mobility in the system. The problem with mobility is that an HO procedure always takes some time. Preliminary measurements must be made, their processing takes some time, and the time taken by the actual HO signaling is not negligible. If a fast-moving mobile station arrives into a small microcell, it is quite possible that this vehicle is already in the center of the cell before an HO from the macrocell to this microcell is completed. This means that the mobile station has transmitted with high power levels directed to the macrocell base station far away until the HO is completed, and the microcell (which uses the same frequency in this example) has been blocked from all traffic. An attempt to fix the situation can be made with a low antenna installation and antenna tilting [10]. But the problem remains that possibly the whole macrocell is in SHO; thus, no capacity gain can be obtained.

When hierarchical cells are used (i.e., pico-, micro-, and macrocells in the same area), there must be a strategy for when to perform HOs between these hierarchy levels. Generally, the mobile station should be directed to the smallest available cell because this has the greatest effect on improving the system's spectral efficiency. However, fast-moving mobile stations should be handled by the macrocells and leave stationary or slow-moving mobiles in the microcells. A fast-moving mobile station served by microcells generates successive HOs and lots of signaling traffic to the network. It may even block the microcell base station if the HO is not done fast enough. The speed of a mobile station can be determined from its fading characteristics, or even from the number of HOs it has performed over a period of time. Some sort of time hysteresis is also needed. A car at a red traffic light should continue to be served by the macrocell, as it will soon start moving quickly again.

Another parameter to be taken into consideration is the capacity requirement of a mobile station. "Heavy users" should be handled by the smallest available cell; that is, a picocell when possible. As explained previously, the capacity and coverage parameters interact strongly in WCDMA, and high-bit-rate services in a macrocell would reduce its coverage area considerably.

Picocells usually do not provide continuous coverage; it may be that there is only one picocell in an office, so any HO from such a cell is obviously an HO to a different hierarchy level and to a different frequency.

9.5.5   Microcell Deployment

The fundamental problem with microcells seems to be fast-moving mobile stations. Microcells increase the system capacity substantially, but they cannot be used to handle high-speed mobiles. As shown in Section 9.5.4, a fast mobile could block the whole microcell because of the delay in HO execution. An obvious solution is to only use macrocells for fast mobiles.

Another solution to this problem is the use of directed cells. A directed cell could deploy a long and narrow beam along a busy road (Figure 9.6). This cell handles the traffic generated by the moving mobiles on the busy road.

Distributed antennas can transmit several small cell beams, which logically form only one cell (Figure 9.7). This means that there is no need for SHOs in subcell boundary areas. However, the mobile station power control must function quickly so that the mobile's transmissions don't block distributed receivers if it passes them with high speed.

Another problem inherent to microcells is the so-called corner effect. This is depicted in Figure 9.8. Here we see two base stations deployed on crossing streets. They both use the same frequency. If a mobile station moves west along Pembroke Street, it will have a connection with base station A. Because this base station is quite far away, the received signal strength in the mobile station is low. On the other hand, the mobile has to transmit using lots of power so that it can be heard by base station A's receiver. When the mobile station arrives at the junction of Pembroke Street and Trumpington Street (point 1), two things happen very quickly. First, base station B is very near the junction, so the signal strength of its transmission received by the mobile station is very high. It will easily block everything base station A attempts to send to the UE. Second, the UE is transmitting with so much power that it will likely block the receiver in base station B.

FIGURE 9.6
Directed cells.

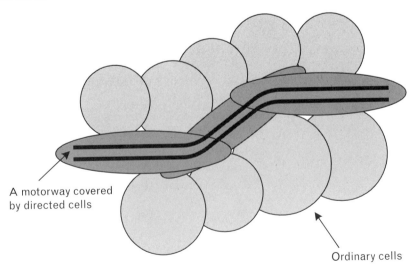

A motorway covered
by directed cells

Ordinary cells

FIGURE 9.7
Distributed antennas.

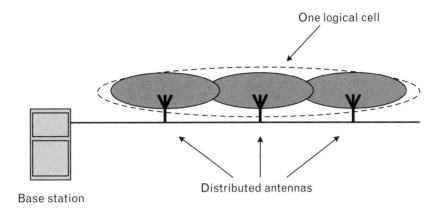

This unfortunate scenario can be avoided if the base stations and cell beams are deployed so that an SHO situation is already achieved before the mobile reaches the street corner. In practice, because of reflections and diffractions, the transmission from base station B can most probably also be received in Pembroke Street, at least in the area near the corner. The power control must be very fast, but note that even fast power control doesn't help if the connection is already lost. Power control commands do not get through to the peer entity if there is no connection.

If the call connection survived at point 1, then there are new problems ahead when the UE moves farther west along the street. In a few seconds (point 2) it loses base station B and must start using A again. Network planning in town centers is not an easy task.

9.5.6 Picocell Deployment and Indoor Planning

Picocells are cells with very small coverage areas. The cell range is only a few tens of meters. These cells could be used in traffic hot spots, and typically

FIGURE 9.8
Corner effect problem.

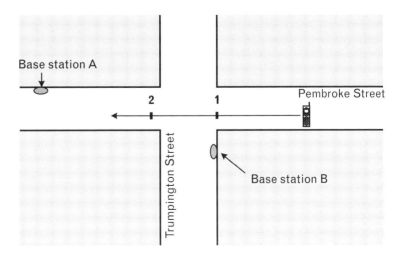

indoors. The problem with outdoor usage is handset mobility. The same problems microcells have with fast-moving mobiles are even greater with picocells. Typical sites for picocells will include offices, shopping centers, airports, and generally all buildings with a large number of slow-moving mobile telecommunications service users.

The TDD mode is an attractive technology for picocells. There are some problems with the TDD mode usage outdoors, but these can be avoided in indoor usage. The TDD mode can provide asymmetric data connections efficiently (spectrum-wise), and heavy downlink traffic is most probable indoors. Indoor wireless data traffic is more likely to be highly asymmetric to the extent that the users tend to use the service as a wireless local-area network (WLAN). In practice, the problem with the TDD mode will be that dual-mode FDD/TDD handsets could be somewhat more expensive than single-mode handsets, and it is not certain whether consumers want to pay extra just to get better service indoors.

While indoors, path loss calculations become a much more complex issue. Walls and floors may cause highly varying path losses depending on the particular construction material. A light textile wall may cause only 3 dB of attenuation, but a concrete wall can bring up to a 20-dB path loss. See [11] for an overview of penetration losses for various building materials.

If the TDD mode is used, then the planning process must take care of some additional issues. In TDD, all HOs will be hard handovers. This means that in the cell border areas, one cannot have the additional gain provided by SHOs. Thus, it would be useful to produce a map showing the equal power boundaries; that is, the areas where the reception power levels from two base stations are equal. This boundary also functions as a "trip wire" for HOs (assuming that no hysteresis parameter is used to delay HOs). The placement of this boundary should be such that it minimizes the amount of HOs between the two cells.

Asymmetric transmission is the strength of the TDD mode, but it also contains some problems. It can create nasty interference situations in which the same timeslot/frequency pair is used for downlink traffic by one base station and for uplink traffic by another (nearby) base station. The situation will be especially troublesome in the boundary area of these two base stations. UE 1 in Figure 9.9 may try to receive something from base station 1, while at the same time UE 2 transmits with high power in the same timeslot to base station 2. If both UEs are close to each other, UE 1 probably cannot receive anything except interference during its designated timeslot.

A solution to this problem is to do some resource planning in advance. Timeslots can be allocated into mutually exclusive groups. This idea was presented in [12]. It is similar to frequency planning in TDMA systems, except that here it is timeslots that are being sorted. There are altogether 15 timeslots in one TDD frame. These are sorted into groups in which each group typically has an equal number of timeslots, but this is not necessary.

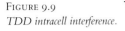

FIGURE 9.9
TDD intracell interference.

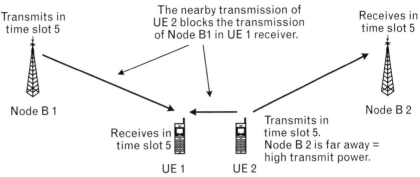

The groups are combined into clusters so that the same timeslot will not be allocated twice to different groups within the same cluster. And finally, these groups are mapped into base stations, or base station sectors.

As a result of this clustering, each base station is free to allocate the timeslots in its group at will between the downlink and uplink. This will not cause any interference to other base stations and their users within the same cluster. The drawback is of course that a base station can use only a portion of its full capacity.

An alternative would be to define a fixed allocation pattern for all timeslots within a TDD system. For example, timeslots 1 to 3 would always be used for uplink traffic, and timeslots 4 to 15 would be allocated for downlink. Here all timeslots would be in use, but the system is no longer as flexible and dynamic in its timeslot allocations between the uplink and downlink.

Notice that in some cases, the adjacent channel interference between operators in different frequencies may cause problems.

As picocells are used only in traffic hot spots, they will not provide continuous coverage. Therefore, HOs to and from picocells will quite often be interfrequency HOs. For example, if there is one picocell per office floor, the staircase could be covered only by a micro- or macrocell. A mobile user walking from one floor to another would have to perform two HOs in this case, first from a picocell to a microcell, and then from the microcell to another picocell. See also Section 8.6.3 on small base transceiver stations.

9.5.7   Sectorization and Adaptive Antennas

Sectorization can be used to increase the capacity of a WCDMA system. Sectorization means that a cell is divided into several subcells, typically into three or six sectors. If the radiation pattern of a sector is ideal, then a cell with N sectors should have N times more capacity than a cell without sectors. However, this is not the case in practice. In real cells the sectors always overlap somewhat. This reduces the capacity increase with increasing N.

For example, for a three-sector base station the capacity increase is generally approximated to be only 2.5. This is called the sectorization gain. The sectorization gain with nonideal radiation patterns is investigated in [13].

Sectorized cells have areas in which the sectors overlap. While a mobile station is in these areas, it can have a connection with both sectors simultaneously (see Figure 9.10). This is a special case of an SHO, called a softer HO. In this case, the signal combining is done in Node B, and not in the RNC, as with an ordinary SHO. This saves transmission capacity in the Iub interface, and also makes it possible to use more efficient combining technologies.

Adaptive antennas (also called smart antennas) are another technology that makes it possible to increase a cell's capacity. Here the base station can form a narrow beam toward a mobile station, tracking the movements of the mobile station. This feature has two desirable properties: (1) increased effective cell range, and (2) increased cell capacity. A dedicated beam exists for each communicating mobile.

A multibeam antenna is a smart antenna that transmits several fixed beams (not adaptive beams) within a sector. These beams cannot track the user.

The difference between sectors and adaptive antennas is that sectors are minicells. They have fixed radiation patterns, and each sector transmits its own control channels. Adaptive antennas are a method to accommodate more traffic channels into a cell. The radiation pattern of an adaptive antenna beam is very dynamic; it can follow a mobile station and adjust the signal strength according to radio path conditions. Sectors and adaptive antennas can coexist in a base station. Adaptive antennas are discussed more in Section 12.11 and in [14–16].

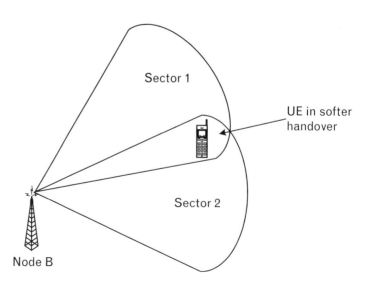

FIGURE 9.10
Softer HO.

9.5.8 Other Network Elements

So far the network planning discussion has concentrated on radio related issues on the air interface. However, an operator also has to plan the other parts of its network. This includes base station controllers (RNC), switches, registers, and the transmission media. It is difficult to give any general advice on how to plan and acquire these because every operator faces a different situation. Moreover, cellular network infrastructure is not bought from your nearest hardware store, but after detailed negotiations with (possibly several) equipment manufacturers. However, this section contains some observations that might be useful.

The physical placement of various network elements has to be considered. The various network architecture figures in this book show the logical architecture of the system, not the physical one. For example, some RNCs may be colocated with their MSC/SGSN. It all depends on how the available transmission resources can be best utilized. The physical network structure does not have to be hierarchical; for example, RNCs and their switches may be connected to a loop cable.

The cost of the network equipment is not the only factor in network planning. An operator must also think about running costs and reliability. Transmission costs can be a substantial expense if the operator has to use leased lines. Reliability is important as the network should be able to stand faults. A misguided excavator cutting a wrong cable in a construction site should not silence the whole cellular network. There should be inbuilt redundancy in the system so that local problems remain local. Note that in times of emergency, mobile phones are likely to be used extensively, but only if they work.

Building a new cellular network is expensive, especially if a completely new radio access network with thousands of base stations is needed. In terms of technology, 3GPP standards allow the sharing of a common radio access network between operators. There would only be one physical RAN, but two or more logical ones. It is even possible to share the frequency carrier between operators, as the specifications allow a list of PLMN identity codes to be broadcast on a carrier instead of the earlier case when only one PLMN code was broadcast. Each operator would most probably have its own core network because this is not so expensive to build, and moreover, the registers in the core network contain, for example, subscriber user profiles, and this is information operators will not want to share. The user subscription controls which core network handles this user's traffic. However, in some countries the 3G licence conditions may stipulate that each 3G operator has to build its own radio access network. Also note that sharing a network means that not only the costs are shared, but also the capacity of the network. More capacity needs more hardware, even if it is all located in the same base station tower.

The capacity and placement of interfaces also needs to be planned. Generally, the traffic in interfaces should be minimized whenever possible. For example, the traffic in Iur interface can be minimized if the area (i.e., group of Node-Bs) controlled by an RNC is chosen so that there are few handovers over the RNC border.

9.6   Admission Control

Admission control is a process that guards the access to the radio access network. It prevents the overallocation of resources. If not performed properly, this could result in the unnecessary reduction of the QoS of current users and even dropped calls. Admission control is performed when a UE attempts to make a call or create new bearers. The control procedure checks whether the requested resources can be granted without problems later on. Additionally, admission control has to be performed for mobile-terminated calls as these also consume resources. HOs are still one further procedure that must include admission control.

In 2G TDMA systems, such as GSM, admission control is more straightforward than in a WCDMA system. The capacity in a TDMA cell is typically not interference limited (a soft limit), but depends on the number of channel elements in the base station (a hard limit). It is easy to determine the number of unused resources because it is simply the number of free timeslots in the channel elements. Admission control has to set some threshold after which the admission is limited. There might be different admission-control criteria for different services. An incoming call (a mobile-terminated call) may have a slightly higher priority than outgoing calls as it is more probable that a mobile user will answer a call than a fixed-line user (in most markets). Of course, the outgoing call could also be addressed to another mobile user. HOs must have an even higher priority as dropped calls are quite annoying for users. Emergency calls must have the highest priority of all.

The problem of admission control is a more complex one in a WCDMA system because a new call increases the overall interference level in the cell (as well as in neighboring cells); thus, it has a direct effect on the QoS the other users in the cell are receiving. As shown earlier, the increased interference reduces the effective coverage radius of the cell. Thus, it may happen that a new connection in the cell may drop another connection, or even several connections, near the cell boundary. Clearly, this is not the best way to run a network. Users may accept the occasional blocking of their call attempts, but not the excessive dropping of existing calls.

We will consider both the uplink and the downlink admission control algorithms separately. The new resource can be granted only if both algorithms show a green light. These algorithms will have to be different, as the free capacity in the uplink and in the downlink depends on different factors. The downlink uses orthogonal codes; thus, the capacity-limiting factor is mostly intercell interference (and sometimes the number of free codes). In the uplink, most of the interference is intracell, but intercell interference has to be considered as well. Finally, the traffic patterns in the uplink and the downlink can be quite different.

The admission-control algorithm has to estimate the increase in the usage of resources the new user would cause. This estimation may be quite difficult to accomplish, as a new user can also affect the resource consumption of other users, especially in the uplink direction. If the estimation reveals that the new status would exceed the preset threshold, then the request for the resource has to be rejected. Note that while the total load level in a cell is low, an increase to the load level increases the interference level only by a small amount. But once the cell already accommodates a large amount of load, then a load increase of similar size will cause a large increase in the interference level. This is depicted in Figure 9.11.

The admission-control algorithm must reside in the RNC, which has sole access to all the required information to run an efficient admission policy. One can find several different admission-control schemes in the literature. Prasad et al. [17] give the following three examples:

Received-power admission control (RPAC) admits a bearer service i if $Z_k < Z_i^t$, where Z_k is the total received power before the new user, and Z_i^t is the service-vector dependent threshold for bearer service i.

Busy-channel-based admission control (BCAC) admits a bearer service if $N_k < N^t$, where N_k is the number of used channels at a base station k, including the new channels, and N^t is a predefined threshold.

Transmitted-power-based admission control (TPAC) admits a bearer service i if:

$$P_k < P_i^t \text{ for } \forall (k \in B)$$

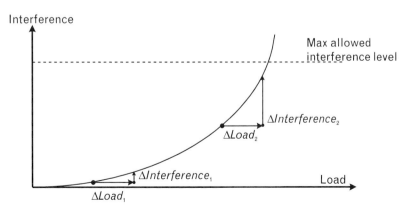

FIGURE 9.11
Estimation of the increased interference level in uplink.

where P_k is the transmitted power from a base station k before the new user, P_i^t is the service-vector dependent threshold for bearer service, and B is a set of base stations. (That is, a new bearer service should be admitted only if P_k is below the threshold at all base stations in set B.)

RPAC can be used in the uplink, which is interference limited. BCAC is used in the downlink when it is code limited, and TPAC also in the downlink in case it is interference limited.

Holma and Toskala [4] present two other examples of admission-control algorithms. The first of these is the wideband power-based admission control strategy. The user will not be given the requested resources if the resulting total interference level is higher than the threshold value:

$$I_{\text{oldtotal}} + \Delta I > I_{\text{threshold}} \tag{9.1}$$

This equation can be used in the uplink direction. A similar kind of strategy can be applied in the downlink. The user is granted the resources if the new total base station transmission power is smaller than the threshold power level:

$$P_{\text{oldtotal}} + \Delta P_{\text{total}} > P_{\text{threshold}} \tag{9.2}$$

The two threshold values are set in the radio-network-planning phase.

It is rather difficult to estimate the ΔI because the increased interference is due not only to the new user, but also to all other users in the cell whose transmission power levels will increase as well. In the second equation, the value ΔP_{total} is easier to determine because, at least in principle, the codes in the downlink are orthogonal; thus, they should not interfere with each other. Thus, the power increase should only be due to the new user, and this value can be estimated by the open-loop-power-control algorithm. In practice, however, various diffractions and reflections will also cause intrasystem interference in the downlink; thus, the value acquired from the open-loop power control may not be accurate in all environments. Note also that the second equation assumes that the downlink is interference limited and not code limited. If the number of orthogonal codes is the limiting factor, then this equation cannot be used.

The second example from [4] is the throughput-based admission-control strategy. Here the new user is granted the requested access if the following load equations are true:

$$\eta_{\text{UL}} + \Delta L > \eta_{\text{UL\_threshold}} \tag{9.3}$$

$$\eta_{\text{DL}} + \Delta L > \eta_{\text{DL\_threshold}} \tag{9.4}$$

where η_{UL} and η_{DL} are the uplink and the downlink load factors, respectively, before the new user is granted access. The calculation of the load factor ΔL of the new user is given in [4]. Both the uplink and the downlink have to be considered separately, and both cases must allow the access before it can be granted. Note that this scheme is a modification of the interference-based scheme presented in Figure 9.10. Instead of the maximum interference level, this scheme considers the maximum allowed load level as depicted in Figure 9.12. Interference as such is not considered at all in this scheme, even as increased load also means increased interference.

Admission-control schemes are something that will not be specified in 3GPP technical specifications. The various measurements needed for admission control, and the signaling procedures to support them, will be specified, but the actual control algorithm is an internal function of the RNC. It has to be defined either by the operator or by the equipment manufacturer.

What makes this task exceptionally difficult is the fact that 3G can provide services that have never been used before in any system. The traffic will be packet-based, which is also a quite new concept for commercial mobile radio networks. GPRS can provide some valuable experience in managing packet traffic to operators. It will be difficult to predict the usage of various services and, thus, the volume and the pattern of traffic [18]. Moreover, there is the issue of QoS. In the circuit-switched networks, the delay can be guaranteed, but in the packet-switched world the situation is a more complex one. The admission-control algorithm also has to take the delay into consideration. For example, if delay requirements for a certain application are relaxed, then the network can provide more throughput.

However, once an operator sets up the 3G infrastructure, it will probably be very underutilized at first. This gives the operator some time to measure and analyze the user traffic and its development with the new services. Even a crude admission-control mechanism is probably sufficient at first, and then it can be adjusted according to the accumulated experience.

FIGURE 9.12
Load-based admission control.

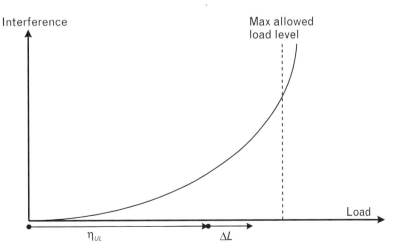

Dimitriou and Tafazolli show the importance of admission control in [19]. See also [8] for admission control in the UTRAN.

9.7 Congestion Control

In some sources, congestion control is also called load control. Like admission control, congestion control attempts to manage the resource usage in the telecommunication network so that it will not become overloaded. The difference between these control procedures is that admission control is a one-off procedure, but congestion control a continuous one. The admission-control process considers each resource request as a separate case, whereas congestion control manages the network as a whole. Admission control decides whether a connection can be set up or not; congestion control then monitors and manages the existing connections.

In principle, a really efficient and clever admission-control scheme should render congestion control unnecessary. If the admission-control process could predict the changes in the amount of traffic and interference levels, it could decide whether a new connection/bearer could be set up or not. However, in practice this is not always possible, particularly with packet-based applications and traffic. There are services that can generate highly varying data rates, and the actual amount of traffic cannot be predicted when the service is requested. The interference level in the cell can change over time. The mobility status of the UEs in the cell may change. A connection from a sports car, for example, may cause more interference than a stationary user, and there may, in any case, be new HOs to the cell.

Congestion control must be performed separately for both the uplink and the downlink, as they may have different traffic load levels. The control procedure has to compensate for the possible overload situations, and bring the network back to within the operating parameters defined during the radio-network planning phase. It has several means to reduce the load in the radio access network [4]:

- Deny downlink power-up commands from the mobile;

- Reduce the uplink E_b/N_0 target used in the uplink fast-power control;

- Reduce the throughput of packet-data traffic;

- Move part of the users to other WCDMA carriers;

- Move part of the users to other radio access technology (RAT) carriers;

- Reduce the bit rates of services that have adjustable bit rates;

- Drop calls.

As seen, power control is an important tool for congestion control. It should be possible to handle most of the overload situations. The through-put of packet data traffic can be reduced by the packet scheduler in the RNC. It can reduce the data rates of non-real-time applications, thus releasing the capacity for the congested real-time users.

HOs to other carriers are also an efficient way to remove excessive load from a carrier. This requires that the other carriers must have free capacity. Note, however, that interfrequency HOs are not always possible, even if there is free capacity in the other carrier. For example, it is not useful to move fast-moving vehicles from a motorway macrocell to local microcells as the interference would only block the microcells without bringing any new capacity. In the case of intersystem HOs, we should take care that the new system can provide the same services as the UTRAN; that is, voice service users could probably be transferred to GSM, but interactive video users cannot.

Reducing the bit rates of adjustable bit rate services can be applied, for example, to speech call users with the AMR codec. It is possible to reduce the speech codec bit rate slightly without losing the call, but the QoS may be also reduced. Dropping calls should obviously be the last method considered. However, it is better for the RNC to drop some chosen calls in a controlled way than to let an overload situation cut many more calls. Congestion control is further discussed in [4,7,8,17].

References

[1]   Ranta, P., and M. Pukkila, "Interference Suppression by Joint Demodulation," in *GSM —Evolution Towards 3rd Generation Systems*, Z. Zvonar, P. Jung, and K. Kammerlander (eds.), Norwell, MA: Kluwer Academic Publishers, 1999, pp. 153–186.

[2]   Lister, D., et al., "UMTS Capacity and Planning Issues," *3G Mobile Communication Technologies Conference, IEE Conference Publication, No. 471*, London, March 2000, pp. 218–223.

[3]   Irons, S., et al., "Supporting the Successful Deployment of Third Generation Public Cellular Technologies-System Dimensioning and Network Planning," *3G Mobile Communication Technologies Conference, IEE Conference Publication No. 471*, London, March 2000, pp. 156–160.

[4]   Holma, H., and A. Toskala, *WCDMA for UMTS: Radio Access for Third Generation Mobile Communications*, New York: John Wiley & Sons, 2000.

[5]   ACTS Project AC016, Software Tools for the Optimization of Resources in Mobile Systems [STORMS], STORMS Project Final Report, http://www.infowin.org/ACTS/RUS/PROJECTS/FINAL-REPORTS/fr016.pdf, April 1999, p. 58.

[6]   Correia, L.M., *Wireless Flexible Personalised Communications*, Chichester, U.K.: John Wiley & Sons, 2001.

[7]   Ojanperä, T., and R. Prasad, *Wideband CDMA for Third Generation Mobile Communications*, Norwood, MA: Artech House, 1998.

[8]   3GPP TR 25.922, v 5.0.0, Radio Resource Management Strategies, 2002.

[9]   Johnson, C., and J. Khalab, "The Coexistence of WCDMA Macrocell and Microcell Radio Network Layers," *3G Mobile Communication Technologies Conference, IEE Conference Publication No. 489*, London, May 2002, pp. 81–85.

[10] Shapira, J., "Microcell Engineering in CDMA Cellular Networks," *IEEE Trans. on Vehicular Technology*, Vol. 43, No. 4, November 1994, pp. 817–825.

[11] Rappaport, T. S., *Wireless Communication Principles and Practice*, Upper Saddle River, NJ: Prentice Hall, 1996.

[12] Koshi, V. "Radio Network Planning for UTRA TDD Systems," *3G Mobile Communication Technologies Conference, IEE Conference Publication No. 471*, London, March 2000, pp. 161–165.

[13] Yang, X., S. Ghaheri, and R. Tafazolli, "Sectorization Gain in CDMA Cellular Systems," *3G Mobile Communication Technologies Conference, IEE Conference Publication No. 471*, London, March 2000, pp. 70–75.

[14] Mogensen, P. E., et al., "Antenna Arrays and Space Division Multiple Access," in *GSM—Evolution Towards 3rd Generation Systems*, Z. Zvonar, P. Jung, and K. Kammerlander (eds.), Norwell, MA: Kluwer Academic Publishers, 1999, pp. 117–151.

[15] Morrison, A., et al., "An Iterative DOA Algorithm for a Space-Time DS-CDMA Rake Receiver," *3G Mobile Communication Technologies Conference, IEE Conference Publication No. 471*, London, March 2000, pp. 208–212.

[16] Boukalov, A., "System Aspects of Smart Antennas Technology," Presentation 6.5.1999 at Radio Communication Systems Department/School of Electrical Engineering and Information Technology (EIT) at the Royal Institute of Technology (KTH), Stockholm, Sweden, at http://www.s3.kth.se/radio/seminars/sa.pdf. Accessed on 1.25.2001.

[17] Prasad, R., W. Mohr, and W. Konhäuser, *Third Generation Mobile Communication Systems*, Norwood, MA: Artech House, 2000, pp. 243–248.

[18] *GSM Quarterly*, Issue 17, pp. 28–30, Mobile Communications International, June 2000.

[19] Dimitriou, N., and R. Tafazolli, "Resource Management Issues for UMTS," *3G Mobile Communication Technologies Conference, IEE Conference Publication No. 471*, London, March 2000, pp. 401–405.

CHAPTER 10

Network Management

10.1   Telecommunication-Management Architecture

Telecommunication-management architecture in UMTS is largely based on the Telecommunications Management Network (TMN) as defined in [1].

Note that while other areas of UMTS are specified by the 3GPP in quite a detailed way, the telecommunication-management specifications contain definitions only for general management frameworks and concepts. There are no detailed definitions of messages or signaling-flow diagrams to be used in network management, nor do we find a clear definition of the list of tasks to be performed under this banner. Even the physical entities performing these tasks are not specified.

The reason for this lack of specificity is that it is very difficult to define a common telecommunications-network-management scheme suitable for all the various networks. Even if the telecommunication networks are strictly defined, as in UMTS, their implementations can vary considerably. The exact implementation of the same network entities, such as the particular hardware components and the software code, are different with each equipment vendor. The 3G specifications contain numerous optional features that may or may not be implemented in a particular network. The network may consist of different radio access networks-for example, operator A may have only UTRA-type access networks, while operator B uses UTRA, GSM, and DECT access networks. The transmission networks may be different. The physical size of the UMTS network can vary considerably; an operator with millions of subscribers has entirely different management needs than an operator with only 5,000 customers.

This all means that if a strictly defined common TMN does not exist in GSM, then one will hardly exist in UMTS. However, as UMTS networks will be multivendor networks, network management will be a logistical nightmare if every vendor uses a completely different management approach to its equipment. The 32-series specifications from the 3GPP aim to provide a common framework for designing the management networks, but the work on these specifications is likely to take a long time. Still, the various management entities will never be plug-and-play connectable. Some extra work will be needed when combining the management systems from different manufacturers.

The tasks the UMTS management functions have to perform include the following:

- Fault management;

- Configuration management;

- Performance management;

- Roaming management;

- Accounting management;

- Subscription management;

- QoS management;

- User equipment management;

- Fraud management;

- Security management;

- Software management.

Some of these tasks are discussed in the following paragraphs. We start with fault management.

10.1.1   Fault Management

The purpose of fault management is to detect a fault in the network as soon as possible after it occurs, locate it, make a notification of the fault, and if possible fix the fault automatically and test the fix.

There are many things that can go wrong in a complex telecommunication network:

- A physical hardware component can break down.

- All software contains bugs, some more, some less, but no large software system is totally free of bugs. If a bug is serious, it may cause a system malfunction.

- Overloading a network entity may either block the entity altogether or reduce the capacity of the entity.

- The worst failures are functional failures in which no clear reason for the problem can be found. The problem cannot be fixed; thus, it is likely that the fault will happen again.

- Communication failures between network entities are also possible.

Once a fault is detected, it must be reported to the network management system. This report must contain enough information so that the management system can make the right analysis and react correctly.

A fault report (also called an alarm) should contain (at least) the exact location of the fault, its type and severity, its probable cause, the time stamp of the fault detection, and the nature of the fault. All alarms generated by a network entity and not yet fixed and cleared must be stored in a list of active alarms in that entity.

Most faults can be both detected and cleared automatically by the fault-management system. There are clear conditions for the fault occurrence and fault clearing. These faults are referred to as automatically detected and automatically cleared (ADAC) faults in 3GPP specifications. An automatically detected and manually cleared (ADMC) fault is one for which an occurrence condition exists, but there is no clearing condition. This implies that an ADMC fault cannot be cleared automatically, but a manual input from an operator is needed.

Once a fault has been detected and reported, it must be repaired. There are no common rules about how to fix a fault; each case depends on the type of the fault and the particular network element implementation. Hardware faults are fixed by changing the faulty unit. Software faults can be fixed by reboot, by a partial software upgrade (patch), by an updated software release, or with totally new software. Communication faults may be fixed by changing the transmission equipment or by removing the cause of the interfering noise. A fixed fault is then reported by raising another alarm, this time with the severity field set as "cleared."

Fault management is specified in [2].

10.1.2 Configuration Management

Configuration management provides the operator the means to control and monitor the configuration of the network elements. There are two distinct areas in configuration management: one that targets single network elements and another that considers the network as a whole.

3GPP is making an attempt to at least partly standardize configuration-management in UMTS, as these networks can be expected to be multivendor networks, and configuring such a network without any kind of common configuration-management rules would present a logistical puzzle.

Configuration management must ensure that changes in the network configuration can be done in a controlled and efficient way. A network life cycle contains three phases:

1. Network installation and service launch;

2. Network operation stage, in which only short-term dynamic reconfiguration is needed;

3. Network upgrades to meet long-term requirements.

The network installation is usually done with the help of equipment vendors. Software is loaded to the network elements and initialized. Configuration-management entities are part of the installation. The system must be thoroughly tested before the network launch to the customers.

In the network operation phase there is no need for a large-scale network reconfiguration. The only configuration operations required are short-term and temporary in character, and related to changes in network operational parameters. It is often necessary to change some parameters to ensure efficient use of network resources, because, for example, of a change in traffic-distribution patterns. A conference may require more traffic capacity locally, just as may the main roadways during rush hours. Office blocks may require a great deal of traffic capacity during working hours but hardly any capacity outside business hours.

The real challenge for the configuration-management system comes with network upgrades. Once the number of subscribers increases, new capacity is needed; thus, new equipment must be bought. As the technology develops or subscribers begin to demand new services, new upgrades are again needed. In addition, the manufacturers will improve their existing equipment and components, and an update may be a sensible thing to do just to improve the performance of the existing equipment.

A network system can be either updated or upgraded. The difference between these terms is that an update improves some existing functionality or features of the system. It does not introduce any new features. An upgrade implements new features or facilities to the network. Upgrades also include changes such as extensions, reductions, or replications of the network entities.

An important property that must be included in the configuration-management system is reversibility. This means that once an upgrade is made and activated, it must be possible for configuration management to quickly restore the old configuration if problems arise with the new configuration. Lengthy breaks in service are not allowed in a modern telecommunication network.

To be able to provide configuration management functionality, a configuration-management system has two service components: the system-modification service component and the system-monitoring service component.

The tasks of the system-modification service component include the creation, deletion, and conditioning of network elements and resources. The system-modification service component issues reports on the configuration of the entire network or parts if it. These reports can be issued spontaneously or on request.

Configuration management is specified in [3].

10.1.3 Performance Management

One of the functions of configuration management is to optimize the performance of the network resources. This is only possible if the managing entity has all the necessary and accurate information on which to base the configuration decisions.

The task of performance management is to collect data that can be used to assess the current configuration of the network. Thus, in a way the name of this management function is a bit misleading. A better name might have been "performance-measurement management."

The type of data to be measured include the following:

- Traffic levels within the network, both user data and control signaling;

- Verification of the network configuration;

- Resource-access measurements;

- QoS;

- Resource availability.

Traffic measurements may include, for example, the traffic load on the radio interface and the usage of resources within the network nodes.

The changes to the network configuration must be verified with the help of performance measurements. It is not enough to believe that a change improved the effectiveness of some functionality in the network. This verification must be based on hard facts. The measurements are needed from the measured target both before and after the configuration.

The resource-access measurements must be performed at regular time intervals across the network.

The QoS measurements measure the QoS attribute as it is experienced by the user. These measurements can also be used for determining the charges of the services used, if QoS is used as one variable when calculating the charges.

The resource-availability measurements gauge the availability of the chosen resources at different phases of the life cycle of the system.

Measurement collection, or measurement administration, includes the following functions:

- Measurement job administration;

- Measurement result generation;

- Local storage of results at the network element;

- Measurement result transfer;

- Performance data presentation to the operating personnel.

Measurement jobs are processes that are executed in the network elements in order to manage measurement data collection there. Their administration includes actions such as creating, modifying, and deleting measurement jobs, as well as scheduling, suspending, and resuming them.

Measurement reports are generated at a particular frequency known as the granularity period. During this period, measurement samples are collected, and at the end of the period, a report is generated. The samples from the previous periods do not influence the new report; only the latest samples are considered.

If the measurement report retrieval is deferred to a later moment in time, then the network element has to provide temporary local storage for these results. The storage capacity and the storage duration depend on the implementation.

Measurement results can be transferred using two different methods. They can be sent forward as soon as they are generated, as notifications, or they can be stored in the local network entity as files and transferred only after they are requested by the network manager.

Finally, the performance data must be presented to the operator and/or the configuration-management application. This task includes the storage and preparation of the data.

Performance management is specified in [4].

10.1.4 Roaming Management

Roaming is a process whereby a mobile subscriber uses the resources of a network other than his or her home network. This provides users with much better service, as they can use their mobile phones outside the home-network service area. Technically this process introduces new challenges; for example, routing of a mobile-terminated call to the user requires the cooperation of the HLR and the VLR.

Roaming management handles issues related to the roaming agreements between operators. A roaming agreement is a contract between the home-network operator and the serving-network operator. There will be a great number of 3G operators, and the number of required roaming agreements will be even larger. In theory, n operators could have a maximum of $(n-1)!$ roaming agreements between them. The actual number of roaming agreements will be smaller than this maximum.

In this agreement, the participants must agree on many items, including

- Tariffs and pricing;

- Signaling and traffic interconnection;

- Call detail record (CDR) exchange format and schedule;

• Problem handling;

• Service-level agreements.

The roaming agreement can be an ordinary direct agreement between both service providers or it can be established by the means of a clearing-house. Roaming management is discussed in [5].

10.1.5   Accounting Management

Accounting management concerns the management of charging and billing of services provided by a 3G system. These items are discussed in their own sections: charging in Section 10.2 and billing in Section 10.3.

Accounting management must be in close cooperation with fraud management in order to prevent the dishonest usage of system resources. For local (HPLMN) users this is relatively straightforward, but in the case of visiting users, interactions between the two network operators (between the home and the serving networks) are required.

Cost control is also part of accounting management. It must be possible to provide the user with an indication of the charges a requested service will generate. The user may also want to set a limit on the charges per time interval for the usage of certain services, and accounting management must be able to monitor this limit and notify the user if it is reached.

10.1.6   Subscription Management

In the very first cellular systems the only service provided was voice, and there was no difference between users on what could be provided for them. However, in modern systems the user subscriptions are highly customized for each user. The network has to access the user's subscription data (or as the jargon goes, a user's service profile), and decide what kind of services can be provided for the user.. Subscription management is a process that maintains the user subscription data in the home subscriber server (HSS), and also in USIM.

In 3G networks the subscriber information is stored in HSS, which is basically an extended HLR. 3GPP aims to standardize the interface to HSS for customer care activities. The use of proprietary interfaces is inconvenient for those operators using multiple vendors' equipment since their systems have to accommodate multiple proprietary interfaces that perform essentially identical functions. Moreover, a standard interface would make it easier to generate customer self-care applications that allow customers to amend their own subscription data.

Subscription management is discussed in [6].

10.1.7   QoS Management

QoS management in 3G networks primarily consists of two functional areas: QoS policy provisioning and QoS monitoring. QoS policy provisioning is the process of configuring and maintaining selected network elements with QoS policies that are created based on customer subscription and observed network performance. In 3G networks, multiple networks must interwork to provide the end-to-end QoS required by end-user applications. Furthermore, there are many classes of network elements from many network infrastructure suppliers, each of which require configuration in a consistent manner to fulfill network operator's QoS objectives. QoS monitoring is the process of collecting QoS performance statistics and alarms. This data is then used to generate analysis reports for making changes or upgrades to the network.

QoS management is discussed in [6].

10.1.8   User Equipment Management

This is a feature that allows a network operator to activate and deactivate the tracing of a particular subscriber within the network. Once activated, the trace activity is reported back to the network management system. This feature is called user equipment management (probably because user tracing has a bad connotation). The activation/deactivation and reporting interface for trace management will be standardized.

User equipment management is briefly discussed in [6].

10.1.9   Fraud Management

Fraud management includes all those activities designed to detect and prevent fraud in mobile telecommunication networks. Note that the fundamental feature of this network—mobility—makes fraud prevention much more difficult. A fraudulent customer cannot always be found as easily as in fixed networks. The services of a mobile network can be used from anywhere within the coverage area. This does not necessarily have to be the home-network coverage area, as roaming provides the services via other networks.

Fraud prevention aims to prevent fraud from happening at all. Fraud detection aims to detect fraud that has already taken place. These tasks include at least the following functions:

- Classification of customers according to the levels of fraud risk;

- Revision of fraud risk levels;

- Detection of fraud patterns;

• Taking appropriate action to suspend service provision;

• Contacting visiting (roaming) users' home service provider for a credit check.

The first item in this list can be accomplished with a credit information check. If the credit check result is not satisfactory, the operator can refuse to provide a subscription altogether, or provide only a limited subscription with no international roaming privileges. A rejected subscription application does not prevent the customer from using mobile network services. It is also possible to buy a prepaid SIM, whereby the customer pays for airtime beforehand and then may use the services as long as there is credit left.

The second item suggests that it is necessary to continuously monitor the customer's behavior and usage patterns; if bills are not paid on time or credit information changes, the operator may reconsider the subscription agreement.

Fraud attempts usually follow some set pattern. It is useful to build a "database" of fraud patterns and then look for these patterns. If fraud is detected, then the service must be suspended from this user. There must be mechanisms to also suspend the service if the user is roaming.

Visiting customers represent a bigger risk for an operator, as it cannot make the same tests for them as for the home subscribers. However, the operator can consult the home service provider and international registers such as the Central Equipment Identity Register (CEIR) to detect known fraudulent users.

Fraud detection must be done as much as possible in real time. This way the damage caused by the fraudulent activity will be smaller, and it will also be easier to identify and catch the fraudulent user.

Fraud management is discussed in [5].

10.1.10 Security Management

The security-management architecture of the TMN is divided into two layers. These layers are the O&M IP network layer and the application layer.

The O&M IP network can be built as a private trusted network, but in practice this might be difficult due to the sheer size of the network. In that case, the only security provided is that the network is logically separated from the Internet. O&M traffic can be encrypted, however.

The application layer provides authentication and authorization services. The authentication of users ensures that every party involved is securely authenticated against every other party. Authorization means access control; that is, checking that a user is authorized to perform an operation upon a specified target at a given time. This second item includes the logging of events, which makes it possible to check who did what.

Security management is discussed in [6].

10.1.11 Software Management

Software management can be divided into two processes: the main software-management process and the software-fault-management process. The main software-management process handles the management of new software releases and correction patches. The software fault-management process then takes care of network monitoring and handles faults that are caused by problems in the installed software.

The main software-management process can contain the following stages:

- Delivery of software from the vendor;

- Forwarding of the software to network elements or element managers;

- Validation of the software to ensure that it is not corrupted;

- Activation of the software to an executable state;

- Validation of the software to check that it runs properly;

- Accepting or rejecting the software based on outcome of the previous stage;

- Returning to the previous version of the software if the new software is rejected.

This is only one example of the software-management process. There can also be other valid steps and sequences. For example, in the beginning of the process, the operator and the vendor may have a lengthy negotiation procedure, defining the requirements of the software to be delivered.

The software fault-management process focuses on the detection of the actual fault and how to solve the problem. Thus, the steps in this process include the following:

- Monitoring of the software problems ("bugs") in the network;

- Problem solving, which includes locating the reason for the problem and then performing one of the following three corrective actions:

1. Revert to an earlier software version;

2. Activate a new corrected version of the software;

3. Reactivate the current software ("reboot").

Reverting to an older version of the software may sometimes help, but it may also contain the same bug (it is possible that the bug simply wasn't found before). Also, an earlier version may contain some other bugs that were corrected in the later version.

The best solution is, of course, a new corrected software version (that has been properly tested). But a new version of software takes time to produce. It may take weeks or months for the software vendor to locate the error, fix it, test the fix, and make a new software release. If the problem is severe enough and prevents the normal operation of the network element, then the operator must fix the problem temporarily using other methods while waiting for the new version.

Reactivation of the current software is the easiest and quickest solution to the problem. The reason for the problem is still there, but with luck it will take some time before the fault occurs again.

Software management is discussed in [6].

10.2   Charging

Charging is a special case in telecommunications management, in that it is specified with considerable detail. This is understandable because charging for services is indeed quite important for operators. Furthermore, UMTS will be a global standard, and roaming will be widespread. The home public land mobile network (PLMN) and the visiting PLMN must have mechanisms to exchange charging data so that the user can be billed correctly and the participating networks get their remittances. These charging mechanisms have been specified by the 3GPP.

A prerequisite for call charging is the collection of the call and event data. The collected data is not only used for charging purposes but for the statistical analysis of service usage and as evidence of customer calls. It can be used to provide itemized bills for customers. Also, in case of disputes and complaints, the operator has to be able to show that the charged calls have actually taken place.

The following new requirements are given in [7] for UMTS charging and accounting:

- Provide a call detail record (CDR) for all charges incurred and requiring settlement between the different commercial roles;

- Allow fraud control by the home environment and the serving network;

- Allow cost control by the charged party;

- Provide at the beginning of a chargeable event an indication to the charged party of the charges to be levied for the event (i.e., advice of charge [AoC] supplementary service);

- Allow itemized billing for all services charged to each subscription;

- Enable the home environment to provide a prepay-service and enable the serving network to support that prepay service;

- Allow interoperator charging, including mobile-operator to mobile-operator, mobile-operator to fixed-operator, and mobile-operator to IP-network-provider;

- Allow network operator to make use of third-party supplier charging;

- Provide details required for customer care purposes.

A central item in the call-charging process is the CDR. It is a data record that contains all the necessary information about the resources and services used by a subscriber. Each CDR has a unique identity within a certain time span. CDRs are generated by network elements in the serving network and they are sent to the billing system in the home network. They are also used in the intra-PLMN billing; that is, where the serving network is the home network. There is no single CDR format, but the actual contents of the CDR depend on the used service and on the network entity that generated the CDR.

The CDR information may include, for example:

- User identity;

- Terminal identity;

- Identifier of the service requested;

- Resource requested (e.g., bandwidth);

- QoS parameters;

- Quantity of data transferred;

- Serving network identity;

- Time stamp at the instant the resources were provided;

- Time duration covered by this call record;

- Unique CDR identity.

A CDR is always generated when a service is used, even if the service might be free for a subscriber. As indicated earlier, a CDR can also be used for purposes other than charging, and the network element generating the

CDR cannot (and should not) know about the billing issues. A CDR does not contain information about the costs generated; this will be calculated by the billing system. Note also that the calling party is not necessarily the charged party.

There are some special cases in charging that are worth mentioning here. If a call is a very long one (this is quite possible with low-volume packet data connections), then it must be possible to generate a CDR while the call is still ongoing. This "breakpoint" can occur after a certain amount of data has been transferred, or after some duration of time. A CDR generated in this way is called a partial CDR.

If the call is a multimedia call, then it may generate several CDRs, one for each component of the call. The billing system must be able to combine these CDRs and indicate to the user on the bill that these charges belong to one call.

E-commerce will be an important class of applications in the new UMTS network. Customers will be able to buy goods and services via telecommunication networks. The service providers and merchants will of course want payments for these goods, but generally these payments are not done via the UMTS charging system but via a separate electronic payment system. The UMTS charging will be used only for those charges that have occurred because of the use of UMTS services or UMTS transmission resources. It is also possible for the network operator to act as an e-commerce merchant, in which case the charges can be handled via the UMTS charging. However, regardless of the merchant's identity, all usage of UMTS network transmission resources will generate CDRs. Electronic payment systems are not part of UMTS specifications. These are, however, briefly discussed in Section 14.4.5.

CDRs are part of the data-collection-management task for charging. These data flows go from network elements to the charging gateway function (CGF), or to some other network management element. There are also data flows going in the opposite direction. Tariff management must ensure that tariff information is distributed to all applicable network elements, so that the AoC service can be supported. AoC is described in [8] and [9].

The charging of circuit-switched and packet-switched services differs considerably, so they are discussed separately below.

10.2.1 Charging of Circuit-Switched Services

The charging of circuit-switched calls is relatively straightforward, because there will be only one mobile services switching center (MSC), the so-called anchor MSC, that manages of a circuit-switched call for its whole duration. Even after HOs have been carried out, all signaling is still routed via the anchor MSC. The charging of circuit-switched calls is typically based on the call duration and possibly on the QoS and the use of any supplementary services. These are all managed by signaling; thus, the anchor MSC knows about them. At the end of the call, the MSC can generate and dispatch a CDR. In case of a circuit-switched call, the CDR can be sent directly to the billing system.

In case the call is an intersystem call, (i.e., it is either originating or terminating in another fixed or mobile network) and then the home PLMN, the situation is a bit more complex. Here it is the gateway MSC (GMSC) that generates CDRs for call routed from or into other networks.

If a user is visiting (roaming) another network, then the VPLMN generates CDRs for this subscriber in the usual way and sends them to the billing system of the VPLMN. The billing system stores the information from CDRs into the transferred account procedure (TAP) records, which are sent regularly to each HPLMN. Thus, CDRs as such are not sent into other networks, but TAPs are used for this purpose. TAPs are not a 3GPP-specific method, but they are already used by the GSM Association.

MSCs are not the only entities that generate CDRs. Even though CDRs stands for charging data records, they are also used for other purposes than charging, such as accounting and statistics. Thus, all call attempts, both originating and terminating, generate CDRs. This also happens if an HLR or VLR is interrogated, or if some supplementary service is used, for example.

10.2.2 Charging of Packet-Switched Services

The charging of packet-switched traffic is a more complex issue. There is no fixed "anchor" entity that exists for the entire duration of a packet "connection." If the UE moves into the service area of another serving GPRS support node (SGSN), all call-handling functions move to the new SGSN. Therefore, CDRs can be generated by more than one SGSN. Furthermore, the gateway GPRS support node (GGSN) has to generate its own CDR to relay some of its call information to the billing system.

The problem in the billing system is how to combine the separate CDRs, which refer to the same packet session. This is solved by adding a unique charging ID number (C-ID) to all CDRs. It is generated in the GGSN when the PDP context is activated, and then transferred to the SSGN. It is further relayed to new SSGNs if routing-area updates are performed. Because the C-ID is only unique within a GGSN, the billing system has to check both the GGSN identity and the C-ID when combining CDRs.

Figure 10.1. demonstrates how the charging data is collected in a typical GPRS session. The charging gateway functionality (CGF) transfers the charging information from the SGSN and GGSN to the billing system. Each GSN (an SGSN or GGSN) may have its own CGF, or there can be one centralized CGF in a separate network element called the charging gateway (CG). This example includes a CG. The CGF has to collect the CDRs from the GPRS nodes, possibly store them temporarily, and then

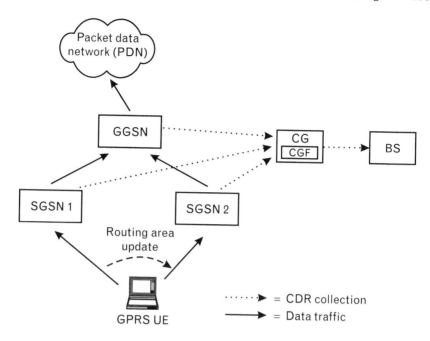

FIGURE 10.1
*CDR collection in a
packet-switched call.*

transfer them to the billing system. It may also perform some preprocessing of CDRs for the billing system, but the internal functionality of a CGF (or a billing system) will not be specified by the 3GPP.

The charging of packet-switched traffic is typically based on the volume of the traffic. This is easiest to calculate at the SGSN. There it can be implemented in a centralized way as opposed to distributing the task to radio access network (RAN) nodes. SGSN counters have the knowledge of how many packets have been transferred in both the uplink and downlink. However, the SGSN cannot know the exact amount of packets delivered to a UE in the downlink direction. Some packets may have gone missing in the RAN because of a weak radio connection. Therefore, the RNC has to indicate to the SGSN the amount of data packets not successfully delivered to the UE. Partially transferred packets are also counted as failures. This number is then deducted from the downlink counter value at the SGSN. In the uplink direction this kind of problem does not exist, as all packets arrived at the SGSN can be regarded as transferred to their destination.

Charging issues are specified in [7] and [10–13]. The charging in IMS domain will be discussed in 3G TS32.225, but this specification was not yet available at the time of this writing.

10.3 Billing

Billing as an issue is often overlooked in technical discussions. This is probably because it is not, and will not be, included in the specifications.

Furthermore, billing systems are typically not provided by the major telecommunication-infrastructure vendors, but by special billing system manufacturers. Thus, the marketing and press material from the big telecommunications companies does not include billing system items.

Billing is, however, an important issue for the operators and service providers. First, they get their money from it. Second, the bill is often the most important contact point between the customer and the operator or service provider. A confusing billing statement may cause irritation for the customer, and in the worst case force him or her to change the service provider.

In principle, the billing process is simple. The billing system gets the charging data in the form of CDRs from the network elements, calculates the cost of the service used based on current tariffs, and dispatches the bill statements to customers. The bills may be sent directly by the operator, or by the service provider, if they are used in the country in question.

In practice, preparing a billing statement will become an increasingly complicated task with UMTS. In early 2G systems the only major service provided was voice, for which billing was relatively easy. There were probably only a few tariffs in use, and even an itemized statement was easy to compose. In 3G, however, the customer will have a large selection of services to choose from. Pure voice will still be charged based on duration, but for data services other criteria have to be used. The amount of data is one criterion, but it is not sufficient in every case. Different services have different requirements for the QoS. An e-mail service does not have to be strictly real-time, but an interactive game application does not allow even for short delays. Different tariffs are therefore needed. Also, how do we measure and discount the shortfalls of the QoS? A customer may have subscribed to a certain service with a certain QoS. What happens if the service is provided with lower QoS because of congestion? A billing system must be able to solve this kind of problem automatically.

Setting up these tariffs will be difficult, but educating the customer about them will be even more so. An average customer (let's call him John Smith) does not know about packet-switched technologies and average delays, and probably doesn't want to know. If John's billing statement is full of telecommunications jargon, he will not understand it and may have a bad feeling about his service. The billing statement has to be as simple as possible, but also provide detailed information about the services consumed.

In the future it is possible that a customer will have several subscriptions with the same telecommunications operator for different services. John Smith may own a fixed-line telephone and have a subscription for it with Telegiant Ltd. He may also own a PC and have an Internet connection via Telegiant Ltd. The cable TV in his living room is well supplied with sex and violence by Telegiant Ltd., and his mobile telecommunications operator happens to be a certain Telegiant Ltd. It would probably make sense for

Telegiant to combine these four bills into one statement and send only that one to John Smith. This is called convergent billing. For operators, it would mean reduced billing costs, and provide a way to offer the customer special tariffs according to his needs. On the other hand, a combined "superbill" could scare the customer. If all telecommunications costs are combined into one statement, the resulting sum could be uncomfortably large for the customer to receive at one time. It could encourage the customer to considerably reduce his service usage the next month.

A billing statement from Telegiant Ltd. to Mr. John Smith is depicted in Figure 10.2. The reverse of the statement could contain tariff information and customer service contact information.

The phenomenon wherein an existing subscriber leaves an operator to join another one is called *churn*. This is a big problem in some markets. New customers can be lured to join through attractive offers, but at the same time, existing customers are lost to other operators who are giving similar offers. Changing the operator is made easy in many markets in that the customer is allowed to keep the old telephone number in case he or she changes the operator. The result is a very unsettled market. One way to combat this problem is to offer attractive contracts with some minimum mandatory duration. On the other hand, this may not solve the problem but only worsen it. Once the minimum contract duration has elapsed, the customer is again free to change operators and will probably do so if there are similar attractive offers from other operators. The new non-voice services in the 3G systems could provide another solution to the problem. If an operator can develop a suite of products that can capture and hold subscribers, then there is no need for long contracts or equipment subsidies. This is, however, an extremely difficult task, and other operators will surely provide tough competition.

Prepaid

The prepaid services form a special case in billing. In GSM, prepaid services have found a lot of success lately. The prepaid concept means that a customer does not have a subscription with an operator, but buys airtime for a SIM card in advance and then uses the phone as long as there is unused credit in the account, after which new airtime must be purchased.

Prepaid service has some advantages over the traditional postpaid concept. Because all phone usage is paid in advance, a credit check is not needed for new customers. Thus, those customers with imperfect credit information can use mobile services under the prepaid scheme. In fact, the user remains anonymous in a prepaid plan. The handset and SIM can be bought from a superstore together with some airtime, and the mobile is ready to be used. The operator does not know the identity of the handset buyer. Also, because there is no contract with the operator, the customer is free to change the operator at any moment, although the unused credit on the prepaid SIM is forfeited.

FIGURE 10.2
*Hypothetical
telecommunications billing
statement.*

TeleGiant Limited			
Monthly Statement: 1.2.2004 – 1.3.2004			
			Euros
GiantCable TV Service			
Monthly subscription			19.90
GiantSport service packet			9.90
Pay-per-view movies:			
Batman XIV		3.90	
Eurocop		3.90	
The Return of Rambo		4.90	
Total movies			12.70
(The GiantCinema service packet would cost you only 14.90 per month)			
GiantTalk Phone Service			
Monthly subscription			5.20
Local calls			13.20
National calls			5.32
International calls			
Italy	12.2/20.12–20.23	11 min. 2 sec.	3.20
Finland	12.2/20.25–20.45	20 min. 23 sec.	5.10
Canada	23.2/21.03–21.39	35 min. 45 sec.	15.56
Total international calls			23.86
GiantNet Internet Service			
Monthly subscription			7.99
On-line games:			
Aces High	4 hours 34 min.	6.80	
(10 000 points: 40% discount & promotion to Group Captain)		−2.72	
Grand Prix Circuit	10 hours 5 min.	8.92	
Total on-line games			13.00
GiantMobile Service			
Monthly subscription *(inc. 300 min. off peak air-time)*			4.90
Calls made	1 hour 16 min.	15.20	15.20
Emails sent	56	@0.09	5.04
Multimedia postcards sent	10	@0.25	2.50
JamBuster traffic info service packet			2.90
		Subtotal	111.81
TeleGiant Preferred Customer Discount		−5%	−5.59
		Total	106.22

There is no monthly subscription charge in prepaid service. This makes it attractive for people who will use the mobile services only occasionally.

For example, a prepaid SIM is attractive to a customer who needs a phone only for emergency use but is not going to use it otherwise. Here the mobile handset does not generate any costs if it is not used. Another good use of prepaid service is for the international vacationer from a country with one kind of cellular technology who is going to a country with another incompatible technology; for example, North Americans visiting Europe, or Europeans visiting North America.

Prepaid service also offers an efficient way to control the cost of the usage, even though there are also ways to do this with a postpaid subscription. The children in a family can be given prepaid mobile phones, and this automatically limits their phone service usage to a certain amount.

For the operator, prepaid service means no need for billing. The customer has already paid for the services beforehand. It may even happen that the customer never uses the services he or she has paid for.

But the prepaid system also includes some problems. For the customer, the most obvious problem is that only some operators allow international roaming for prepaid cards. This is because real-time credit checks are not available to check the prepaid-card credit status. This limitation will be removed little by little in the future; for example, with the implementation of the CAMEL concept. There are more and more prepaid card users, and it is important to provide international roaming services for these users. Also, the calls are typically more expensive with prepaid than with postpaid SIMs. And prepaid service also demands an added task for the customer, as new airtime vouchers must be bought more or less regularly. Of course, the prepaid airtime will run out just when you need it most.

For the operator, the biggest problem is probably the anonymity of the customer. This is a security problem, as it is very difficult to catch fraudulent users. Airtime vouchers can be counterfeited and misused. Stolen or duplicated vouchers can be used without a risk, as the operator does not know to whom the handset belongs. This also means that prepaid phones are excellent way for criminals to communicate because they are difficult to trace.

Also because of this anonymity, the operator cannot use efficient marketing techniques. It does not know what kind of services this customer uses or how much he or she uses them. The customer cannot be provided with new tariff or service offers because contact information is unknown. One possible solution could be to encourage prepaid customers to send their addresses and usage details to the operator, for example, by offering some benefits such as extra airtime vouchers or participation in a contest.

It is also possible for a customer to buy a prepaid SIM only to receive calls. Most countries use the calling party pays (CPP) pricing structure for mobile communications. The receiving party does not pay anything for incoming calls after the initial card purchase (there are no monthly fees in a prepaid scheme). This problem can be partly avoided by limiting the validity of the airtime vouchers to a certain period of time. Obviously, churn is also quite high among prepaid users.

An interesting detail with prepaid service is that although it is increasingly popular in most GSM countries, it is used very little in Finland, which has one of the highest mobile phone penetration in the world. The reason for this may be the fact that handset subsidies are illegal in Finland. Because handsets must be purchased at the current market price, the network usage is then relatively inexpensive, as the operator does not need to collect the subsidy cost from the customers. The monthly subscription charges can be kept low; thus, prepaid service with no monthly fee would not make much difference for the customer.

On the other hand, in some other countries where the mobile boom started later, the prepaid scheme has been very popular. In Italy and Portugal, some operators have more than 80% of their subscribers on prepaid schemes. And it is the new subscribers especially who opt for prepaid schemes. This makes the prepaid system very important for operators.

Prepaid cards, and the pricing of mobile cellular systems in general, are given a thorough treatment in [14].

10.4 Service Providers Versus Operators

A telecommunication network is operated and managed by an operator. The operator has an operating license for the network, and it owns all the infrastructure needed: base stations, base station controllers, switches, registers, management equipment, and so on. The transmission network, however, could be leased and not owned. The operator could make roaming agreements with other operators.

A service provider is a company that sells the services of telecommunication networks to customers. They can combine various services into service packages, price and market them, sell subscriptions to customers, and handle all the billing. A service provider does not own any kind of network infrastructure (except maybe the billing service equipment), but it buys airtime from an operator. One service provider may sell subscriptions to several operators. It has been argued that service providers increase the competition in the telecommunications market by offering a wide variety of services.

However, the use of service providers is not a global phenomenon. For example, in Finland, service providers are unknown, and the customers make their subscriptions directly with the network operators. One could argue that this provides customers with less to choose from. On the other hand, it has been said that, in the United Kingdom, for example, a large number of service providers offering many different contracts makes it virtually impossible for customers to evaluate all of the offers. Depending on the point of view, it can be argued that service providers increase the competition in the market, or add an extra management layer between the customer and the operator and, thus, increase the cost of using the network services.

Mobile Virtual Network Operator
The latest development in this area of business is the concept of mobile virtual network operator (MVNO). As a concept it can be located between service providers and operators. An MVNO would not have a license to use radio spectrum itself; thus, it would have to buy its airtime from one or several "nonvirtual" (real) operators. Therefore, an MVNO does not need any kind of RAN infrastructure. It could have its own core network (CN), or it could hire the services of some other operator's CN. The user's SIM, however, would always be issued by the MVNO.

The most likely type of MVNO would be one with its own CN. The connection to the rented RAN would be via the "host" CN (the mobile network operator or MNO) through a mobile services switching center (MSC) as depicted in Figure 10.3. The MNO-MSC handles the outgoing

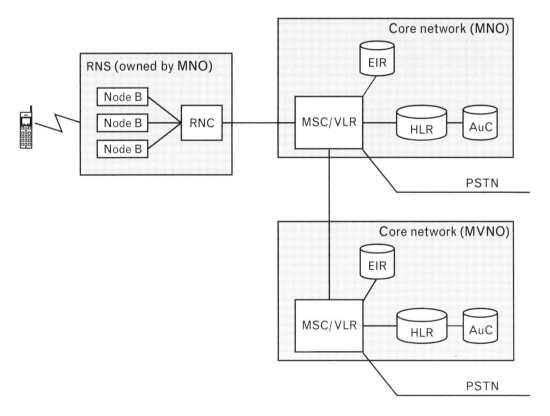

FIGURE 10.3 *MVNO concept.*

call routing to MVNO-MSC. Incoming calls would be routed from the MVNO-MSC to the appropriate MNO-MSC.

However, there are some problems with this arrangement, both technical and commercial. For example, the MNO's capacity planning would be difficult if it doesn't know how much traffic an MVNO is going to generate via the MNO's network. It could be that an MVNO has contracts with several MNOs, and it could relay the traffic via the network with the cheapest tariff at that moment. Thus, the traffic patterns could change considerably depending on the time of the day. On the other hand, this situation is in a way somewhat similar to international roaming. An MVNO can be seen as just another foreign PLMN, with its customers just happening to roam in the MNO's network.

Commercial questions are also difficult, as an MVNO will be at the same time the MNO's customer and its competitor. The telecommunication regulators will have a difficult task defining the rules for MVNO and MNO relationships.

It is almost certain that MVNOs will become a reality in 3G networks. In 3G spectrum bidding contests, many well-known telecommunications companies have been left without operating licenses, but these companies will still want to participate in 3G somehow. The MVNO concept provides a way to do just that. Moreover, the companies with successful license bids have very heavy debt burdens, and they may be willing to sell airtime to other parties so that they can bring their entire network capacity into use as soon as possible. A customer probably would not notice any difference between an MVNO and an MNO. MVNOs would have their own PLMN network codes, SIMs, and so on.

What does the MVNO concept mean for service providers? Probably it is just a new layer in the service hierarchy. Service providers could buy airtime from the MNOs and MVNOs alike. In countries where service providers are not used, this concept might bring some new competition into the market. Big MVNOs would be quite powerful negotiators against MNOs. MVNOs could also provide a way for new companies in the mobile telecommunications business to learn CN management and operator business without costly RAN construction. These companies could then bid for their own nonvirtual operating licenses in the later phases of 3G. So MVNOs could provide a path for some service providers to become operators. On the other hand, some MVNOs could set up their own marketing network with service outlets, and bypass the service providers altogether.

A good presentation on MVNOs can be found in [15].

References

[1]   ITU-T Recommendation M.3010 (1996): Principles for a Telecommunications Management Network.

[2]   3GPP TS 32.111-1, v 5.0.0, Telecommunication Management; Fault Management; Part 1: 3G Fault Management Requirements, 2002.

[3]   3GPP TS 32.600, v 4.0.0, 3G Configuration Management (CM); Concept and High-level Requirements, 2002.

[4]   3GPP TS 32.401, v 5.0.0, Telecommunication Management; Performance Management (PM); Concepts and Requirements, 2002.

[5]   3GPP TS 32.101, v. 3.4.0, 3G Telecom Management: Principles and High Level Requirements, 2001.

[6]   3GPP TS 32.101, v. 5.0.0, 3G Telecom Management: Principles and high level requirements, 2002.

[7]   3GPP TS 22.115, v 5.2.0, Service Aspects; Charging and Billing, 2002.

[8]   3GPP TS 22.024, v 4.0.0, Description of Charge Advice Information (CAI), 2001.

[9]   3GPP TS 22.086, v 4.0.0, Advice of Charge (AoC) Supplementary Services - Stage 1, 2001.

[10]   3GPP TS 32.200, v 5.0.0, Telecommunication management; Charging management; Charging principles, 2002.

[11]   3GPP TS 32.205, v 5.0.0, Telecommunication management; Charging management; 3G charging data description for the Circuit Switched (CS) domain, 2002.

[12]   3GPP TS 32.215, v 5.0.0, Telecommunication management; Charging management; 3G charging data description for the Packet Switched (PS) domain, 2002.

[13]   3GPP TS 32.235, v 4.1.0, Telecommunication management; Charging management; Charging data description for application services, 2002.

[14]   "Cellular Mobile Pricing Structures and Trends," OECD Report (DSTI/ICCP/TISP(99)11/FINAL), 2000, http://www.oecd.org/pdf/M00020000/M00020419.pdf, accesssed on 12/7/2002.

[15]   OFTEL Statement on Mobile Virtual Network Operators, October 1999, http://www.oftel.gov.uk/publications/1999/competition/mvno1099.htm, accessed on 12/7/2002.

CHAPTER 11

Procedures

We return to the details of the UMTS in this chapter. Many of these procedures were already presented at a high level in the discussion of RRC in Chapter 7. Those examples included only two communicating entities, UE-RRC and UTRAN-RRC, but here some other participating entities are discussed. In this chapter the most common signaling procedures are shown as signaling flow diagrams. These diagrams help to illustrate how the various network entities and protocol tasks interact to perform a function. The UTRAN air interface procedures are further discussed in [1]; this chapter is rather UE-centric. Some examples that are more UTRAN-centric are given in [2].

Note that these diagrams do not show all the signaling needed to complete the procedure in question. Typically, part of the internal signaling of an entity (i.e., in the UE) has been left undefined in the specifications (or shown in only a general form) and its exact implementation is manufacturer-specific. The specification may imply what kind of data is necessary to transfer over an internal interface, but the format of this data is not stipulated. Therefore, the author has invented some of the signal names in these diagrams, and they are not based on any specification. Moreover, the procedures described here do not generally show any confirmation primitives. In real implementations these are often necessary to confirm that the requested action is possible or was successfully carried out in the applicable protocol task.

11.1 RRC Connection Procedures

The UTRAN separates the concepts of a radio connection from a radio bearer (RB). A radio connection is created first, and then the network can create one or more RBs independently of the radio connection. An RB can also exist without a dedicated radio connection. In this case, the RB uses the common channels, such as the random access channel (RACH) and the forward access channel (FACH). Therefore, the concept of an RRC connection is used here. An RRC connection implies that a radio connection exists, but this connection can use either dedicated or common resources. So an RRC connection is a logical concept, and a radio connection is a physical concept. The physical reality implements and enables the logical concept. A dedicated connection allocates the resource exclusively to one user, so common channels should be used whenever possible.

11.1.1   RRC Connection Establishment

The establishment of an RRC connection is always initiated by the UE. This is true even with a mobile-terminated call. In this case the network informs the UE about the incoming call via a common (paging) channel, and the UE then performs a normal RRC connection establishment procedure.

The UE initiates this procedure, but the UTRAN controls it. It may decide that no radio resources can be allocated for the UE, and respond with an RRC connection reject message. The UTRAN may also direct the UE to another UTRA carrier or to another system with this message. See Figure 11.1, and also Section 7.5.2.6.

11.1.2   Signaling Connection Establishment

The following procedure shows how the RRC connection-establishment procedure is used by the higher layers; that is, by the nonaccess stratum (NAS). All higher-layer signaling messages, including the initial message (in this example it is a CM service request, but it could be a location update request or some other NAS message) are relayed through the radio interface encapsulated in direct transfer messages. The direct transfer message itself is defined in the RRC layer specification [3]. The initial message is included in a radio-connection-establishment request primitive sent to the RRC layer.

The connection establishment has to be indicated to higher layers, with a –CNF (confirmation) primitive in the originating side, and with an –IND (indication) primitive in the other side. If the establishment fails, then the UE-NAS must receive a negative confirmation. See Figure 11.2.

11.1.3   RRC Connection Release

This is the normal RRC connection release when there is a dedicated channel (DCH). Note that the PDUs here are sent in unacknowledged mode. To improve the reliability of the RRC connection release complete message, the UE can send it repeatedly to the network. If this message is still lost, then the network will clear the connection once the UTRAN-L1 is found to be out of sync. See Figure 11.3.

It is also possible within the UTRAN to have a logical RRC connection without a DCH; for example, packet data transfer is handled via shared channels. In this case the radio interface signaling, including the RRC

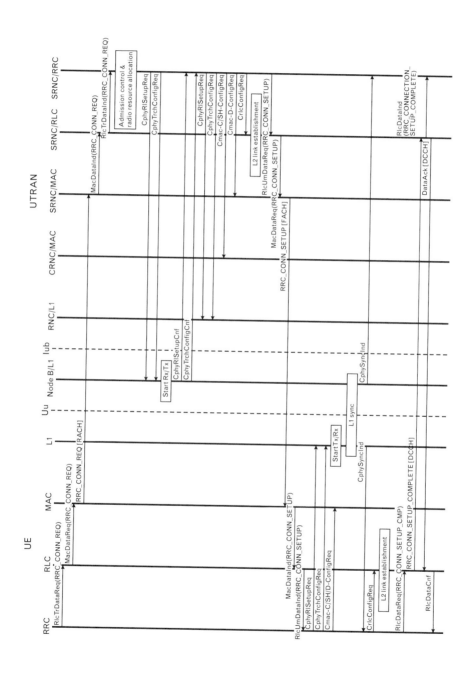

FIGURE 11.1   *RRC connection establishment.*

FIGURE 11.2
*Signaling connection
establishment.*

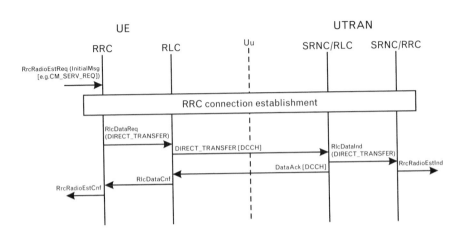

connection release signaling, is handled via RACH and FACH channels
(Figure 11.4).

11.2   Radio Bearer Procedures

As previously explained in connection with the RRC connection establishment, a radio connection and an RB are two separate concepts in UMTS.
The radio connection is a static concept. It is established once, and survives
until it is released. There is only one radio connection per terminal. On the
other hand, the RB defines what kind of properties this radio connection
has. There may be several RBs on one radio connection, each having different capabilities for data transfer. The capabilities are based on the requested
QoS parameters. RBs are also dynamic: they can be reconfigured as necessary. This is not to say that radio connections cannot be configured; they can
be reconfigured in many ways. The most common radio connection reconfiguration is probably the handover (HO) procedure.

It is also possible to have an RB without a dedicated radio connection.
Circuit-switched bearers or bearers using real-time services typically need
dedicated radio channels to meet their strict delay requirements. Packet-
switched bearers or bearers using nonreal-time services, however, often do
not need a permanent association to a dedicated radio resource, but they can
use common or shared radio channels.

11.2.1   Radio Bearer Establishment

An RB establishment is always initiated by the UTRAN. This is because
each RB uses some radio resources, and only the network knows what kind

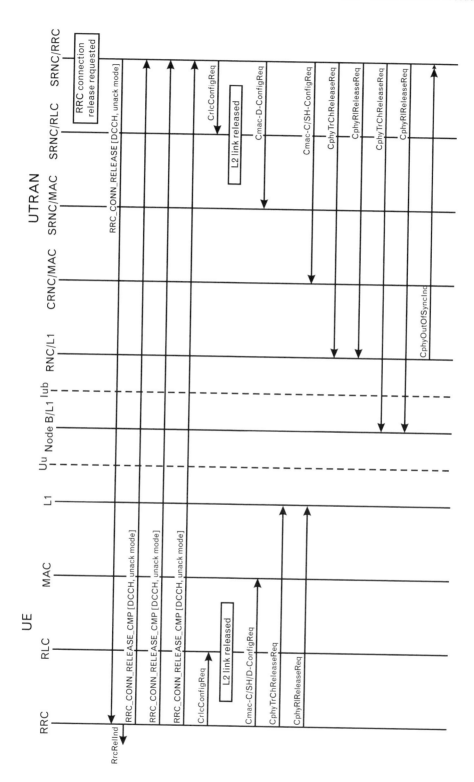

FIGURE 11.3   *RRC connection release from dedicated channel.*

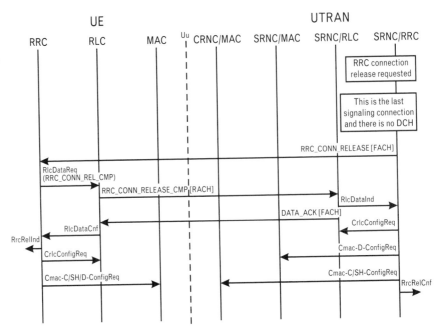

FIGURE 11.4
*RRC connection release
without dedicated channel.*

of resources it can grant to a UE. The UE can make a request for an RB to be established, but this request is signaled to UTRAN using higher-level signaling (CM); thus, it is not part of the access stratum RB establishment procedure.

The UTRAN RRC gets the request for a new bearer from higher layers. At the RRC level the signaling is simple: the UTRAN sends a radio bearer setup message, and the UE responds with a radio bearer setup complete. However, the interlayer signaling can be quite different depending on the requested QoS parameter and whether there is already a suitable physical channel in place.

The following cases are presented below:

1. A radio bearer setup with a dedicated physical channel activation;

2. A radio bearer setup with an existing physical channel modification —an unsynchronized modification;

3. A radio bearer setup with an existing physical channel modification —a synchronized modification;

4. A radio bearer setup without an associated dedicated physical channel.

11.2.1.1 Radio Bearer Setup with Channel Activation

The first case (with a dedicated physical channel activation) is quite straightforward (Figure 11.5). The UTRAN-RRC starts by configuring the

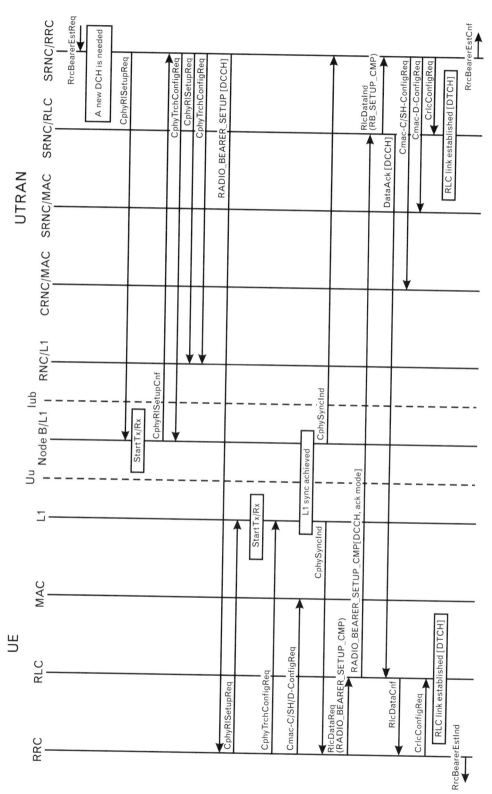

FIGURE 11.5 *Radio bearer establishment.*

physical layers in both the RNC and Node B (note that there could be several communicating Node Bs if the UE is in soft handover [SHO]). Only after this is done is a radio bearer setup message sent to the UE. The UE-RRC then configures the lower layers according to instructions in this message. Once layer 1 synchronization is obtained, the RRC sends a radio bearer setup complete message back to the UTRAN, and the UTRAN-RRC then configures the lower layers in the RNC. Once the radio link control (RLC) connection (L2) is established, both communicating RRCs indicate the RB establishment to their higher layers. Note that the RLC link establishment is a local procedure in both the UE and the UTRAN; it does not require any signaling over the air interface.

11.2.1.2   Radio Bearer Setup with Unsynchronized Channel Modification

The second case presents a scenario in which it is possible to modify the current physical channel to meet the requirements of a new bearer (Figure 11.6). This modification is unsynchronized, meaning that the old and new physical layer configurations can coexist at the same time in the peer entities; that is, the UTRAN can use one configuration and the UE another, and they can still continue to communicate. This also implies that no activation time is required.

In practice, all this means that the procedure is quite similar to the first case, except that setup primitives are now replaced by modify primitives. The UTRAN RRC modifies all the participating physical layers and sends a radio bearer setup message to the UE. The UE RRC modifies its lower-layer protocol tasks and sends a radio bearer setup message to the UTRAN. The UTRAN then modifies its lower protocol layers (the MAC and RLC) and once the RLC link is established, both parties inform their higher layers.

If L1 cannot modify the physical channel as requested in CphyRl-ModifyReq, it sends a negative acknowledgment back to the RRC. This may happen if the physical layer cannot support the requested configuration.

11.2.1.3   Radio Bearer Setup with Synchronized Channel Modification

The third case is an RB establishment with a synchronized physical channel modification (Figure 11.7). In this case it is necessary that physical channels are reconfigured in both peer entities at the same time because the old and new configurations are incompatible and cannot coexist. This problem is solved by including an activation time in the radio bearer setup message.

The procedure starts when the UTRAN-RRC checks the availability of the requested configuration from all the participating Node B physical layers. If they all respond with a positive acknowledgment, then the RRC

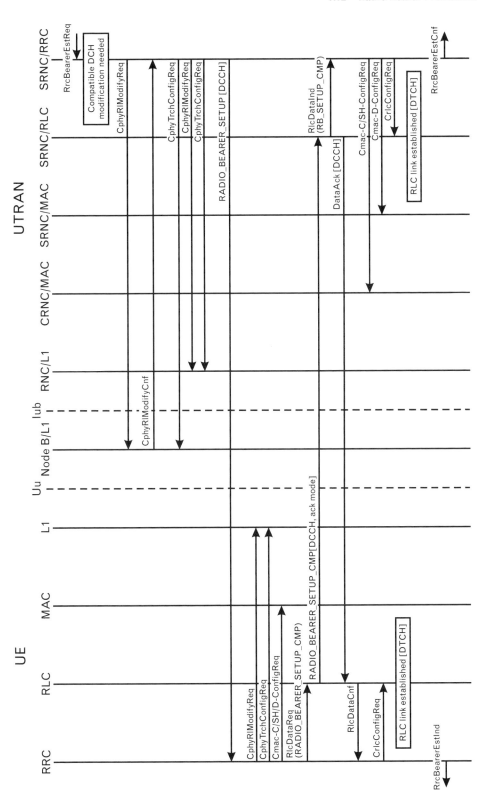

FIGURE 11.6 *Radio bearer establishment with unsynchronized dedicated channel modification.*

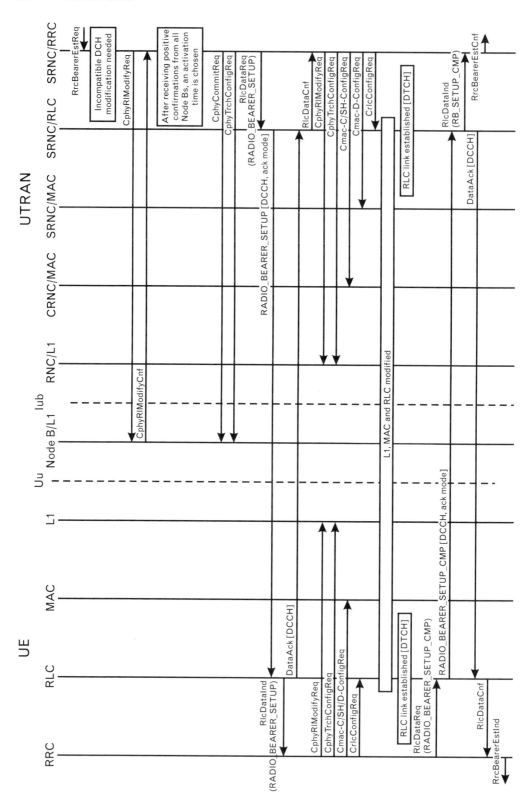

FIGURE 11.7   *Radio bearer establishment with synchronized dedicated channel modification.*

chooses an activation time and signals this to the UE and to all the lower layers in the RNC and Node Bs. The UE similarly signals the new configuration and the activation time to all the lower-layer entities. At the activation time all entities start using the new configuration, and the UE-RRC sends a radio bearer setup complete message to the UTRAN.

11.2.1.4 Radio Bearer Setup Without Dedicated Channel

In the fourth case the new bearer doesn't need a permanent DCH (Figure 11.8). Therefore, this procedure doesn't contain any physical layer modifications at all, but just RLC and MAC configurations.

11.2.2 Radio Bearer Release

When an RB is released, the physical channel can be modified or released altogether depending on whether it can be "reused" after the RB release. Also these procedures can be synchronized or unsynchronized depending on whether the old and the new configuration can coexist. In this example the physical channel is reconfigured and the procedure is unsynchronized (Figure 11.9). Note that RB release and radio connection release procedures are separate concepts. A radio connection release may follow an RB release, but they are separate procedures from the UE's point of view.

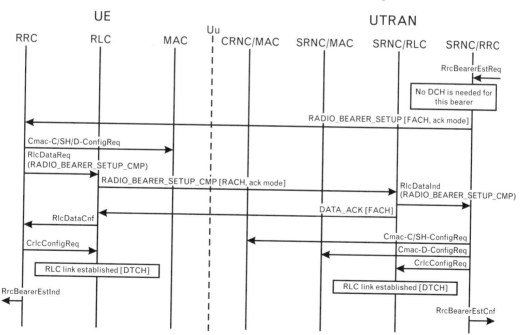

FIGURE 11.8 *Radio bearer establishment without dedicated channel.*

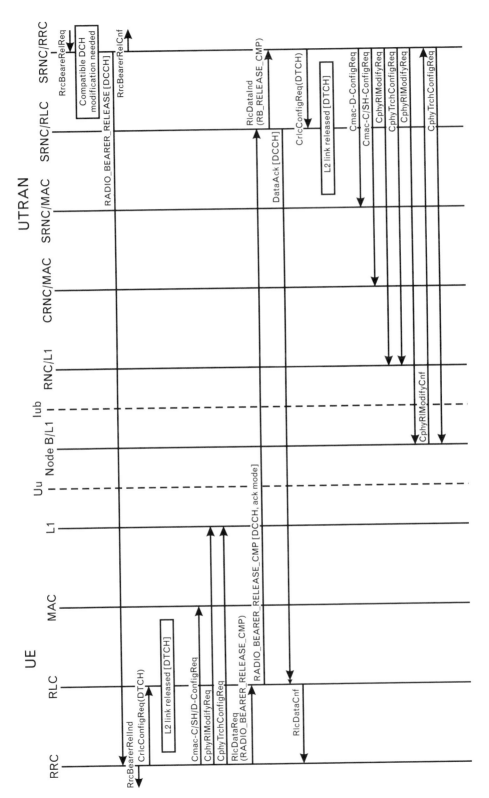

FIGURE 11.9   Radio bearer release with unsynchronized physical channel modification.

11.2.3 Radio Bearer Reconfiguration

If the RB needs to be reconfigured, this can be accomplished using the next procedure. This example shows an unsynchronized RB reconfiguration, where the old and the new configuration can coexist. A synchronized reconfiguration procedure would use an activation time parameter to make sure that the reconfiguration takes place simultaneously in both in the UE and in the UTRAN because the old and the new configuration cannot exist at the same time.

The basic principle in the unsynchronized procedure is simple (Figure 11.10). The UTRAN modifies the L1 in the Node B, and if it is successful, sends the configuration information to the UE in a radio bearer reconfiguration message. The UE-RRC reconfigures all its lower layers (RLC, MAC, and L1) and sends a radio bearer reconfiguration complete message back to the UTRAN. After receiving this message the RNC-RRC reconfigures the lower layers in the participating RNCs. Finally the old configuration, if one existed, is released from layer 1 in Node B.

If the requested modification is not possible in L1, it responds with a negative acknowledgment to the CphyRlModifyReq primitive. This, as well as any other failures in the UE, is further indicated to the UTRAN with a radio bearer reconfiguration failure message.

See also Section 11.2.6 for the use of the radio bearer reconfiguration procedure.

11.2.4 Transport Channel Reconfiguration

The transport channel is a new concept not used in GSM. Transport channels are used in the interface between the L1 and L2 layers; that is, between the physical and MAC layers, so in that sense they are the equivalent of GSM's service access points) (SAPs) that also exist in the interface between layers 1 and 2. Transport channels define how and with what characteristics the physical layer immediately below them transfers the data. The transport format defines the attributes a transport channel is currently using, for example, block size, transmission time interval, and error protection. The transport format set defines all transport formats associated with a transport channel. The used transport format within the given transport format set can be dynamically changed for each transmission time interval (TTI). The selection of the transport format is done by the transmitting MAC immediately above the transport channels.

The transport channel reconfiguration procedure can be used to reconfigure transport channel parameters (the transport format set). The procedure is always initiated by the UTRAN, which sends a transport channel reconfiguration message to the RRC in the UE. As with other modification and reconfiguration procedures, this procedure can also be either synchronized or unsynchronized.

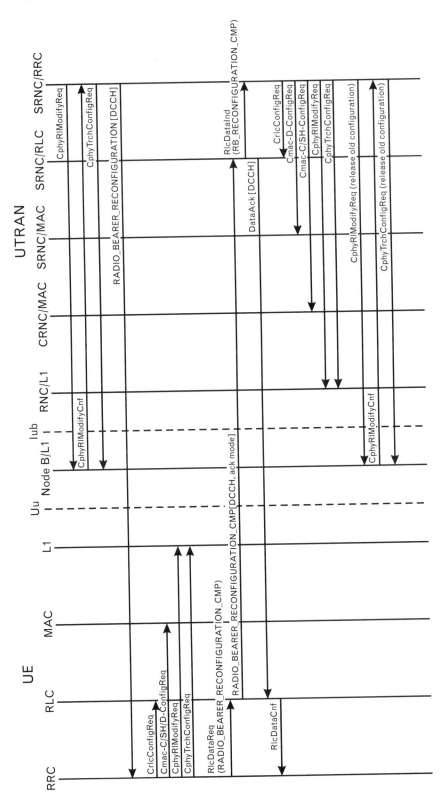

FIGURE 11.10   *Radio bearer reconfiguration.*

11.2.4.1   Unsynchronized Transport Channel Reconfiguration

An unsynchronized transport channel reconfiguration takes place when both the old and new transport channel configurations can coexist. This is implemented so that the UTRAN first sends the request for the transport channel reconfiguration to the UE, waits for it to reconfigure its transport channels, receives the acknowledgment message and then reconfigures the transport channels in the UTRAN correspondingly.

11.2.4.2   Synchronized Transport Channel Reconfiguration

A synchronized transport channel reconfiguration procedure has to be used when the old and new configurations are incompatible and cannot coexist. The UTRAN chooses the activation time, and sends this to the UE within a transport channel reconfiguration message. The RRC decodes the information and sends it to the MAC layer and to layer 1. However, the lower layers will not implement the required changes immediately, but only after the activation time has elapsed.

The next example (Figure 11.11) shows how the network modifies the transport format set (TFS) used by the UE. This is an unsynchronized example, where the old and new configurations can coexist. Notice that the transport format reconfiguration procedure does not impose any changes on the RLC configuration on either side.

See also Section 11.2.6 for the use of the transport channel reconfiguration procedure.

11.2.5   Physical Channel Reconfiguration

This procedure can be used to establish, reconfigure, and release physical channels. The decision to initiate this procedure is made by the UTRAN. The decision may be based on the measurement information received from the UE, or it may be purely a UTRAN internal decision (e.g., because of low network resources). The UTRAN sends a physical channel reconfiguration message to the RRC in the UE. The continuation depends on whether the procedure is synchronized or unsynchronized. Both of these cases are explained next.

11.2.5.1   Unsynchronized DCH Modify

An unsynchronized physical channel reconfiguration takes place when the old and new physical channel configurations can coexist. This is implemented so that the UTRAN first sends a request for a physical channel reconfiguration to the UE, waits for it to reconfigure its physical channel, receives the acknowledgment message, and then reconfigures its own physical channel(s) in the UTRAN correspondingly.

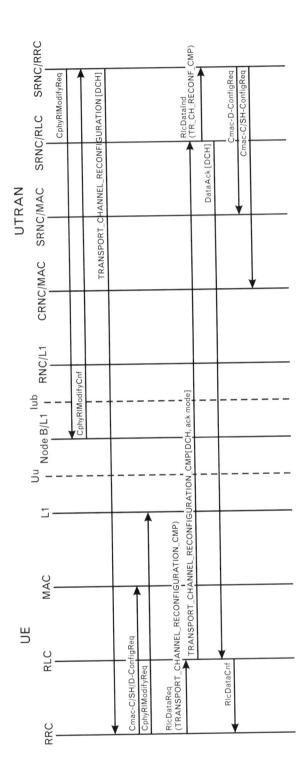

FIGURE 11.11 *Transport channel reconfiguration.*

Note that the coexistence of the old and the new configurations does not mean that a UE has to maintain two configurations at the same time. It means that a UE can use one configuration (the new one) and the network another (the old one), and they can still continue to communicate.

11.2.5.2   Synchronized DCH Modify

A synchronized physical channel reconfiguration procedure must be employed when the old and new configurations are incompatible and cannot coexist. The UTRAN chooses the activation time, and sends this to the UE with a physical channel reconfiguration message. The RRC decodes the information and sends it to its lower layers in various signals. However, the lower layers will not implement the required changes immediately, but only after the activation time has elapsed.

The activation time is given in the form of a connection frame number describing the instant the operation or changes caused by the related message should be executed. Note that if the activation time is missing, then this means that the modification must be carried out immediately. However, any changes to channel configurations while there is active communication ongoing must take place on TTI boundaries. Figure 11.12 depicts a synchronized physical channel reconfiguration.

See also Section 11.2.6 for the usage of the physical channel reconfiguration procedure.

11.2.6   Control of Requested QoS

The UTRAN air interface is very flexible, which allows for the dynamic allocation of system resources. The UTRAN allocates, for each UE, the minimum amount of resources so that the negotiated QoS can be maintained. This chapter contains several subsections that explain individual procedures related to the QoS and system resource control. The overall picture (i.e., how the individual procedures are related to each other) is explained in this section.

In the connected mode, the UE may be required to perform traffic volume measurements in its MAC layer. These are done to determine whether the current channel configuration can cope with the current traffic. If the UE suspects that the present configuration is not the optimal one, it sends a measurement report to the network. The rules explaining when and what to report are given in the measurement request message; see Section 7.5.2.14. The network may then decide to trigger a channel-reconfiguration procedure. The network may also trigger a channel reconfiguration without any UE input at all. Possible reasons include, for example, channel inactivity and low network resources.

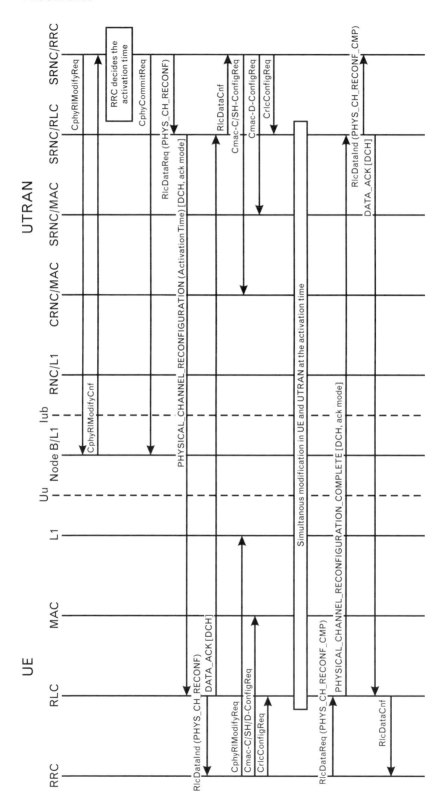

FIGURE 11.12 *Synchronized physical channel reconfiguration.*

The individual procedures discussed in the following sections include the following:

1. Physical channel reconfiguration,

2. Transport channel reconfiguration,

3. Radio bearer reconfiguration,

4. Traffic volume measurements, and

5. Transport format combination control.

11.2.6.1 Increased Data

The initial resource allocation (radio bearer setup) was discussed earlier. Once data transmission is ongoing, the RLC task in the UE monitors the amount of unsent data in the RLC transmission buffer. If the buffer fills over a predefined threshold, a measurement report is sent to the UTRAN. This message acts as a request message for additional resources.

The UTRAN may allocate additional resources to the UE, depending on the availability of the resources and the negotiated QoS of the UE. There are several ways it can do this.

Small amounts of data are typically transferred over common transport channels, but not necessarily. Even small amounts of data could be sent via DCHs, if the data has strict delay requirements because we cannot easily guarantee the delays in common channels. If the buffer fills up while the common channels are in use, the UTRAN may assign a DCH to the UE (i.e., the case of a RACH/FACH to a DCH/DCH). This will be done using the physical channel reconfiguration procedure; see Section 11.2.5.

If the UE has a DCH already assigned, any increase in data transmit capacity must be handled by reconfiguring the existing channels. For this purpose there are three possible procedures that could be employed. A physical channel reconfiguration is used when there are already suitable transport formats (TFs) and transport format combinations (TFCs) defined for a UE, and these can be used in the new configuration. If the TFs and TFCs must also be reconfigured, then the transport channel reconfiguration (Section 11.2.4) must be used. The most powerful procedure of these is the radio bearer reconfiguration (Section 11.2.3). It can change the QoS parameters as well as the multiplexing of logical channels in the MAC, reconfigure transport channels, and alter physical channels. Note that it is the UTRAN that decides the procedure to be used. The UE simply has to follow the orders given to it via the RRC signaling.

So in the order of the amount of change these procedures cause, the most disrupting first, we can list them as follows:

1. Radio bearer reconfiguration;

2. Transport channel reconfiguration;

3. Physical channel reconfiguration.

Thus, the radio bearer reconfiguration procedure can be used to change more parameters than the physical channel reconfiguration. A higher procedure may also include all the configuration parameters of a lower procedure.

It is also possible that new radio bearers are allocated or old radio bearers are released, but these actions are not due to traffic measurements in the RLC, but because a new application or service needs one (or in case of release, does not need an RB any longer).

The issues explained above apply also to downlink data, except that the transmission buffer to be monitored is now in the UTRAN rather than in the UE.

11.2.6.2   Decreased Data

If the amount of transmitted data decreases, then the underutilized resources must be deallocated. The triggering event is, again, the transmission buffer in the RLC. If the buffer occupancy remains under a threshold for a certain time, then the traffic volume monitoring process sends a measurement report message to the UTRAN.

It is up to UTRAN to decide what kind of actions (if any) will be taken. As in the increased data case, there are three procedures to choose from: (1) physical channel reconfiguration, (2) transport channel reconfiguration, and (3) RB reconfiguration. If the amount of transmitted data is low enough, then the UTRAN may command the UE to release its DCHs and use common channels instead (i.e., the case of a DCH/DCH to a RACH/FACH). This change can include any of the three reconfiguration procedures mentioned above.

11.2.6.3   Other Considerations

The UTRAN must consider the RLC buffer occupancy in both the UE and the UTRAN when it decides which procedure to use. For example, even if there is very little uplink traffic (the buffer in the UE-RLC is empty), it cannot move the UE from a DCH/DCH to a RACH/FACH if there is a large amount of downlink traffic. The UE cannot have a DCH in only one direction (i.e., combinations of RACH/DCH and DCH/FACH are not allowed).

The UTRAN can also "fine-tune" the UE's data transfer resources with the transport format combination control procedure (Figure 11.13).

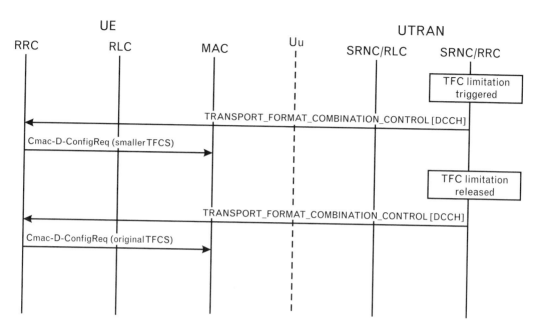

FIGURE 11.13   *Transport format combination limitation.*

This procedure simply modifies the allowed uplink TFCs within the transport format combination set (TFCS). If the network is temporarily congested, it may restrict the data flow from a UE by changing its TFCS, which defines the capabilities of the allocated transport channel. Once the congestion is over, the network can return to the old parameters.

11.3   Data Transmission

The example in Figure 11.14 presents a downlink data transfer, in which two channels are used in the air interface. Here a downlink shared channel (DSCH) is associated with a DCH. This shared channel can be used to help the DCH with traffic peaks. If the downlink traffic is generally of low volume, but high peak rates exist, then a DCH + DSCH combination offers a good solution to the problem. Allocating some DCH capacity (channelization codes) to meet these rare peaks would be a waste of resources. It is better to let the DCH handle the constant part of the data traffic and to transfer the traffic peaks via the DSCH. The DCH will carry the management data (power control, transport format indicator [TFI]) for the shared channel. The receiving UE knows from this DCH when there is data on an associated DSCH to be decoded.

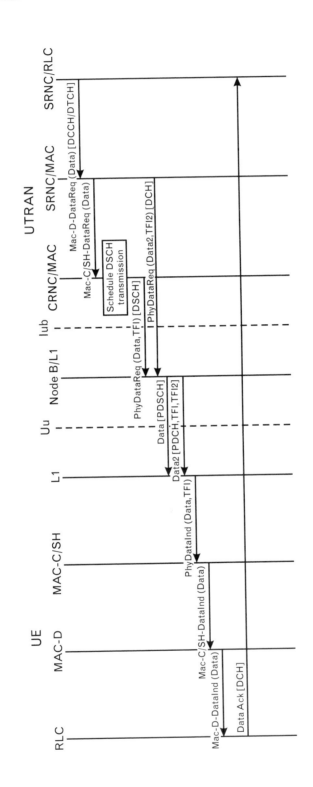

FIGURE 11.14   *Downlink data transmission on DCH + DSCH.*

Note that because the DSCH uses common channels, it might not be available at all times. Therefore, any data transferred via the DSCH may suffer delays. If the data stream is from a multimedia application, the DCH can transfer the real-time component, and DSCH the nonreal-time component.

In the next example (Figure 11.15), the data transfer is done in the uplink direction using only the shared packet channel. Note that there are no DCHs allocated, so all signaling must be done via the CPCH/FACH pair. The common packet channel (CPCH) is defined only for the frequency division duplex (FDD) mode.

The CPCH is a contention-based RACH-like channel, so a UE starts transmission by first probing the channel with access preambles. These are sent using increasing power levels until a matching response is received from the network or the maximum number of preambles has been sent, whichever occurs first.

The actual data transmission begins with a power control preamble, and continues with data packets without any delay. Acknowledgments are sent via FACH, as there is no DCH. The transfer of data packets is power controlled.

A CPCH is suitable for low-to-middle-volume nonreal-time packet data traffic. If the traffic is of a high volume, or it has tight real-time requirements, then a DCH must be allocated instead.

Figure 11.16 presents the uplink packet data transfer in the time division duplex (TDD) mode. The uplink shared channel (USCH) is defined only for the TDD mode, but its usage is similar to other shared channels. It is suitable for nonreal-time low-to-middle-volume packet traffic. This channel may or may not be associated with a DCH. Note that in this example the RRC handles the management of the channel, but the actual data to be sent arrives at the RLC from the user plane PDCP task (not shown). The physical uplink shared channel (PUSCH) capacity request message can be sent both on the RACH and on the USCH. The RRC in the UTRAN schedules the usage of the USCH. It sends a physical shared channel allocation message to the UE specifying the allocated resources and the period of time for which they are allocated. The acknowledgment messages can be sent on a FACH or on a DSCH.

Figure 11.17 shows the usage of a DSCH in the TDD mode for downlink data transmission. In the TDD mode the DSCH can also exist without an associated DCH. In the UTRAN the RRC monitors the RLC transmission buffer level, and may decide to allocate shared channel resources for the UE in question. Again a physical shared channel allocation message is used to indicate to the UE the allocated resources and the period of time for which they are allocated. The acknowledgment messages on the uplink can be sent either via a RACH or a USCH.

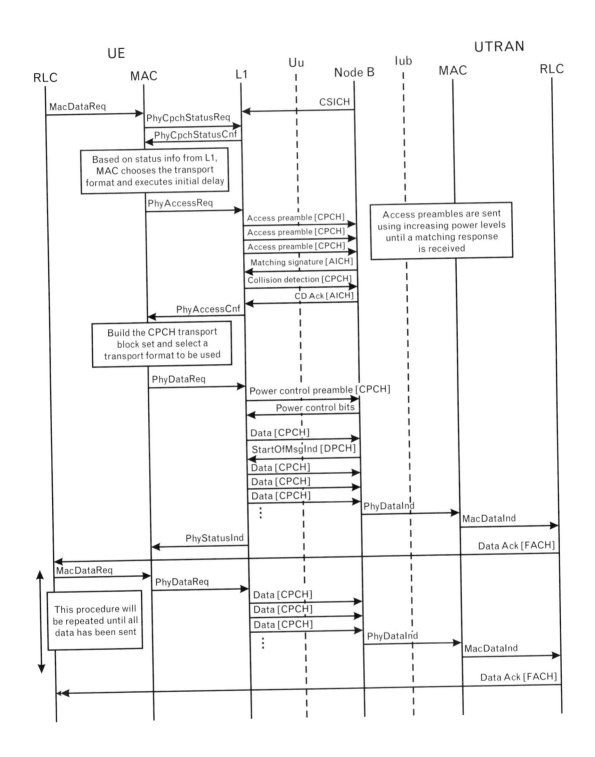

FIGURE 11.15   *Uplink data transmission on CPCH/FACH.*

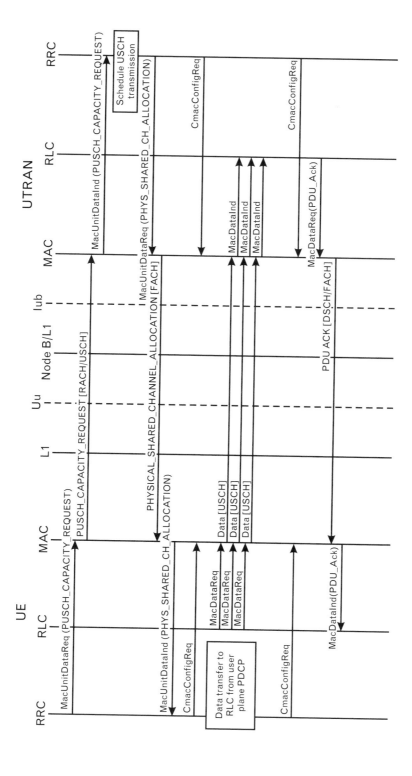

FIGURE 11.16 *Uplink data transmission on shared channels in TDD mode.*

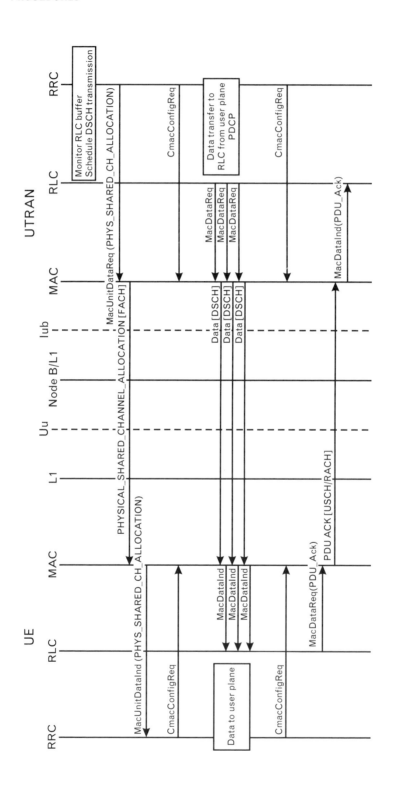

FIGURE 11.17 *Downlink data transmission on shared channels in TDD mode.*

11.4 Handovers

This paragraph contains examples from three types of HOs (soft, hard, and intersystem HOs). An HO means that the radio connection is transferred from one base station to another. HOs are discussed more in Section 2.5. The preceding measurement procedures discussed in Section 7.5.2.14 are not shown in the following procedures, even though they are typically carried out before HO decisions are made.

11.4.1 Soft Handover

An SHO takes place when a UE receives data from more than one base station at the same time. We need to make a distinction between the UE's receiving information from more than one transmitter and receiving information from a single base station through two or more different antennas. The latter case is a special type of SHO, namely the softer HO. The group of base stations communicating with the UE is called the active set. SHOs are widely used in CDMA systems because all base stations use the same frequency in this technology; thus, implementing an SHO is relatively straightforward. It is mostly employed in areas near cell boundaries where several cells may overlap.

The UTRAN must both receive and send via all base stations belonging to the active set on behalf of a UE. Note that the uplink SHO does not require any special transmitter support from the UE, as it can continue transmitting as before. The same signal is merely received by several different base stations instead of one. Changes to the active set are only done according to commands from the network; the UE should not update the active set by itself.

Site-selection diversity transmission (SSDT) is a special form of SHO that can be employed by the network. In this method the SHO is used only in the uplink direction; that is, the UTRAN receives the uplink signal via several base stations and then chooses one of them to handle the downlink transmission. This method does not cause any additional interference to the radio interface, as only one transmitter is used in both the uplink and downlink.

Figure 11.18 shows a procedure in which a new Node B is added to the active set. In CDMA jargon we call this procedure "adding a branch" to the active set. The RNC first configures the corresponding Node B(s). Only after the configurations have been acknowledged does it send an active set update message to the UE-RRC. This message contains all the necessary information the UE needs to start receiving the transmission via the new Node B(s).

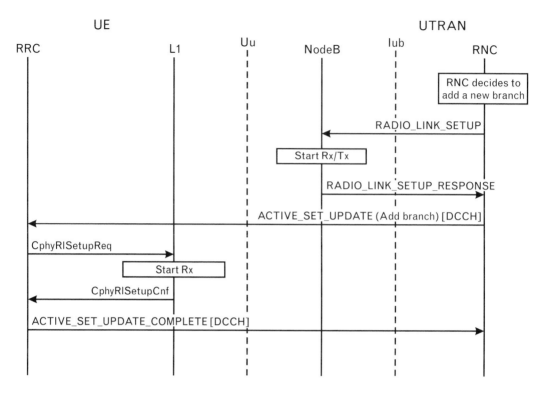

FIGURE 11.18   *Radio link addition.*

The next figure (Figure 11.19) depicts how a branch is removed from the active set. In this procedure the UE is first advised to stop the reception via the deleted branch, and only after that is the Node B in question configured to stop the transmissions and receptions for this UE.

11.4.2 Hard Handover

An HHO is technically a more complex procedure for the UTRAN than an SHO. An HHO is another term for an interfrequency HO. The problem with the interfrequency HO is making the required measurements on the new channel, which are difficult to carry out in a CDMA system. In connected mode a CDMA handset is transmitting and receiving continuously; thus, there are no spare slots available for making measurements on other frequencies. The HHO procedure and its problems are further discussed in Section 2.5.3.

Once the network decides to perform an HO procedure, it sends a handover command via the old cell. Note that there is no message called the "handover command" in the RRC specification. Instead the HHO procedure can be initiated by several other messages, such as physical channel

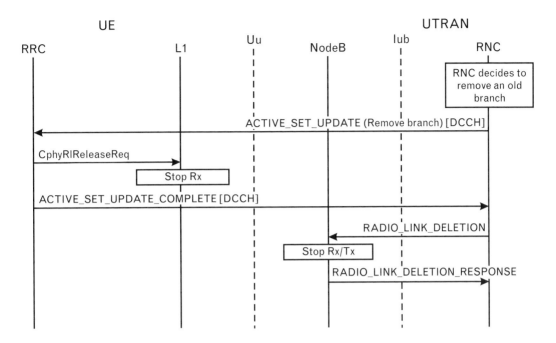

FIGURE 11.19 *Radio link removal.*

reconfiguration, radio bearer establishment, radio bearer reconfiguration, radio bearer release, or transport channel reconfiguration. An HHO is performed as a side effect or consequence of these procedures because the assigned radio frequency channel may change as a result of the procedure. An HHO is just a series of normal radio link reconfigurations. If the reconfiguration happens to include a new frequency, then an HHO has taken place.

The example here is of the nonseamless HO type because the UE first stops its transmissions in the old channel, and then starts them again in the new channel. A nonseamless HO means that the user can probably notice the HO from the temporarily reduced QoS (audible clicking sound) during the procedure. There are some methods to make the HHO a more seamless one, such as overlapping the compressed mode slots in both frequencies while the HO takes place. Also, if dual radio parts are available in the UE, then an HHO does not cause any perceived problems.

The following figure (Figure 11.20) presents a rather simplified example of a basic HHO. The UTRAN and the core network (CN) signaling is actually much more complex, and there are several different variations of this procedure depending on whether the HO is an intra-Node B, intra-RNC, intra-MSC, inter-MSC, or an intersystem HO. This list also gives the order of the complexity of these procedures. An intra-Node B HO is a quite simple procedure because only one network component has to

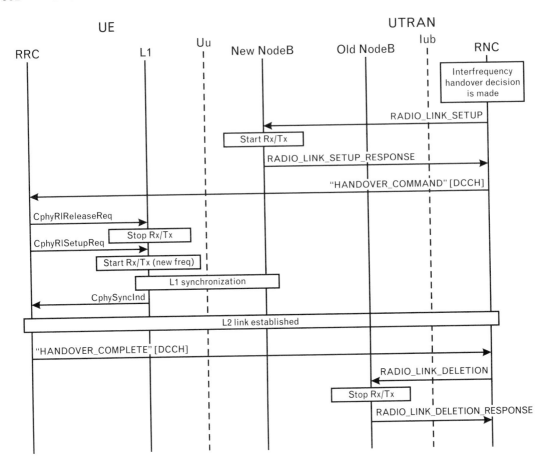

FIGURE 11.20 *Hard handover.*

take part in it, whereas an intersystem HO requires the cooperation of several entities from two different systems. Consult [4] and [5] for more information.

An intermode HO between FDD and TDD modes is similar (at a signaling level) to the "ordinary" intramode HHO. The biggest difference is in how to perform the preceding interfrequency measurements, which must precede the HO itself. See [6] for a description of intermode FDD-TDD HOs.

11.4.3 Intersystem Handovers

Intersystem HOs between GSM and UTRAN networks will be quite important, as many 3G operators will provide their wide-area coverage with the GSM network and use 3G in relatively small traffic hot spots. Every time a dual-system UE moves between the two types of coverage areas in these networks, an intersystem HO must be performed. The 3GPP specifications make wide use of the term *RAT* when they refer to the applicable network technologies. Thus, an intersystem HO is often called an inter-RAT HO.

11.4.3.1 Intersystem Handover, GSM → UTRAN

From the UE's point of view the intersystem HO procedure can be divided into three phases:

1. Preliminary UTRAN neighbor cell measurements in GSM mode;

2. Reception of the handover command message in GSM mode;

3. Synchronization to the UTRAN.

The first problem of how the GSM layer 1 actually performs the inter-RAT measurements is beyond the scope of this book. If the handset is an expensive high-end model, then it may have two receivers; one receiver can be used to monitor the UTRAN while the other receiver handles the normal GSM reception duties. Normally, however, there is only one receiver part in a dual-system mobile, and the UTRAN neighbor cell monitoring must be done during the idle GSM slots (non-Tx/Rx slots). The reports are eventually delivered to the GSM-BSS in a measurement report. The GSM-BSS may decide to perform an HO based on analysis of the results.

The second problem with the inter-RAT HO in the UTRAN is that synchronization to the UTRAN requires a large amount of information; information such as spreading and channelization codes and frequency information. Relaying so much information to a UE using an extended (and possibly segmented) GSM HO command would be impractical. The solution is to use predefined UMTS radio configurations. The UE should download up to 16 predefined radio configurations via the UTRAN system information broadcast (SIB 16). Once the HO takes place, the network only indicates the identity of the predefined configuration to be used, and possibly some additional parameters, in the handover to UTRAN command message. This solution requires that the UE has already been using the services of the UTRAN prior to entering the GSM network. If the UE has not downloaded any suitable preconfigurations, then the network has to ask the UE to use one of its default configurations. These are fixed, defined-in-the-standard configurations permanently stored in the UE. They can provide a temporary connection, during which the UTRAN can reconfigure the connection into a more suitable one.

Synchronization to the UTRAN is the third problem. The UE has to know the UTRAN's frame timing before it can start communicating with the network. How to do this is explained in Section 2.5.4. Note that if the

UE has already made measurements from this carrier, it probably has at least some of the required synchronization information already.

The actual signaling for the intersystem HO is simple. The SRNC-RRC configures the appropriate Node B and the lower layers of the RNC. Then the GSM-BSS sends a handover to UTRAN command to the GSM part of the UE. The message is decoded and the relevant information is relayed to the 3G part of the handset. This information includes the identification of the selected predefined configuration. The RRC then configures the lower layers and establishes a DCCH link. The handover to UTRAN complete message is sent to the UTRAN using the new link in the acknowledged mode. The signaling flow diagram is given in Figure 11.21. Note that the CN's internal signaling is not shown, but it can be studied in [7].

The QoS in the used application should not cause any problems because a UMTS system should be able to provide at least the same services and quality as a GSM system because its service palette is larger and the data rates are higher.

11.4.3.2   Intersystem HO, UTRAN → GSM

An HO from the UTRAN to GSM is a more difficult task than the other way around (Figure 11.22). In principle, a CDMA mobile station is receiving continuously in the dedicated state, which makes it difficult to measure other systems, like GSM cells, at the same time. An additional problem is that GSM data rates are generally lower than in UMTS, so the active applications must be able to adapt to the new lower rates after the HO to GSM.

The problem of measuring other systems can be solved with the use of compressed mode. The UTRAN base station leaves "gaps" in its transmissions and informs the UE about the pattern of the gaps. During these gaps the UE can monitor other frequencies and other systems by "looking through" the gaps. Alternatively, the UE can use a dual-receiver approach, where the UE employs two receivers, one receiving the FDD resource and another handling the GSM resource.

The HO from the UTRAN TDD mode to GSM can be implemented without the simultaneous use of two receivers because there are already some gaps in the downlink transmission in the TDD mode. A UE can do the measurements either by using idle slots or by getting assigned free continuous periods in the downlink by the UTRAN. The measurement results are reported back to the network, which may decide to perform an intersystem HO based on its evaluation of the results. The handover from UTRAN command message is received by the RRC, and the contained information is relayed further to the GSM part of the UE. The UE then performs a normal GSM HO to the specified GSM cell. Once the new connection has

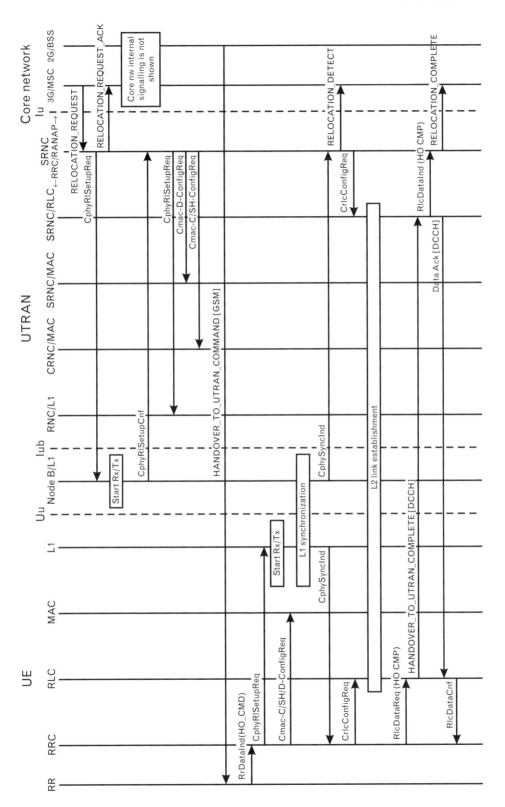

FIGURE 11.21 *Intersystem HO, GSM →UTRAN.*

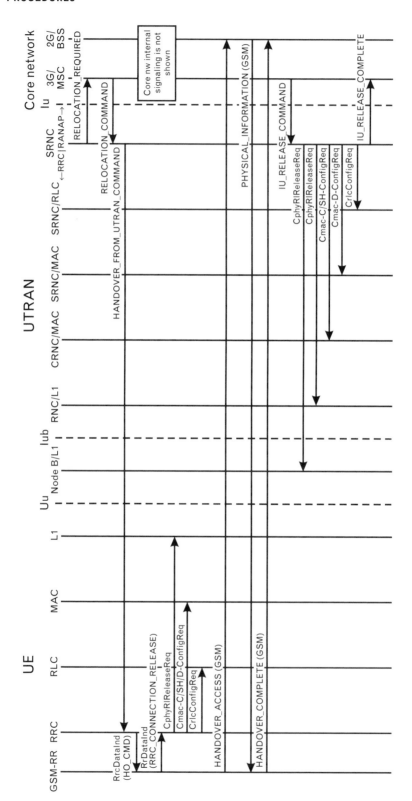

FIGURE 11.22 *Intersystem HO, UTRAN → GSM.*

been successfully set up, the UTRAN can release the resources that were allocated for the connection in the UTRAN.

See [8] for the complete description of the UTRAN to GSM HO and [9] for the description of the compressed mode.

11.4.3.3   Intersystem Change, GPRS → UTRAN

An intersystem change takes place when a UE that is GPRS-attached and supports both UTRAN and GSM systems makes a cell reselection to another RAT. The intersystem change from GPRS to UTRAN in particular takes place when a GPRS-attached UE changes from a GSM cell to a UTRAN cell (Figure 11.23).

If the UE is of the dual-system type, it may be ordered to also carry out intersystem measurements while in connected mode. If the intersystem measurements indicate that a cell reselection should take place, then the UE suspends the packet data transmission to the 2G network and sets up an RRC connection with the UTRAN. The UE has to receive and decode SIB 3 or SIB 4 from the UTRAN candidate cell so that it knows whether it is allowed to use the cell at all. After a successful connection establishment the UE can release the GPRS resources.

Once the UE has decoded the new routing area identity, it can initiate a routing-area update procedure, which tells the UTRAN the new location of the UE. The CN must recover the unsent data packets from the 2G-SGSN buffer and forward them to the 3G-SGSN.

The radio bearer setup message from the network will contain a list of PDCP-SNU acknowledgment numbers. They identify the mobile-originated data packets successfully transferred via GPRS before the start of the update procedure. The missing packets must be resent via the UTRAN once the packet transfer is resumed.

11.4.3.4   Intersystem Change, UTRAN → GPRS

This procedure takes place when a UE connected to a packet-switched CN selects a new cell belonging to the GSM-RAT (Figure 11.24). The rules for intersystem measurements are defined in [10]. The network can set a threshold value Ssearch$_{RAT\_m}$, and if the serving-cell signal quality is better than this threshold, then no intersystem measurements from RAT$_m$ have to be carried out.

If the intersystem measurements indicate that a cell reselection should take place, then the UE will start the process by receiving system information from the GSM cell. Once the new routing-area identity is known, the UE initiates a routing-area update procedure. It will suspend transmission to the UTRAN cell and send a routing area update message to the 2G-SGSN via the new GSM cell. The GSM network will include the receive N-PDU

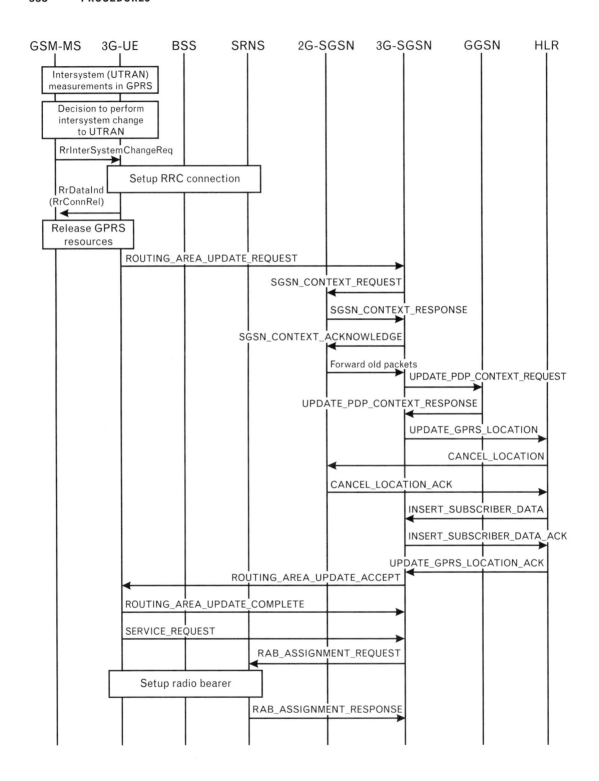

FIGURE 11.23 *Intersystem change, GRPS → UTRAN.*

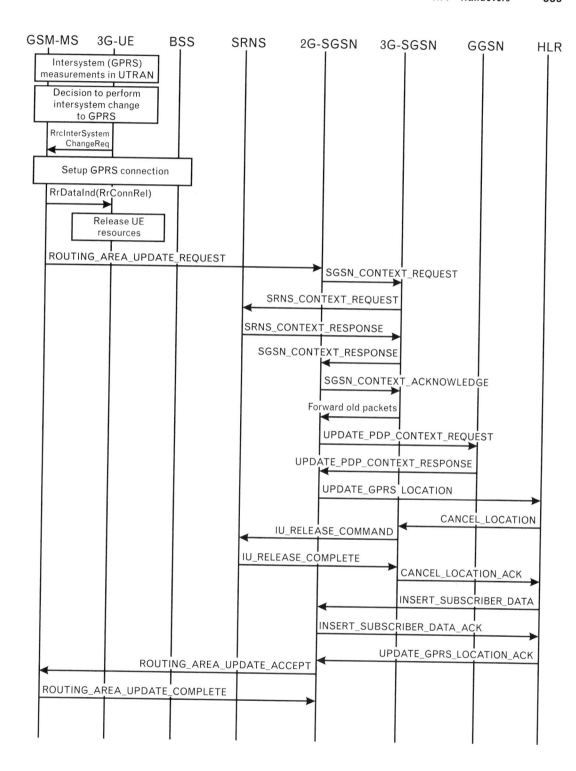

FIGURE 11.24 *Intersystem Change, UTRAN → GPRS.*

number parameter to the routing area update accept message. The receive N-PDU will contain the acknowledgments for each PDCP-SNU sent by the UE and successfully received by the UTRAN before the start of the intersystem change procedure. This value is used to determine which packets should be resent via the 2G-GPRS system.

The network has to carry out several tasks before the packet data transfer can be resumed via the 2G system. First, a new radio access bearer must be set up. Second, the unsent (or unacknowledged) downlink data packets must be recovered from the old 3G-SGSN and transferred into the new 2G-SGSN for retransmission (unless the old 3G-SGSN and the new 2G-SGSN are actually the same equipment). Third, the location of the UE must be updated in the HLR registers, and the subscriber's data must be copied into the new 2G-SGSN.

Intersystem change procedures are further discussed in [11].

11.5   Random Access Procedure

The random access procedure, or RACHing, is a process that is used by the UE to request a DCH allocation or to send small amounts of data to the network (see Figure 11.25).

The RACH transmissions are controlled by the MAC layer. The MAC receives the following RACH transmission control parameters from the RRC:

- A set of access service class (ASC) parameters;

- The maximum number of preamble ramping cycles M_{max};

- The range of the backoff interval for timer T_{BO1}, applicable when a negative acknowledgment on the acquisition indicator channel (AICH) is received.

When there is data to be transmitted, the MAC selects the ASC to be used from the available set of ASCs, which consists of an identifier, i, of a certain physical random access channel (PRACH) partition and an associated persistence value P_i. The procedure for the ASC selection is presented in [12]. How the RRC derives i and P_i is given in [13]. The mapping of the access classes to the ASCs is explained in [14].

Based on the persistence value, P_i, the MAC decides whether to start the L1 PRACH transmission procedure in the current transmission time interval or not. If transmission is allowed, then the PRACH transmission procedure is initiated by sending a PhyAccessReq primitive. If the transmission is not allowed, then a new persistency check is performed in the next

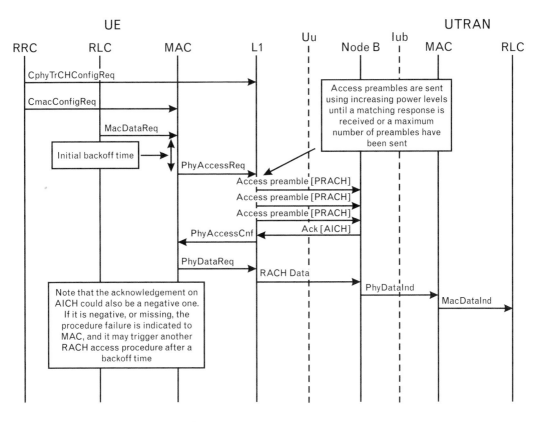

FIGURE 11.25   *Random access in FDD mode.*

transmission time interval. The persistency check is repeated until transmission is permitted. The persistence value (transmission probability) is a value between 0 and 1, and the persistency check simply means that the MAC draws a random number between 0 and 1, and if it is smaller than or equal to the persistency value, then the check was successful.

Note that the purpose of this complex procedure is to make sure that the random access attempts of individual UEs do not happen at the same time but are as evenly spread in the time domain as possible.

Layer 1 repeatedly sends RACH access preambles at the same time, ramping up the transmission power until it receives an acknowledgment or it has sent the maximum number of preambles, whichever occurs first. The purpose of this procedure is to find the lowest possible power level that can be used for transmission on the PRACH. When the preamble has finally been acknowledged to layer 1 on the AICH, this is indicated to the MAC with a PhyAccessCnf primitive. The MAC then initiates the actual data transmission (the RACH message part) with a PhyDataReq primitive. When L1 indicates that no acknowledgment on the AICH has been received when the maximum number of preamble retransmissions is reached, then a new persistency test is performed in the next transmission time interval. If a negative acknowledgment is received on the AICH, then a backoff timer (T_{BO1}) is started. After expiration of the timer, the persistence check is performed again.

See also [15] for the timing relations between the PRACH and the AICH.

REFERENCES

[1] 3GPP TS 25.303, v 5.0.0, Interlayer Procedures in Connected Mode, 2002.

[2] 3GPP TR 25.931, v 5.0.0, UTRAN Functions, Examples on Signalling Procedures, 2002.

[3] 3GPP TS 25.331, v 5.0.0, RRC Protocol Specification (RRC), 2002.

[4] 3GPP TS 23.009, v 5.0.0, Handover Procedures, 2001.

[5] 3GPP TR 25.832, v 4.0.0, Manifestations of Handover and SRNS Relocation, 2001.

[6] Prasad, R., W. Mohr, and W. Konhauser, *Third Generation Mobile Communication Systems*, Norwood, MA: Artech House, 2000, pp. 269–292.

[7] 3GPP TS 23.009, v 5.0.0, Handover Procedures; Section 8.2, 2001.

[8] 3GPP TR 25.922, v 5.0.0, Radio Resource Management Strategies, 2002.

[9] 3GPP TS 25.215, v 5.0.0, Physical Layer - Measurements (FDD), 2002.

[10] 3GPP TS 25.304, v 5.0.0, UE Procedures in Idle Mode and Procedures for Cell Reselection in Connected Mode, 2002.

[11] 3GPP TS 23.060, v 5.1.0, General Packet Radio Service (GPRS); Service Description; Stage 2, 2002.

[12] 3GPP TS 25.321, v 5.0.0, MAC Protocol Specification; Section 11.2.1 (Access Service Class Selection), 2002.

[13] 3GPP TS 25.331, v 5.0.0, RRC Protocol Specification; Section 8.5.12 (Establishment of Access Service Classes), 2002.

[14] 3GPP TS 25.331, v 5.0.0, RRC Protocol Specification; Section 8.5.13 (Mapping of Access Classes to Access Service Classes), 2002.

[15] 3GPP TS 25.211, v 5.0.0, Physical Channels and Mapping of Transport Channels onto Physical Channels (FDD); Section 7.3, 2002.

CHAPTER 12

New Concepts in the UMTS Network

This chapter explains some of the new concepts that will appear with the 3G networks. The overview offered in this chapter is quite short, but as before, references to other sources are listed for those who want to study the subject more thoroughly. The list of new items is not by any means exhaustive, nor is it UMTS specific. Some of these items will also appear in the enhanced GSM networks, and some items such as location services (LCS) will appear in some other cellular networks as well. Note that the CN-related items are especially easy to implement in both the GSM and UMTS systems because the CN can be the same in both cases.

Some of the technologies discussed here support applications that are explored in Chapter 14, together with their unique terminals if required; notably LCS and multimedia messaging service (MMS).

12.1 Location Services

Location services (LCS) is a common name for all services that make use of the knowledge of the (more or less) exact position of the mobile station. The location of the UE can be the offered service by itself, but it can also be used to provide certain value-added services, for which the UE's location is an important input parameter. A subscriber can, for example, request a list of nearby restaurants after arriving in a foreign town.

The applications that will be enabled by LCS are further discussed in Section 14.8.4.

LCS can be used by the operators, mobile phone subscribers, and third-party service providers. The reported location information includes, in addition to the location itself, a time stamp and a location error estimation.

LCS services began in the United States, where it was mandated by FCC that the location of a mobile phone user placing an emergency call had to be known. This is an important service, as more than half of all emergency calls in the United States come from mobile phones. Once the necessary technology for positioning was in place, it was easy to invent other applications that took advantage of this positioning capability. The FCC requirement is also an important incentive to develop user positioning technology, as the penalties for failing to provide emergency call user positioning with specified accuracy by the set deadline are heavy.

Thus, LCS is not something being invented exclusively for UMTS. For example, GSM networks will use these services as well. Note, however, that although the CN LCS elements and the offered services are roughly similar in both UMTS and GSM, the actual techniques that estimate the location of a UE may be different in these networks. The following discussion describes the UMTS LCS system.

The LCS architecture is depicted in Figure 12.1. Note that this is a logical architecture description; it is up to equipment vendors to decide how this logical architecture is implemented in physical equipment. An LCS client is an entity that makes a request for the location information of a target UE. After receiving this information, it can be used for a variety of applications. An LCS client may be located in the public land mobile network (PLMN), it may be external to the PLMN, or it may even be in the target UE, in which instance the UE could include a navigation application that requires the current location of the UE as an input.

A gateway mobile location center (GMLC) receives the location requests from LCS clients. These requests are processed, and some checks are made to make sure that the LCS client is authorized to request the location information of the UE in question and confirm that the UE is capable of supporting LCS. The approved and processed requests are routed to the correct radio network subsystem (RNS).

It is the radio network controller (RNC) that typically performs most of the LCS functions in the network. It must support the calculating function that processes the measurement results received from the UE or from Node Bs, depending on who makes the location measurements.

FIGURE 12.1
LCS architecture.

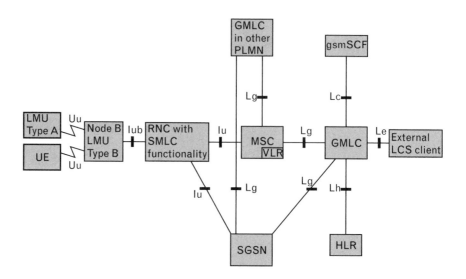

The location measurement unit (LMU) performs the actual measurements and reports the results to an RNC. Depending on the positioning method used, it may also make location calculations based on these measurements.

The LCS requirements for the target UE depend on the LCS method. If a network-based positioning method is used, no support is required from the UE. Here the UE can be located without any active support from the UE itself. If the method is mobile-assisted or mobile-based, then the UE must have additional software and possibly also some hardware to support the scheme. In mobile-assisted methods, the mobile must perform location measurements and report the results to the network. In mobile-based methods, it also must analyze these measurement results and calculate its own location. The fourth method type is network-assisted, in which case the UE does the location measurements and calculations, but the network provides additional information to make these tasks easier. In the current specifications this method is employed in network-assisted GPS.

At the time of this writing, there are three basic methods in the specifications for locating a UE:

1. Cell-coverage-based method;

2. Observed time difference of arrival (OTDOA) (only in Release 4 onwards);

3. Network-assisted GPS methods (only in Release 4 onwards).

Note that in Release 99, only the cell-coverage-based method is supported. Although you may find some aspects of the other two being discussed in some Release 99 specifications, officially they are not part of Release 99. In addition to the three methods discussed here, some additional methods and applications are considered in Section 12.1.4. Some of those may be added to the specifications at some later releases.

12.1.1   Cell-Coverage-Based Method

This is the easiest method to locate a UE as it does not require any measurements at all. Neither the UE nor the network infrastructure requires any modifications to support it. In this method, only the last known cell in which the UE performed some kind of location-update procedure is necessary. This information is obtained by means of normal signaling, for example, by paging, or by some more direct location-update procedure. The location can be indicated as the cell identity or as the coordinates of the actual Node B. The achieved accuracy is rather poor, and the estimated error includes the whole cell area. However, it is possible to improve the estimation in some cases. For example, from the signal propagation delay the network can estimate the UE's distance from the Node B, and in case of a sectored cell this may result in quite a good location estimation. A typical cell size is from a few hundreds of meters to a few kilometers. See Figure 12.2.

Even this kind of rough accuracy is useful for many applications. The operator can launch a scheme whereby the call tariffs are based on the location of the UE (location-sensitive billing). For example, the calls originating from the subscribers home cell can have a lower tariff. This can be used to lure customers away from the fixed telephone network. Another example could be fleet tracking services and traffic information services. Both services can work quite well with a few kilometers' granularity.

12.1.2   Observed Time Difference of Arrival

The observed time difference of arrival (OTDOA) method is based on the measurements made by the UE from the Node B pilot signals. The pilot signal is transmitted on the common pilot channel (CPICH), and it carries the primary scrambling code that is unique to a cell. This means that the cell can be easily identified from the decoded pilot signal, which makes it a suitable channel to be measured. The measurement the UE performs is the standard "SFN–SFN observed time difference" physical layer measurement.

The principle of the OTDOA method is explained in Figure 12.3. The UE carries out the required measurements and reports them to a position calculation function. This function can be situated either in the serving RNC or in the UE itself. To be able to calculate the location of the UE, it also needs to know the exact locations of the transmitters and the relative time differences (RTDs) of the transmissions. Because the obtained component measurements may contain some errors, the UE should perform OTDOA measurements on as many pilot signals as possible to minimize the error in the final result.

The OTDOA between Node B1 and Node B2 is

FIGURE 12.2
Cell-coverage-based location
method.

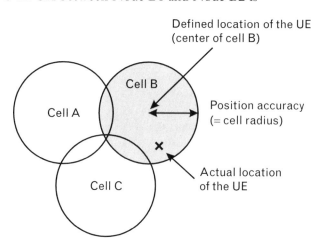

Defined location of the UE
(center of cell B)

Position accuracy
(= cell radius)

Actual location
of the UE

FIGURE 12.3
OTDOA location method.

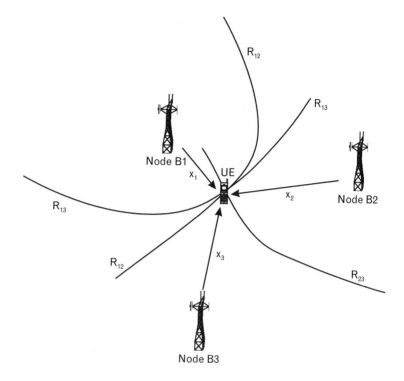

$$R_{12} = x_2 - x_1$$

where x_1 is the pilot signal propagation delay from Node B1 and x_2 is the pilot signal propagation delay from Node B2.

R_{12} forms a hyperbola, depicted in Figure 12.3. If the time difference is measured from only one pair of base stations, then the location of the UE is not accurately known; the UE can be anywhere along the hyperbola R_{12}.

The more time differences are measured on different pilots, the more exact the position approximation can be. In theory even two time-difference measurements should give an exact position, but signal diffractions and interference will introduce errors into the measurements, so more measurements should be made if possible. Therefore, the following measurements have also been made:

$$R_{23} = x_3 - x_2$$

$$R_{13} = x_3 - x_1$$

The UE should be near the intersection of the three hyperbolas R_{12}, R_{13}, and R_{23}. Note that this method gives the best results when the Node Bs measured equally surround the UE.

There are two modes in the OTDOA method: UE-assisted and UE-based.

12.1.2.1 UE-Assisted OTDOA

In the UE-assisted mode, the UE performs the time-of-arrival measurements itself and then reports them up to the network, which makes the location calculations in the RNC.

12.1.2.2 UE-Based OTDOA

In the UE-based OTDOA mode, the UE first performs the time-of-arrival measurements, just as in the UE-assisted mode, and then also performs the calculations for the location determination. In the UE-based mode, the network has to transfer the data needed by the position calculation function to the UE because this function is now located there, rather than in the RNC. Therefore, UE-based OTDOA terminals are probably a bit more expensive than UE-assisted terminals.

12.1.2.3 Enhancements to Basic OTDOA

If a UE is located very close to a base station, then the strong transmissions from this transmitter may block the receiver of the UE so that it cannot measure the other base stations for positioning purposes. This is known as the hearability problem. It can be solved by using the idle period downlink (IPDL) method. Each base station must cease its transmissions for short periods of time, and during the idle period of the interfering base station, the UE can perform its location measurements by measuring the pilot signals of other base stations. Obviously, the network should not schedule all base stations to have their idle periods at the same time because then there would be nothing left for the UE to measure.

However, in some cases it may be beneficial for the operator to time align the idle periods from different base stations. If the amount of traffic is high or pilot signals cannot be heard easily for some other reason, as would be the case with low-level pilots, then common idle periods may ease the problem. During these common idle periods only pilot signals are being transmitted, and then only by some of the base stations. This kind of synchronization technique makes it easier for a UE to find more pilot signals it otherwise could not hear. Furthermore, the power levels of these pilot signals can be increased temporarily during the common idle periods. All this makes the location measurements easier.

It is difficult to achieve a very high accuracy using the OTDOA method. The problem with this method is that the location determination is based on the propagation delay of the pilot signals. In practice these signals will be reflected and obstructed. The distance between the base station transmitter and the UE is then actually shorter than the propagation delay might suggest because the signal has not arrived via the shortest route.

Another factor that reduces the accuracy is the calculation of the RTD. The calculation function should know the exact time when each pilot signal was transmitted. This is not possible because the base stations in the UTRAN are seldom synchronized. This problem can be solved by measuring the RTD regularly and storing the results to a database that is accessible by the location calculation function. One possible method is to plant special location measurement units at known locations and let them measure the time differences of local base stations. If the IPDL method is used in the network, these measurements can be performed during the idle periods.

Note that the RTD does not remain constant because the base stations are not synchronized and their time bases will drift. Therefore, these measurements must be renewed periodically and the calculation function should always use the latest available results. In the TDD mode the base stations are typically synchronized, which makes the calculation of RTD easier. Each operator must decide what kind of error in location accuracy is acceptable, and design the network based on that requirement. Every nanosecond of error in the RTD between base stations increases the location error by 0.3m, and in practice the RTD error can easily be tens of nanoseconds. In some countries the accuracy requirement will be set by law, and this must of course be taken into consideration. The U.S. laws require different levels of accuracy: the more accurate the measurements, the less often (the lower the probability) the accuracy has to be realized. U.S. law also distinguishes mobile-based methods from infrastructure-based methods. The mobile-based methods are seen as more accurate than the infrastructure-based ones.

12.1.3 Network-Assisted Global Positioning System

12.1.3.1 GPS Principles

This location-determination method makes use of the Global Positioning System (GPS). GPS uses a constellation of positioning satellites, 24 altogether, and the calculation of the location is based on the propagation delays of different satellite transmissions. GPS can itself provide an estimation of the UE location, but this estimation can be further improved (or made quicker) with some help from the UMTS network.

GPS is an American system that was originally developed for, and used exclusively by, the U.S. military. The public could buy GPS receivers and use them, but the obtained accuracy was much lower than with military receivers. The U.S. military encrypted part of the GPS transmission, and without the ability to decode this signal, the best accuracy a civilian GPS receiver could provide was around 100m. However, in the spring of 2000, this jamming, which was a form of jittering, was stopped. Without jamming, standard GPS can provide approximately 25m accuracy. GPS can also readily provide the velocity and heading of the measured receiver. This is important for certain types of applications, such as traffic navigation services.

This intentional interference jittering was called selective availability (SA). The inaccuracy created by this method could be corrected by another technique called the differential GPS (DGPS); see Section 12.1.3.2. Now that the GPS encryption has been stopped, the importance of DGPS is reduced, but it can still be used to eliminate some other error sources, such as from multipath and atmospheric propagation errors. We must always remember that the U.S. may start using the SA technique again if it sees it as important for its national security interests.

The Interagency GPS Executive Board (IGEB) (http://www.igeb.gov) presently manages GPS. President Clinton established the IGEB in 1996 to manage the GPS and its U.S. government augmentations as a national asset. Its response to the interesting question regarding the possible return to SA can be found on its Web page: "The United States has no intent to ever use SA again. To ensure that potential adversaries do not use GPS, the military is dedicated to the development and deployment of regional denial capabilities in lieu of global degradation."

12.1.3.2   Differential GPS

DGPS is a method that is used to overcome the errors introduced by SA and other influences, such as atmospheric propagation changes. The principle is simple: Special GPS receivers are planted in known positions. These will receive the GPS signals from satellites and make location calculations. Because these receivers know their positions, they can calculate the needed corrections to the received signals. These correction values are then broadcast so that all other GPS receivers in the area can make use of them. This procedure must be continuous, as the interfering values are also changing continuously.

DGPS is already widely used by various organizations. There are DGPS beacons in many coastal areas, as this method is commonly used in the navigational equipment of ships.

Once the SA jamming was removed in May 2000, the importance of DGPS has diminished, but it can still be used to improve the accuracy. There are location applications that require better accuracy than the standard GPS alone can provide. The increased accuracy provided by DGPS reduces as the GPS receiver moves further away from the DGPS reference receiver, but even from a distance of a few hundred kilometers this scheme brings improvement. DGPS provides an accuracy that is typically better than 10m, but this can be greatly improved with carrier phase tracking techniques, in which case we are talking about millimeters, or a few centimeters.

12.1.3.3 GPS with UMTS

There are two ways to compute location when using the network-assisted GPS technique. If the UE-based method is used, then the UE must contain a full GPS receiver, whereas in the UE-assisted method, the UE only contains a reduced-complexity GPS receiver. The UE-based method requires more expensive hardware in the UE, but the UE-assisted method increases the signaling load in the air interface.

The UTRAN can provide the UE with both timing and data assistance information. The timing assistance contains an estimate of the GPS signal arrival time as calculated by the UTRAN. This can be used to speed up the GPS signal's reception and recovery in the UE. The data assistance consists of some GPS parameters, such as the visible satellite list and clock corrections. Most of this information could also be decoded from the GPS signal, but it is rather low-rate (slow) information; thus, its decoding would take some time (12.5 minutes altogether). A complete cold start, where the UE-GPS receiver does not have any GPS navigation information stored (either from earlier GPS reception or from the UTRAN), takes at least 13 minutes to perform. Thus, in such a case the network-assisted GPS is really an improvement.

12.1.3.4 Other Global Navigation Satellite Systems

There are also other satellite-based location systems, or global navigation satellite systems (GNSS), as they are called. The Russian GLONASS system [1] is currently deployed. Like GPS, it is also a military-based system, but receivers are available for civilian users too, although the lack of affordable handheld receivers for GLONASS has been a problem. Jamming (i.e., intentional interference on the SA model) is not used in GLONASS.

The European Galileo system is still on the drawing board. This project is backed by the European Union, and the aim is to reduce the European dependence on U.S./Russian location systems. The national interests of these countries may not always be congruent with the interests of the EU. In time of crisis, the services of GPS/GLONASS may not be available. The EU formally approved the program on March 26, 2002. In any case, as it will take several years until the system is running, we can expect to use Galileo receivers in 2007 at the earliest. Galileo is a so-called 2G GPS system, and we can expect it to have much better accuracy (~4m) than the standard GPS.

12.1.4 Other Methods

In addition to the previously discussed methods, several other location-service technologies have been proposed for the UTRAN. These are discussed here only briefly. It is possible that some of them will be implemented in later releases. Some of these may be used to supplement and enhance the previously discussed location methods.

The angle of arrival (AOA) method can be used in the base station to estimate the direction of the UE as seen from the base station. If the base station uses sectorized cells or adaptive antennas, which can form narrow directed beams, then by combining the direction information from several Node Bs, the RNC can estimate the location of the UE. As adaptive antennas will certainly be used in some UTRAN systems, AOA is a strong candidate for those networks.

The observed time of arrival (OTOA) method is based on the measurements made on the time of arrival of signals. These propagation delay measurements are only possible if a common reference time is known; that is, if the exact transmission times of the signals are known. If the common reference time can be provided, then this method could be used both in the UE (from downlink signals) and the network (from uplink signals) to determine the position of the UE.

The reference-node-based positioning (OTDOA-RNBP) is an enhancement to the OTDOA method. The network operator can "plant" special equipment in a cell area that has proved to be difficult for LCS, for example because of the hearability problem. The precise location of the equipment is known to the network. It is then possible to use the signals from this equipment as a reference in the network. However, the specifications do not support this method yet, and most probably will not do so.

OTDOA positioning elements (OTDOA-PE) is another OTDOA enhancement. Positioning elements are placed in the cell coverage area in exact locations that are known by the network. They transmit their own secondary synchronization codes (SSC) at known offsets to the Node B's BCH transmissions. The UE can measure the time difference between the arrivals of these signals. From this difference the network can then estimate the location of the UE. This method is especially useful in locations where otherwise only one base station could be received, as would be the case indoors or in coverage boundary areas.

12.1.5 Comparison of Location Methods

Comparison of these methods is difficult because there are so many variables to compare. Accuracy, cost, and speed are the most important factors. And with the cost factor, one has to see whether the cost comes from handset modifications or from network modifications.

OTDOA seems to be the most favored technology for the 3G air interface. It is relatively quick, requiring only a few seconds to calculate the location of a UE. The accuracy with basic OTDOA is a few tens of meters (average), but can be improved with enhancement techniques to about

10m. This method requires modifications to both the handset and the network, but the cost should be relatively modest: only minor software additions are needed in the UE.

Network-assisted GPS also provides a quick result, as most of the required calculations will be done in the network. The accuracy is good; if SA is on, it will be a few tens of meters, but with SA disabled, the accuracy is enhanced to only a few meters. An important factor here is the radio environment. If there are obstacles between the UE and satellites, then the accuracy is worse. GPS signal penetration is very poor and easily stopped by walls. This means that a GPS location method is useless indoors. A general problem with this technique is that a handset requires a GPS receiver, which increases its cost.

AOA and OTOA are quick methods and do not require any changes to the handset. But the negative properties of these methods could make them unusable. The typical accuracy is quite poor, and can hardly fulfill the accuracy requirement in the United States set by the Federal Communications Commission (FCC) (125m, on average, originally, and even stricter in phase II). A location method that cannot provide accurate location finding is hardly worth implementing. Moreover, the changes required to the networks are quite costly.

The cell-identity-based method is also unusable for many applications because of its inaccuracy. If the UE is in a picocell, then its location can be determined with an accuracy of about 100m. But with macrocells, the mobile could be several kilometers away from the cell center. However, this method does not require changes to either the handset or the networks; thus, it can be used to complement other methods. For example, if an operator has both 3G and GSM networks, it can implement some advanced location service in the 3G network, and use the cell-based method in GSM. While the coverage area of 3G grows, better LCS can be offered to a wider user population.

Pure stand-alone GPS is also one possibility. There is one problem common to all methods introduced earlier, and it is that they only work within UMTS network coverage. Particularly in the early stages of the UMTS service, network coverage will be quite limited. However, user positioning information would be useful everywhere, especially to the emergency services. A stand-alone GPS-equipped UE could also determine its exact location when outside the UMTS network coverage, although in this case the UMTS network would not know it. In fact, the achieved accuracy would probably be better outside urban areas because the satellite signals do get easily blocked and reflected in urban settings. The accuracy without SA and with DGPS is excellent (only a few meters). The UMTS network would not require any changes because it is not used for positioning measurements. But the UE would need a fully integrated GPS receiver, which would be a costly addition.

12.1.6   Service Categories

The LCS can be classified into four categories:

1. Commercial LCS;

2. Internal LCS;

3. Emergency LCS;

4. Lawful-intercept LCS.

The commercial LCS is also known as the LCS value-added services. These services are used by external service providers, not by the network operators. They will provide applications that in one way or another make use of the location of the UE. This could be a navigation application or a list of mobile phone shops in the area. The location of the UE itself is not the service provided by these external providers, as this can be provided by the UE, or by the UE and the UTRAN together. The UE's location is merely an input to the service provider's application.

The internal LCS is used to enhance the operations of the UMTS network. To the extent we can increase the resolution and precision of our knowledge of the UE's location, we can enhance the performance of certain network tasks, such as a location-assisted HO.

The emergency LCS is used to locate a subscriber who is making an emergency call. The location of the user is provided to the emergency service provider by the network. This service will be mandatory in some countries, and it may also include some mandatory accuracy requirement. In the United States, the FCC defined the required accuracy as 125m, but in the new phase II of Enhanced Wireless 911 Services, this accuracy will be defined as follows:

- For network-based solutions: 100m for 67% of calls, 300m for 95% of calls;

- For handset-based solutions: 50m for 67% of calls, 150m for 95% of calls.

It is quite probable that similar rules will also be adopted in other countries once the technology is in place.

The lawful-intercept LCS provides location information for the law enforcement agencies, as required by the local jurisdiction. So, in effect, the police can track the UE user, or the user terminal, when there is some need to do so. This feature is a very powerful tool for the police. But on the other hand, it also means that every UMTS phone user is carrying his or her own tracking device. Of course, the phone can be turned off, but still, this service is a frightening tool if it falls into the wrong hands. It is not difficult to imagine what a suppressive government could do with this kind of power.

LCS is specified in [2–7].

12.2   High-Speed Downlink Packet Access

High Speed Downlink Packet Access (HSDPA) is a major work item for Release 5 in 3GPP. HSDPA is designed to enhance the downlink data transmission capacity in 3G systems. This is seen as a necessary step because the practical maximum downlink data rates in Release 99 are too low for multimedia applications, or at least one cell can accommodate only a very few such users. The HSDPA scheme proposes to add an additional wideband downlink shared channel that is optimized for very high-speed data transfer. It could use the existing downlink carriers, or it could use some new, yet to be assigned, frequency spectrum. HSDPA improves only the downlink throughput, but this is intentional, as it is just where the extra throughput and capacity will be needed. HSDPA is specified for both the FDD and TDD systems. In this section the FDD HSDPA system is described, although the principles are similar in both modes.

The shared HSDPA data channel (HS-DSCH) is a more capable replacement of the old DSCH channel. A UE will also have dedicated channels assigned to it while using HSDPA, both in the uplink and downlink, as was also the case with DSCH. As a minimum, these channels will carry some control information, and may also carry real-time user data. Thus, a UE cannot receive HSDPA channels alone. For example acknowledgment data and quality information for HSDPA is sent back to the Node Bs via an uplink dedicated channel (HS-DPCCH).

HSDPA is a combination of several techniques that all contribute to the enhanced capabilities of the downlink channel. A Release 5 HSDPA-capable handset will include both Hybrid ARQ (HARQ), and Adaptive Modulation and Coding (AMC) functionalities. Later releases may adopt Fast Cell Selection (FCS) and Multiple Input Multiple Output (MIMO) techniques to enhance HSPDA capabilities even further. Let's have a look at the HARQ and AMC enhancements.

HARQ is a link adaptation scheme in which link layer acknowledgments are used for retransmission decisions in the UTRAN. In Release 99, retransmission functionality is part of the RLC layer. However, this kind of high-level retransmission scheme is too slow for high-speed data transmissions required for HSDPA. Here the HARQ retransmission buffers are located closer to the physical layer: in the new MAC-hs logical entity that is just above the physical layer (see Figure 12.4). The ARQ combining is based on incremental redundancy; that is, if a transmission fails, the received

FIGURE 12.4
HSDPA protocol stack.

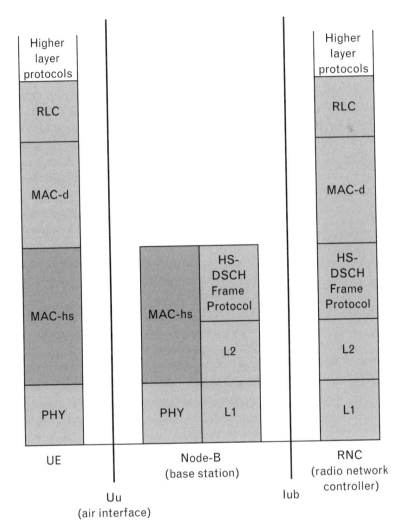

(corrupted) data is still stored to a buffer. Successive retransmissions will include more redundancy, and they are combined with the old data in the buffer. This is repeated until the data in the buffer is considered to have been correctly received, or the maximum number of retransmissions is reached. Moreover, to make HARQ more efficient, a shorter frame length (TTI) is needed. The new shorter TTI will be only 3 timeslots (2 ms), compared to 15 timeslots (10 ms) employed by the other physical channels. When a shorter TTI is used, the UE can inform the UTRAN every 2 ms if the transmission failed. In the old system, 10 ms would have to pass before a failure could have been reported. Shorter frames also mean that the system can have quicker response to changing channel conditions.

AMC (or Link Adaptation) means that the shared channel transport format (i.e., the modulation scheme and the code rate) depends on the channel quality. This is monitored constantly, and the transport format used can be changed in every frame. The quality information is transmitted to the Node Bs via uplink control channels. That is, if the radio channel condition is good, the UTRAN can use higher-order modulation and less redundancy, whereas in poor conditions, a more robust modulation scheme can be employed, and the data packets may also have more redundancy in them. Release 5 employs two modulation schemes, QPSK and 16QAM. Later releases may introduce other schemes, such as 64QAM.

The HSDPA scheme introduces three new channel types, HS-SCCH, HS-DSCH, and HS-DPCCH. The HS-SCCH is the associated shared downlink control channel that indicates when there is data to be received on the actual high-speed downlink shared channel (HS-DSCH). It also carries configuration information for the HS-DSCH, such as the transport format and resource-related information (TFRI), HARQ-related information, and the UE's identity, among others. If the UE identity matches the identity indicated in the HS-SCCH, then the UE knows to receive the next HS-DSCH frame. All HS-DSCH channels use SF=16. However, to increase the throughput of a user, the UTRAN can allocate several such spreading codes for one user. The maximum number of multicodes a UE can support is a UE capability parameter and can be either 5, 10, or 15. Note that if all spreading codes are SF=16 on the HS-DSCH, then 15 multicodes means that the cell could be very close to becoming a code-limited cell. Once the data has been received and processed by MAC-hs, the UE sends an acknowledgment back to the network on the HS-DPCCH channel. HS-DPCCH may also carry Channel Quality Indicators (CQIs) which are based on downlink channel (CPICH) measurements. The HS-DSCH does not support soft handovers.

Note that HSDPA is not suitable for all kinds of services. The HSDPA data channel is a shared channel. It is shared among all active HSDPA UEs in a cell. The shared character of the channel means that maximum transfer delays cannot be (easily) guaranteed, and applications, which have strict real-time requirements, should use dedicated channels and not the HSDPA. On the other hand, the resource allocation in HSDPA channels is very fast. The capacity allocation can be dynamically changed in every frame (2 ms in HSDPA channels). Additionally, the scheduler in the UTRAN may favor higher priority, real-time data streams.

HSDPA provides a suitable tool for an operator looking for a way to increase capacity in data traffic hot spots. HSDPA-capable base stations could be used for pico- or microcells, whereas a normal FDD macrocell provides the overlay coverage. One macrocell could thus contain several smaller HSDPA cells, which do not have to be contiguous (see Figure 12.5).

The HSDPA upgrade requires completely new handsets with HSDPA capability. There will be different UE HSDPA category classes. For example, low-end UEs may conform to the lowest category classes, and high-end UEs can implement the highest classes with better throughput.

FIGURE 12.5
HSDPA microcells.

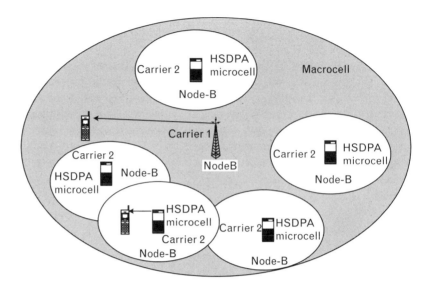

The difference between these classes is mainly the number of multicodes supported, and the length of the inter-TTI interval in the HS-DSCH reception (i.e., the ability of the UE to receive HSDPA data in successive TTIs). However, old handsets can still be used in the network; it is just that they cannot use the new HSDPA-enabled services. Still, even if only some of the users have HSDPA-capable handsets, the overall system downlink throughput will be improved. The theoretical maximum throughput of HSDPA will be 14.4 Mbps in Release 5 (with 16QAM and a code rate of 1), but this can be increased later, for example with the adoption of MIMO antennas. However, note that HSDPA is not so much about increasing the theoretical maximum throughput, but merely about increasing the practical throughput. The theoretical maximum throughput can only be achieved in optimal channel conditions, which are not available in typical usage scenarios. The dynamic and adaptive nature of HSDPA channels means that they will be able to transmit the highest possible amount of data frame after frame even in poor and changing radio channel conditions.

HSDPA specifications include [8], and [9].

12.3 Multimedia Broadcast/Multicast Service

Multimedia Broadcast/Multicast Service (MBMS) is a new Release 6 service. At the time of this writing the technical specification work for MBMS was just starting; thus, this discussion will not go into detail. Only the main principles of this service will be explained.

3G customers can subscribe to many different information services. For example, traffic news and sports news with multimedia content will

certainly attract many customers. This information is typically such that it is
only of interest to anybody when it is new; thus, it should be sent to everybody, or at least to all subscribed users simultaneously. From the operator
point of view, it is clearly waste of resources to duplicate the same information in the network, and then to send it separately to each subscriber using
separate channel resources. If there are several users in the same cell subscribing to the same multimedia sports news, this kind of transmission could
easily consume all the resources of a cell.

A better solution is to send the information as a broadcast or multicast
message using common channels. In this case all users, or all subscribers,
could receive this information simultaneously, but the channel resources
would be consumed only once. The distribution of MBMS information
may be temporarily suspended in some areas if the available radio resources
are running low.

As the name of this service indicates, this is a multimedia service. The
information delivered can consist of several media components. For example the traffic information service could include a map showing the location
of the traffic jam, textual description of the traffic situation, and the same in
audio format because it is dangerous to read the text in a fast moving car. A
sports newsreel could contain a short video with voice showing a goal just
made in a premier league match. This service is not meant for simple text
notifications, as there is already a different service, Cell Broadcast Service
(CBS), for that purpose.

The high-level MBMS network architecture is given in Figure 12.6.
Note that the same service can also be provided for GSM/EDGE as shown
in the figure. CBC does not transmit the multimedia services, but simply
notifications of oncoming broadcast/multicast. All payload data will be
transmitted via SGSN.

MBMS comes in two variations: broadcast and multicast services.

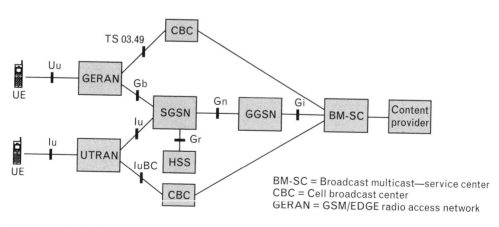

FIGURE 12.6   *MBMS architecture.*

12.3.1   Broadcast Service

The broadcast service is intended for everybody in the network. Thus, its reception does not require an additional subscription from a customer of this network. But the user may also choose not to receive broadcast services, or choose only to receive certain broadcast types. The delivery of broadcast messages is not guaranteed, as their reception is not acknowledged by the UE, but the network may choose to repeat the transmission a few times at certain time intervals to increase the probability of a successful reception by most of the subscribers. A handset could have been out-of-coverage, or switched off, for example; thus, retransmissions can help to reach more users. The receiving handset can recognize duplicate messages and not to show them to the user.

Broadcast services are not charged to the user, but the network may charge a third-party content provider in some cases. For example if the broadcast message was an advertisement, then the advertiser may be charged for it. Broadcast services should be available to all users registered to the providing network regardless of the mode the handset is in, as long as it is switched on and within the network coverage area.

12.3.2   Multicast Service

Multicast services are different from broadcast services in that they are not intended for everybody, but for a specific group of users. In this case the users have to subscribe the service, and they may be charged for it by the operator. On the other hand the operator may choose to charge a third-party content provider, or both the service subscriber and the content provider. The actual charging criteria are not specified. It could be the number of messages, the volume of data, the duration of the transmissions, a flat fee, or something else.

For the user to be able to subscribe to a multicast service, he must first know what services are available. The service announcement mechanism is not specified, but the operator can use a variety of methods. For example, it could employ the ordinary cell broadcast service to advertise the available multicast services. Other possibilities include point–to–point SMS messages, MBMS broadcast mode, or Web URLs that have this information. From the palette of multicast services, the user can select those he wants to receive. By subscribing to a service, the user joins the corresponding multicast group. The same user can belong to several multicast groups, and also receive them even simultaneously if the UE capability allows this.

Because this service is only meant to be received by subscribed users, there has to be a mechanism to prevent other nonsubscribed users from receiving it. A low-level security feature will be provided by the fact that the spreading code for a multicast service data is only known by the subscribers to that service. However, because the number of different spreading codes is rather limited, this kind of security does not provide serious protection. Thus, the operator may choose to cipher the data, and the ciphering key is, thus, distributed only to authorized users when they subscribe to the service. Another aspect in the MBMS security is that the operator must be able to authenticate the content provider and verify the integrity of the data that is supposed to be transmitted to subscribers. This issue also applies to broadcast services.

Typically, multicast services will be transmitted via common point-to-multipoint channels. However, if the number of subscribed users in a cell is small, then the operator may also choose to employ point-to-point channels to deliver the service in this cell. Obviously, if there are no multicast group members in a cell, then there is no need to transmit that particular multicast service into the cell. But the requirement to track the location of multicast subscribers increases the complexity of the MBMS implementation in the network.

The delivery of multicast messages is not guaranteed, and neither is the QoS. If the service is broadcast via common channels, then there is no effective power control, and UEs close to the cell border may have difficulties receiving the service, or at least the QoS they experience is poorer than for UEs close to the base station.

At the time of this writing, MBMS is discussed in two 3GPP specifications. The service aspects are discussed in [10], and the system architecture in [11]. What is still missing is a technical standard that specifies the air interface PDUs, signaling, etc.

12.4 Multimedia Messaging Service

Previously we have discussed Multimedia Broadcast/Multicast Service (MBMS). It differs from Multimedia Messaging Service (MMS) in that MBMS is a point-to-multipoint service, whereas MMS is a point-to-point service. Both of these are non–real-time services, suitable for delivering short multimedia presentations to the users. Real-time multimedia has to be delivered by using other services provided by the IMS domain.

12.4.1 The Service

MMS is one of the sexiest parts of UMTS. This is something 3G marketing staff love to talk about. Video, voice, and text together provide completely new possibilities for building new applications for the mobile communications environment. This section discusses the technical implementation of the service; the applications it enables are discussed in Section 14.8.2.

In GSM systems, the short message service (SMS) has gained a lot of popularity, and it is one of the most rapidly growing segments in GSM business. SMS makes it possible to send short text messages (usually a maximum of 160 characters) to other GSM mobile phones. SMS can be extended into other services via conversion functions into other communication media such as e-mail. The service is a non-real-time store and forward service, meaning that the messages are stored in the network and delivered to the destination address once it is possible to do so, for example, when the target user turns on his mobile or returns to network coverage. The charge is fixed; that is, it is typically not dependent on the destination address or the size of the message, and the sender pays for it. For operators, the SMS business is an excellent source of income because the amount of data transferred is really small. If we estimate that the customer pays 0.1 EUR for an SMS (160 bytes), and typically an SMS costs more than that in 2002, then 1 Mb of SMS data costs over 78 EUR to transmit! Moreover, this data is non-real-time, so the operator can send it when there is unused capacity in the network. Subscribers like SMS as well: The concept is easy to understand, and it is nice to have an address, phone number, part number, or a meeting time in text form.

The problem, to the extent it is a problem, with SMS is that it can only deliver text, and even for text, it can only deliver a limited amount of text at a time. For GSM, some improvement will come in the form of the enhanced messaging service (EMS), which can transfer simple pixel images and music in addition to text. But with UMTS data transmission speeds, a much better service could be provided. This SMS-service "with steroids" is referred to as MMS.

MMS includes only non-real-time services. UMTS can and will also be used for other kinds of multimedia, such as conversational real-time multimedia, but these are not really MMS. Multimedia as a whole is discussed separately later in this book (see Section 14.5).

MMS can include several multimedia components, such as text, (still) images, voice, and video. An MMS message can contain multiple components, which are then combined in the user interface to produce a multimedia presentation. The components are called message elements.

The non-real-time character of MMS makes things much easier for the network. Because the data transfer delay is not an issue, it is possible to use protocols and techniques that make the data transmission more efficient with the network's resources. The protocols can use packet retransmission, long interleaving, and other similar techniques that would be impossible to use with real-time applications because of the inherent variable and long delays caused by these methods. Moreover, the network may postpone the MMS transmission during peak load times.

3GPP MMS specifications do not specify the new applications and services in UMTS, but merely the underlying technology that handles the data transmission needs of these new services. It is impossible to predict all possible applications to be built on MMS. However, these could include electronic postcards (i.e., enhanced SMS service), electronic newspapers, news, traffic information, music on demand, advertisements, online shopping, maps and driving instructions, and so on.

12.4.2 MMS Elements

The architecture of a multimedia messaging service environment (MMSE) is shown in Figure 12.7. The individual elements are explained in the following paragraphs. Note that MMS is not purely a 3G issue; this service could also be provided over the 2G air interface. Of course, in that case the limitations of the 2G data-transmission capability and the user terminal capability must be taken into consideration. The basic GSM air interface can only provide 14.4-Kbps circuit-switched connections, but with the EGDE + GPRS combination (EGPRS), it is possible to reach well over 100-Kbps data-transmission rates. This provides a good basis for MMS applications, especially as MMS is a non–real-time service. The required data for an application can even be loaded using a slow data connection and buffered for playback. Clever MMS applications, which have a unique value to mobile and nomadic users, could make GPRS/EDGE network operators appear to be successful "3G" service providers in the eyes of subscribers. To the extent clever EDGE-enabled MMS applications appear, these represent threats to true 3G businesses.

The MMS user agent is located on a UE, or on some external device connected to the UE. It is a function in the application layer, and it must handle tasks related to the presentation of the arriving multimedia message.

FIGURE 12.7
MMS elements.

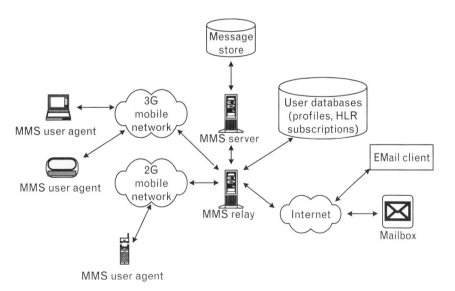

The MMS user agent may also be the initiating party, in which case it has to handle tasks such as MM composition and submission. The agent also has to negotiate terminal capabilities with the MMS Relay/Server. This is because there will be a wide variety of different terminal types, and the network has to know what kind of multimedia messages it can deliver to this particular terminal.

The minimum set of the various media formats that must be supported by all MMS user agents is as follows:

- *Text*
  - Any character set that contains a subset of the logical characters in Unicode.
- *Audio*
  - AMR/EFR coded speech.
- *Image*
  - JPEG.
- *Video*

    ITU-T H.263.

Additionally, the following optional formats have been suggested:

- *Audio*
  - MP3;
  - MIDI;
  - AAC.
- *Image*
  - GIF 89a.
- *Video*
  - MPEG4;
  - ITU-T H.263 profile 3 level 10.

The MMS server is a network entity that must store and process messages. It can use a separate database for message storage. There could be a separate MMS server for each message type: e-mail, fax, and MMS. In practice, an MMS server can be combined with an MMS relay.

The MMS relay is a network entity with a long list of functions. It is a central entity that connects to MMS user agents via various networks (2G, 3G, Internet), to databases, and to MMS servers. The 3GPP MMS specification [12] lists the following mandatory functions for an MMS relay:

- Receiving and sending multimedia messages;

- MMS message deletion based on user profile or filtering information;

- Conversion of multimedia messages into/from formats supported by interworking legacy messaging systems;

- Message content retrieval;

- Message notification to the MMS user agent;

- Generating delivery reports;

- Routing forward multimedia messages and read-reply reports;

- Address translation;

- Temporary storage of messages;

- Ensuring that messages are not lost until successfully delivered to another MMSE element.

Additionally, the specification recommends the following functionalities to be provided:

- Generating charging data records (CDRs);

- Negotiation of terminal capabilities.

There is also a long list of optional functionalities an MMS relay/server can provide.

Note that it is up to the MMS relay to make sure that an MMS user agent receives multimedia messages suitable to its capabilities. The MMS relay must check the MMS user agent profile, which contains information about the capabilities of the terminal, and modify the MMS message to be sent accordingly. It may have to perform media format conversions or delete a message component altogether if the terminal cannot support it. The user profile may also cause modifications to be made to the MMS message. MMS shall support both the MSISDN addresses (i.e., mobile phone numbers; E.164 format) and Internet e-mail addresses (RFC 822 format). Thus, the address translation will sometimes be required, and it will also be done by the MMS relay.

The MMS user database may consist of several separate database entities. These may include a user profile database, a subscription database, and the HLR. The information they have to store includes the following:

- MMS user subscription information;

- Information for the control of access to the MMS;

• Information for the control of the extent of available service capability;

• A set of rules about how to handle incoming messages and their delivery;

• Information about the current capabilities of the user's terminal.

12.4.3   MMS Protocols

The MMS protocol framework is depicted in Figure 12.8. The implementation of "MM1 message transfer protocol" may include WAP, Java, TCP/IP, or other not yet defined protocols.

Wireless Application Protocol (WAP) is an HTTP language derivative that is especially designed for the wireless environment. It will probably be the most widely used protocol in the first phase of UMTS, as it is already used in GSM networks. This head start gains more momentum with GPRS networks and their faster data capabilities (in plain GSM, the WAP applications clearly suffer from the slow data rates). WAP is promoted by the WAP Forum (www.wapforum.org).

Another promising protocol to be used in this interface is Java. The benefit of Java is that it is, in principle, a platform-independent protocol. The user terminals in 3G will contain a wide variety of different operating systems, which will be a problem for application software developers if traditional languages are used. With Java the programmer is free of platform-dependent problems. However, the target operating system must include the Java Virtual Machine to be able to execute Java code. Java is developed by Sun Microsystems.

The possible implementations of MM3 Message Transfer Protocol include a wide variety of different protocols, such as SMTP, POP3, IMAP4,

FIGURE 12.8
MMS Protocol framework.

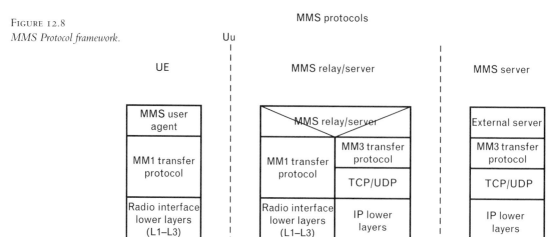

HTTP, or SMPP. Again the selected protocol depends on the particular interface and the application to be executed.

Simple Mail Transfer Protocol (SMTP) is defined by the Internet Engineering Task Force (IETF) [13]. It was originally designed for e-mail transfer via the Internet. Note also that if two MMS relays are connected together, then the interworking protocol between them is SMTP.

Post Office Protocol version 3 (POP3) and Internet Message Access Protocol version 4 (IMAP) are specified in [14] and [15]. HTTP is the standard Internet Hypertext Transfer Protocol. It is used to transfer chunks of HTML across the Internet. HTTP and HTML are being developed further by the World Wide Web Consortium (W3 Consortium) (www.w3.org).

Short Message Peer-to-Peer Protocol (SMPP) is a new protocol developed for sending and receiving short messages via an SMSC. It is used in the interface between the MMS relay and the SMSC. SMPP is specified in [16].

MMS is defined in [12] and [17]. The first specification discusses the service aspects, and the latter specification the technical aspects of MMS.

12.5 Supercharger

The supercharger scheme is a mechanism that could be used to reduce mobility-related network signaling. 3GPP has studied supercharger usage in UMTS networks, but there is nothing preventing this scheme from being used in GSM networks as well. Note that the supercharger is a CN concept, and the UMTS CN is heavily based on GSM/GPRS CNs.

The basic idea behind the supercharger concept is very simple. In current GSM networks, subscriber movement to another MSC/VLR area causes the HLR to provide the new serving MSC/VLR with the subscriber data and to remove the same data from the old MSC/VLR. These signaling procedures produce a significant traffic load in the CN, especially if the MSC/VLR areas are relatively small, as in densely populated urban areas. In supercharged networks the HLR does not remove the subscriber data from the old MSC/VLR; thus, it is already in place when the same subscriber roams back to the old MSC/VLR at some later time.

The reduction in the mobility related signaling traffic can be very significant. Subscribers typically move around within some kind of routine using the same routes every day. An example of this is depicted in Figure 12.9. The subscriber's home and workplace belong to different MSC/VLR areas; thus, in a traditional network two sets of subscriber data updates are needed every day. In the morning the subscriber data has to be downloaded to MSC/VLR 2 and it is removed from MSC/VLR 1. In the afternoon the opposite has to be done. The same data is downloaded to MSC/VLR 1 and removed from MSC/VLR 2.

FIGURE 12.9
*Commuting traffic in a tra-
ditional network.*

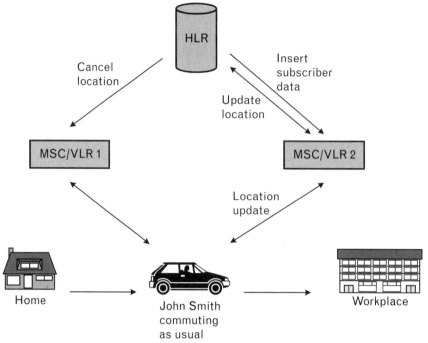

If a supercharged network is used, then the subscriber data-
management traffic can be totally removed. This situation is depicted in
Figure 12.10. The only message necessary in this scenario is the update location message from the new MSC/VLR to the HLR, providing that this is
not John Smith's first day at work, that is, that both VLRs already contain
his subscription data. Then, in the afternoon the only signaling required
between the MSC/VLR1 and HLR is the update location request and the
corresponding response.

The obvious problem with this method is of course that if subscription
data is never deleted from the old VLR, these registers keep on growing
continuously. Clearly there has to be a proper VLR management strategy
for supercharged networks; otherwise, the registers become full and cannot
accept registrations from new mobiles anymore.

The supercharger technical specification [18] presents three schemes to
manage the overgrowth of VLRs:

1. Utilization of a larger database;

2. Periodic audit;

3. Dynamic subscription data deletion.

Utilization of a larger database is not actually a solution to the problem,
as it only delays the moment when the database becomes full. However, a

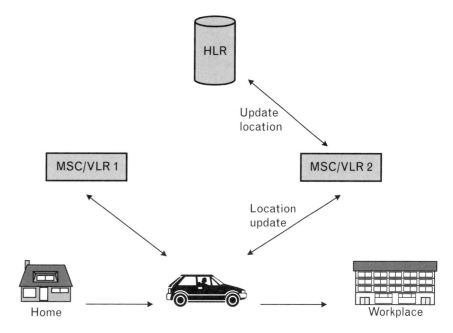

FIGURE 12.10
Commuting traffic in a
supercharged network.

supercharged network would clearly benefit from a large database, as it would contain more subscribers, and this would make the operation of the network more efficient. The larger and more full the database, the more efficient the concept of supercharging, as then it is more probable that a database already contains the subscriber information for a newly arrived mobile. Thus, a supercharged network would need to have larger databases than a traditional network. But this must be combined with some other scheme, which can handle the data deletion for unused data.

The periodic audit is a process that periodically removes the subscription data of inactive users. The amount of data to be removed depends on the estimation of the amount of new users expected during the next audit cycle. If too few entries are removed, the database overflows; if (far) too many are deleted, then the network cannot fully exploit the supercharging functionality.

The dynamic subscription data deletion method deletes a subscription data entry from the register if the register is already full and new subscriber data should be added to it. This is a dynamic run-time scheme performed every time space is needed for a new entry.

In both deletion schemes—periodic audit and dynamic subscription data deletion—the removal algorithm must check that the entry chosen for removal does not have any ongoing activity in the area of responsibility of a VLR. The actual criterion used for choosing the entries to be removed will not be specified. It could be the oldest entry in the register (this scheme is easy to implement), or it could be the entry that has been unused for the longest time.

Note that these management strategies are only presented as examples. The management mechanisms for register databases will not be specified, and are implementation options.

There are also other smaller problems that have to be taken care of when a supercharging scheme is used. For example, if the subscriber data in the HLR is changed, then all the copies of this data in various VLRs will have to be updated. There is no point in the HLR sending updated subscription data to all VLRs that have a copy of the old data. This would destroy the very idea of supercharging, which is to reduce location-management–related CN signaling. Moreover, the HLR does not even know which VLRs have such a copy.

This problem can be solved by attaching an age indicator to the subscriber data entry. If the subscriber data in the HLR is updated, then its age indicator is also updated. All copies of the data in VLRs will now contain an older age indicator. As the HLR knows where the subscriber is currently located, it can also send the new entry (subscriber data and the age indicator) to the serving VLR, but other copies of the data in other VLRs will remain unchanged. Once the subscriber moves to a new location area, a location update request is triggered and the serving VLR adds the age indicator to the update location request sent to the HLR. If the age indicator is the same as the one stored in HLR, then the VLR in question already has the latest subscriber data and no update is needed. If the indicator is older, then the HLR sends the updated data, including the new age indicator, to the VLR.

The reduction of the required signaling with a supercharging scheme can easily be seen from Figures 12.11 and 12.12. Supercharging is further discussed in [18] and [19].

12.6   Prepaging

Consider a GSM CN that will be also used as a UMTS CN. Now consider a mobile terminated call in which the call path is set up in the GSM network before the MS in question has been paged. This means that if the mobile station doesn't respond to paging, then network resources have been allocated unnecessarily. Prepaging is a scheme in which paging is done before the call path is set up in the CN, thus preventing the needless squandering of network resources. They would only be used if the MS actually responds to the paging.

The differences between the current mobile-terminated call (MTC) procedure and the prepaging MTC signaling are shown in Figures 12.13 and 12.14. As the radio connection is set up only after the paging message has been received, the prepaging method actually increases the time the radio resources are allocated to a mobile. However, this is not a major

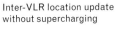

FIGURE 12.11
*Inter-VLR location update
without supercharging.*

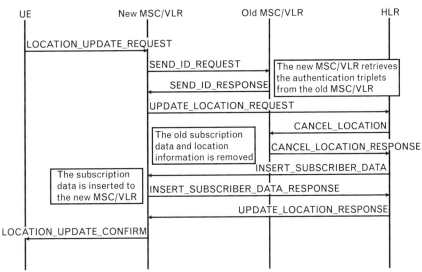

FIGURE 12.12
*Inter-VLR location update
with supercharging.*

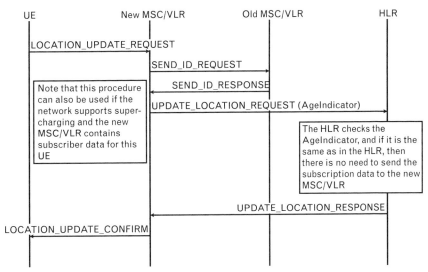

problem, and the total time to set up a mobile-terminated call will not increase. The same messages are sent as before; only the order of those messages changes.

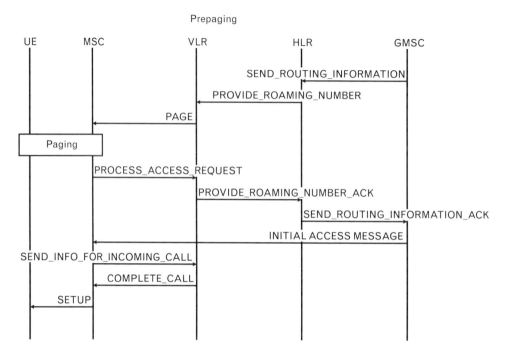

FIGURE 12.13   *MTC setup in a network supporting prepaging.*

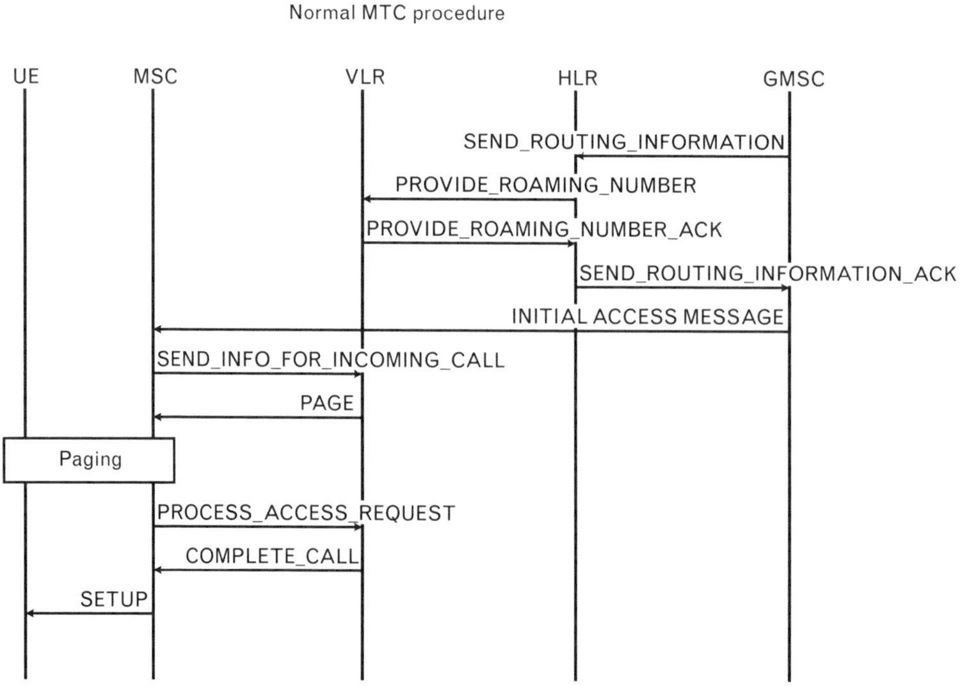

FIGURE 12.14   *MTC setup in a network without prepaging support.*

Note that in practice the MSC and its VLR are always colocated, and the interface between them (B Interface) is thus an internal proprietary interface. The messages used in this interface may not necessarily be the same as specified in the MAP specification.

In prepaging some MAP messages require small modifications. For the radio interface this scheme does not bring any changes, except that the radio resources are allocated a bit earlier.

The case in which the UE is not reachable is shown in Figure 12.15. It is easy to see that the use of prepaging in the CN saves resources in the GMSC, VMSC, and VLR and on internetwork signaling links when the mobile terminal does not respond. In traditional MTC the paging failure would only be noticed at the end of the procedure. Even though this scheme moves some of the burden from the CN into the air interface, which appears counterproductive on first glance, it is justified because of the high proportion of unanswered pages in mobile networks.

Prepaging will be an optional feature in UMTS (and possibly also in upgraded GSM networks). If a network does not support it, then MTCs are always set up using the traditional signaling shown in Figure 12.14. However, note that prepaging is purely an internal issue for the core network. Neither the UTRAN nor the UE knows whether prepaging is used.

Prepaging is discussed in [20].

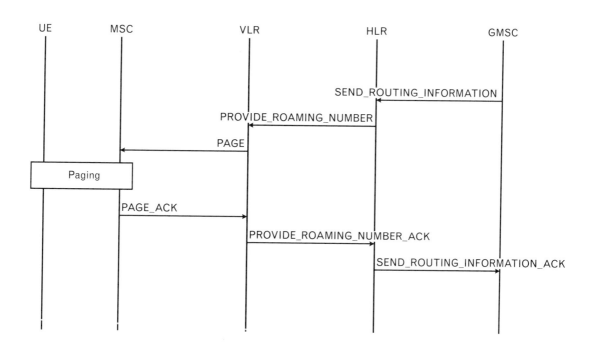

FIGURE 12.15 *Prepaging when the UE is not responding.*

12.7   Gateway Location Register

The gateway location register (GLR) is an entity that handles the location management of a roaming subscriber in a visited network without involving the HLR.

The aim of the GLR scheme is to reduce inter–PLMN signaling traffic, which can be quite expensive, especially in the case of long-distance international links. UMTS will be a truly international system, which means that a large number of roaming users will have to be supported by the UMTS networks. On the other hand, the number of mobile phone users is increasing rapidly, which means that MSC/VLR equipment will probably serve a much smaller geographical area than earlier. When a UE roams in a visited public land mobile network (VPLMN), the network has to inform the HLR about every VLR area change the UE makes. If VLR areas are relatively small, a rapidly moving UE can generate a considerable amount of inter–PLMN traffic while it roams.

The solution to this problem is the GLR. The GLR is a "pseudoHLR" that fools VLRs and SGSNs into thinking that they are communicating with the HLR, when in reality they are exchanging signaling messages with a GLR located in the same PLMN as they are. A GLR acts like the HLR toward the VLR and SGSN in the VPLMN, and like the VLR or SGSN toward the HLR in the home PLMN.

The location of GLRs in the UMTS CN architecture is depicted in Figure 12.16. There is only one GLR per PLMN, and one GLR cannot serve more than one PLMN, at least not in the first phase of UMTS.

The GLR handles the location management of a roaming user in a visited network. For this purpose, it stores the subscriber profiles and location information of users roaming in this PLMN. The subscriber information is downloaded from the HLR during the first location update from this

FIGURE 12.16
GLR in the CN architecture.

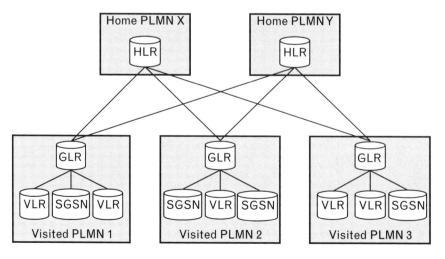

VPLMN. The subsequent location updates from this VPLMN are then invisible to the HLR. The GLR intercepts the update location message addressed to the HLR, updates its internal registers, and returns an acknowledgment message to the VLR. The user information is stored in the GLR until a cancel location message is received from the HLR indicating that the user has moved to another PLMN (either HPLMN or another VPLMN).

The working principle of the GLR is best described with a series of signaling flow diagrams. Let's see what actually happens in the first diagram (Figure 12.17). Once the UE registers for the first time in the VPLMN, the

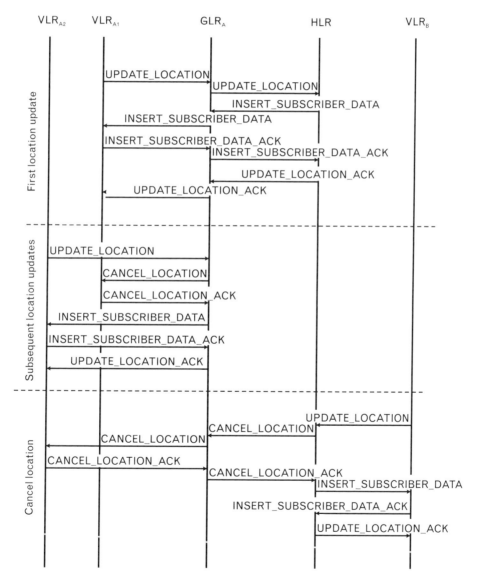

FIGURE 12.17   *GLR signaling.*

location of the UE is disclosed to the HLR by the VLR by the sending of an update location message. The GLR in the VPLMN intercepts this message, and checks whether its registers already contain an entry for this subscriber. If not, this location update is the first one for this user. The GLR creates a register entry and fills it with information from the update location message. This data includes the VLR and the serving MSC numbers, and the subscriber's data. Thereafter, the GLR modifies the message and replaces the VLR number with its own GLR number, and the MSC number with a logical IM-MSC number. The modified message is sent to the HLR.

The HLR stores the received GLR and IM-MSC numbers as genuine VLR and MSC identification numbers. It does not know anything about the GLR's existence. To the HLR, this VPLMN seems to contain only one VLR (the GLR), and its identity is the GLR number. If the HLR wants to send a message to the VLR, it uses the GLR address, and the GLR then checks the actual VLR address in its registers and forwards the message to the correct VLR. The GLR is transparent to both the HLR and the VLRs. This property makes the addition of a GLR into a network rather easy, as neither VLRs nor HLRs need any modifications. The MAP signaling messages do require some changes though; these are explained in [21].

The subsequent location update messages within the VPLMN are again intercepted in the GLR, but as the register already contains an entry for this UE, no message to the HLR is needed. The GLR simply updates the corresponding entry with the new VLR and serving MSC numbers. If new messages from the HPLMN are addressed to this UE, they are now forwarded to the new address by the GLR. These subsequent location update procedures do not generate any inter-PLMN signaling traffic.

If the UE roams into another PLMN, either to an HPLMN or to another VPLMN, the VLR in question sends an update location message to the HLR. Actually this VLR can also be another GLR; it does not matter, as the HLR cannot tell the difference. The HLR then sends a cancel location message to the previous VLR. In this case it is addressed to the GLR_A, which forwards it further to the VLR_{A2}. After the acknowledgment message the GLR deletes the entry of this UE from its registers.

The GLR works just as well for improving the packet-switched CN performance. This is depicted in Figure 12.18. $SGSN_{A1}$ and $SGSN_{A2}$ belong to the same VPLMN, which is served by GLR_A. $SGSN_B$ belongs to another PLMN, and it could as well be another GLR.

The signaling shown is exactly equivalent to the circuit-switched case described earlier. Only the reasons that trigger these procedures are different.

Once the UE roams into a new VPLMN and makes the first GPRS attach in this network, the new SGSN ($SGSN_{A1}$) sends an update location message to the HLR. The GLR of this PLMN intercepts the message, and once it notices that this is a new roaming subscriber, it creates a new entry in its registers for this subscriber. The GLR also modifies the message (the

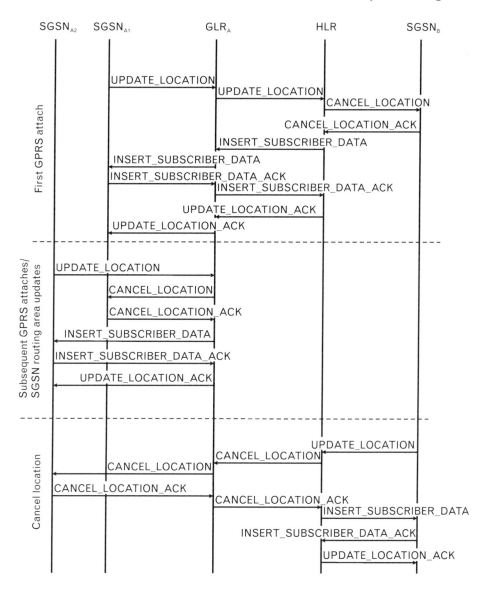

FIGURE 12.18 *GLR signaling with GPRS.*

SGSN number is replaced by the GLR number and the SGSN address by the IM-GSN address), and sends it further to the subscriber's HLR. To the HLR it seems that the subscriber is served by a SGSN, which is identified by the GLR number/IM-GSN address-pair.

Any subsequent intra-PLMN GPRS attach or routing area update messages will be intercepted by the GLR, and they will not be sent further to the HLR. This will not be necessary because, from the HLR's point of view, nothing has changed. The UE is still served by the same pseudo-

SGSN (the GLR). The GLR updates its internal registers with the new SGSN number and addresses it so that it can route inter-PLMN messages to the right SGSN.

Only when the UE roams outside the old VPLMN$_A$, and makes its first GPRS attach in the new network, does the HLR get an update location message. This triggers a cancel location message to the old GLR. The GLR can now delete its entry for this subscriber.

As seen from these examples, the principle of the GLR is very simple, and therefore a practical one. However, in practice it is a bit more complex than what was presented here. The GLR concept does complicate certain issues in the network, like the recovery procedures from the HLR, VLR, or GLR failure.

In these examples there were also some references to the Intermediate MSC (IM-MSC) and the Intermediate GSN (IM-GSN) entities. These are logical nodes, which will be physically implemented in connection with the GLR itself. They are used to fool the HLR. If the HLR has to send something to the serving MSC in the VPLMN, it sends the message to the address of the IM-MSC. Similarly, in the case of SGSN GTP signaling, messages from the HLR are addressed to the IM-GSN. The logical node (the GLR in practice) intercepts these messages and provides them with the address of the real MSC or SGSN.

The use of the GLR reduces the inter-PLMN MAP signaling traffic considerably in typical roaming scenarios, but one has to remember that this applies only to inter-PLMN signaling traffic. Even in our brave new world, the average user very seldom travels abroad and generates this kind of traffic in the first place. Therefore, the GLR concept is an important enhancement to the network infrastructure, but it is not an absolutely essential entity. Indeed, having a GLR in the network is optional for the operator.

The GLR concept is fully compatible with 2G GSM networks. It can be used in those networks without changes to the existing network nodes. The GLR is specified and explained in [22], and [23].

12.8 Optimal Routing

Optimal routing is a feature that makes it possible to route calls directly to a mobile user regardless of his location. Normally, all mobile phone calls are first routed to the HPLMN of the called subscriber, and then further to the visited PLMN, if the called subscriber is roaming. This may result in situations in which the network's resources are wasted. An example of this is depicted in Figure 12.19. Mark is an Orange subscriber in the United Kingdom who tries to call his friend Ulla, who has a subscription in the Radiolinja network in Finland. Ulla is, however, currently visiting the United

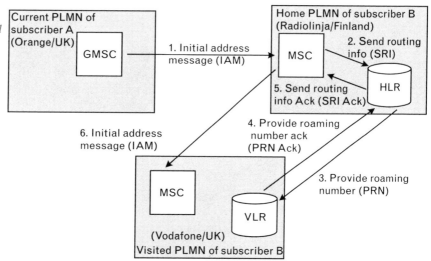

FIGURE 12.19
Basic mobile-to-mobile call without optimal routing.

Kingdom and is roaming in the Vodafone network there. Without optimal routing, this call would be first directed to Ulla's home network in Finland, and then after checking the roaming address from the HLR back to the United Kingdom into the Vodafone network. Thus, the call would require two inter-PLMN legs, Orange → Radiolinja and Radiolinja → Vodafone. Clearly this is not an efficient use of network resources.

A better solution is first to query the location of the called mobile, and then to route the call directly to the PLMN in question. This is depicted in Figure 12.20. After Mark's call attempt, Orange's GMSC sends a Send_Routing_Info message to Radiolinja's HLR. If Ulla is roaming outside Radiolinja's network, its HLR sends an additional message to the VLR in question (Vodafone in this case), requesting a roaming number for Ulla.

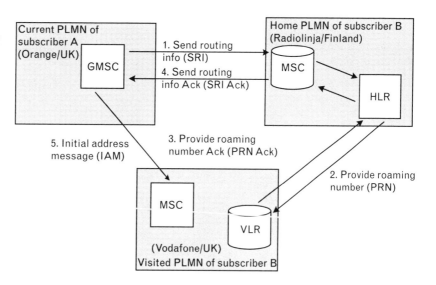

FIGURE 12.20
Basic mobile-to-mobile call with optimal routing.

Once this is received, it is forwarded to Orange's GMSC, and the call can be set up directly from Orange to Vodafone.

The specifications differentiate two cases in which optimal routing can be used. The first one is the basic mobile-to-mobile call, which was discussed in the previous paragraph. The second case is conditional call forwarding. Conditional call forwarding can be either early or late call forwarding. In early call forwarding the call is forwarded before it has been extended to the VPLMN of the called subscriber. In late call forwarding the call is first extended to the VPLMN of the called subscriber, before it is found that forwarding of the call is required (it could be that a call should be forwarded only when the subscriber is busy). In this case, the VPLMN has to return control of the call back to the called GMSC. The GMSC may then query the HLR of the called subscriber to acquire forwarding information and set up the call to this destination.

Figure 12.21 depicts a late call forwarding scenario. It starts with a basic mobile-to-mobile call, in which the GMSC acquires a roaming number from the VLR (1–4). The call attempt is rejected (7), for example, because the called subscriber is busy or not reachable. The visited MSC returns control of the call to the GMSC (8), which can then query the forwarding information from the called subscriber's HLR (9–10). The GMSC can also use the forwarding information returned by the VLR (7–8), if the HLR so indicated in the earlier SRI Ack (4). Finally the call is forwarded to the indicated number (13).

In either case (early or late forwarding) without optimal routing the call would be first extended to the VPLMN of the called subscriber and then from there to the forwarded destination.

Note that optimal routing has implications on charging. In GSM the principle has been that the caller only pays for the first leg of the call, and if

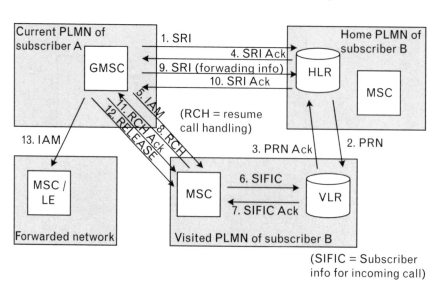

FIGURE 12.21
Late call forwarding with optimal routing.

the called subscriber roams, he pays the call costs from his home network onward. With optimal routing it is possible that the call costs for the caller would be bigger with optimal routing than without it. For example, this may happen if Seamus from Eircell in Ireland is calling Pete from Vodafone in the United Kingdom, and Pete is roaming in Finland. Without optimal routing, Seamus would only pay the call costs from Ireland to United Kingdom, and Pete would pay the costs of the international United Kingdom–Finland leg. However, if optimal routing is used, then Seamus would pay all costs because there would be only one call leg, and this might be more expensive than the Ireland–United Kingdom connection.

Because the caller cannot know for sure where the called person is located, he shouldn't pay more than what he would pay without optimal routing. Therefore, at least in the first phase of optimal routing, this feature is not used if the call costs for the caller would be increased. It is possible that in later phases with more intelligent GMSCs, a call's cost can be calculated by the GMSC and then divided between the parties in a fair manner. Then, optimal routing could be used only when it would enhance the use of network resources.

Note that whereas the optimal routing of basic mobile-to-mobile calls is only applicable between mobile stations, the optimal routing of call forwarding can also be done to fixed network addresses.

As the optimal routing feature is purely a CN functionality, it can be implemented as well in 2G GSM networks.

Optimal routing is discussed in [24] and [25].

12.9   Adaptive Multirate Codec

Adaptive multirate codec (AMR codec) was originally a GSM 2+ feature. Although the name suggests it, it is not yet another new speech codec, but a method that works out how to get the most from a set of existing (or new) speech codecs. The basic idea is that the system switches among available speech codecs, or their modes, dynamically, to maximize the channel capacity at the same time as trying to maintain an acceptable voice quality. The network monitors the voice quality and adjusts the speech codec accordingly. That is, in poor channel conditions, AMR uses a robust coding scheme that can provide acceptable quality, but requires more bits to achieve this. In good conditions, that kind of robust coding is unnecessary as the same quality can be provided with fewer bits by a weaker coding scheme.

The UTRAN multirate speech coder is a single integrated speech codec with several different source rates, called codec modes. There are actually two kinds of AMR codecs in UTRAN. The original AMR codec is employed in Releases 99 and 4, as well as in GSM. The audio bandwidth of this codec is 3.4 kHz. Release 5 introduces a new wideband AMR (AMR-WB) codec that provides a wideband speech service with audio bandwidth extended to 7 kHz. The result is an improved and more natural voice quality, especially in hands-free environments. The following discussion applies to both AMR codecs. The most notable difference between these codecs is that in the narrowband AMR (AMR-NB) there are eight source rates from 4.75 Kbps to 12.2 Kbps (see Table 12.1), whereas the AMR-WB has nine source rates from 6.60 Kbps to 23.85 Kbps (see Table 12.2). In addition, both codecs have a separate mode for background noise. If these silence indicator (SID) frames are transmitted continuously, they need only 1.8 Kbps (1.75 Kbps in AMR-WB). The selected source rate can be switched every 20 ms, which is the length of a speech frame.

In addition to the speech encoder and decoder functions, the AMR codec also needs voice activity detector (VAD), comfort noise insertion, and speech frame substitution functions. The VAD exists at the transmitting side and inspects the speech frames to see if they contain any speech. If not, SID frames can be sent instead of encoded speech frames. At the receiving side, the reception of SID frames triggers the generation of locally available comfort noise. This comfort noise is artificial noise, which is generated to convince the listener that the connection still exists. Users tend to yell into the handset in the absence of comfort noise because they assume that there must be something wrong with a completely silent connection. The SID frames relay noise parameters, which characterize the background noise situation at the transmitting side. These are used at the receiving side to generate the right kind of noise.

TABLE 12.1   AMR-NB SOURCE CODE BIT RATES

AMR-NB CODEC MODES	SOURCE RATE (KBPS)
AMR_12.20	12.20 (= GSM EFR)
AMR_10.20	10.20
AMR_7.95	7.95
AMR_7.40	7.40 (= IS-641)
AMR_6.70	6.70 (= PDC-EFR)
AMR_5.90	5.90
AMR_5.15	5.15
AMR_4.75	4.75
AMR_SID	1.80 (if SIDs were sent continuously)

TABLE 12.2   AMR-WB SOURCE CODE BIT RATES

AMR-WB CODEC MODES	SOURCE RATE (KBPS)
AMR-WB_23.85	23.85
AMR-WB_23.05	23.05
AMR-WB_19.85	19.85
AMR-WB_18.25	18.25
AMR-WB_15.85	15.85
AMR-WB_14.25	14.25
AMR-WB_12.65	12.65
AMR-WB_8.85	8.85
AMR-WB_6.60	6.60
AMR-WB_SID	1.75 (if SIDs were sent continuously)

If an occasional speech frame is lost in the transport network, then the speech frame substitution function can synthesize speech to fill the gap. The substituted speech is based on previously received speech frames. It is not allowed to insert silent frames to replace the lost frames, as this would be quite irritating to the listener. However, if several subsequent frames are lost, then silence must eventually be used to indicate to the user that there are problems in the transmission. This is also because the synthesis of speech frames doesn't work with several successive frames. The result is not speech, but some incomprehensible noise.

The aim of the AMR is to minimize the number of bits sent over the air interface. This reduces the interference level in the system and, thus, increases the available capacity. Of course, the drawback to this advanced technique is its increased complexity. This especially hits the handset equipment, which has to accommodate all these new codecs and the associated management functions.

Two additional concepts related to speech coding are tandem free operation (TFO) and transcoder free operation (TrFO). The problem with speech codecs is that if they are cascaded, the quality of the reconstructed speech after the last decoder might be quite poor. A typical cascaded case would be a mobile-to-mobile call between two networks. The first encoder is in the originating mobile, the first decoder is in the first MSC, the second encoder is encountered when the signal leaves the second MSC, and the receiving UE has yet another decoder. Note that transcoders in the network don't have to be within the MSC, but this is the usual location for them for transmission cost reasons. A simple solution to this problem is to bypass the unnecessary speech codecs. Only those codecs that are in mobile stations are used, and the speech signal is transferred in encoded format along the whole path between the communicating mobiles. This is depicted in Figure 12.22. The use of AMR codecs makes this scheme a bit more complex. Normally each air interface leg is evaluated independently with the AMR, and so the selected codecs may have different rates in different legs. In the TFO/TrFO method this is not possible because the codecs in the network are not used. Therefore, the worst of both air interface legs determines the used common rate.

This scheme of transcoder bypassing is called TFO if the required control signaling is in-band signaling. TrFO uses out-of-band signaling instead. In a Personal Digital Cellular (PDC) system this same concept is known as transcoder bypass.

The AMR codec principles and TFO are discussed in [26] in an easily understandable way.

There are several 3G specifications related to AMR. Start your exploration with the general description, [27] for AMR-NB and [28] for AMR-WB. Note that the AMR will also be used with just minor modifications in GSM. TFO and TrFO are discussed in 3GPP specifications [29–31].

12.10   Support of Localized Service Area

Support of localized service area (SoLSA) is a mechanism that can be used as a platform for providing special tariffs or a special set of service features for certain subscribers within a regionally restricted area or areas. The motivation for this concept is to create a means for network operators to build new service and tariff packages, which take into account subscriber groups and their needs. This method is optional for both the network and the UE. Usage and implementation of the different SoLSA service features may vary

FIGURE 12.22
TFO/TrFO scheme.

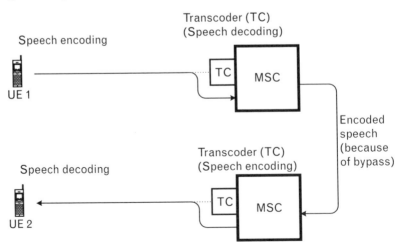

according to operator's service packages. In the GSM world, this concept is known as the Phase 2+ item "support of home area priority."

A localized service area (LSA) consists of a number of cells (or even of one cell) within a PLMN. Each LSA is identified by an LSA identity. The service subscriber can also define a name for the LSA, and this is then displayed to LSA users on their handsets. There can be several overlapping LSAs, and one user can simultaneously belong to more than one LSA. Cells belonging to an LSA do not have to provide continuous coverage. Also, if hierarchical cell structures are used, the overlapping cells in different hierarchical levels do not have to belong to the same LSA. See Figure 12.23.

When a UE makes an initial cell selection or a cell reselection, it will favor cells belonging to its LSAs. Similarly, when an HO is performed, the network will favor the cells belonging to the UE's LSAs.

The SoLSA service also includes the concepts of *exclusive access*, *LSA-only access*, and *preferential access*. Exclusive access means that it is possible for the operator to define cells in which only the members of one (or more) LSA is allowed access. In effect, these cells are then private cells. Other users cannot make HOs to these cells, but they can camp on them to initiate location updates. Note that emergency calls are also allowed for everybody.

LSA-only access means that users belonging to a certain LSA are not allowed to access the network outside this LSA. Again, emergency calls are an exception; they must be allowed to be initiated everywhere.

Preferential access means that the network can allocate a cell's resources to LSA users if required. For example, the last traffic channel in a cell is allocated only to LSA users or in case of congestion when an ongoing call of a non–LSA user is released and replaced with a call of an LSA user.

FIGURE 12.23
Localized service area.

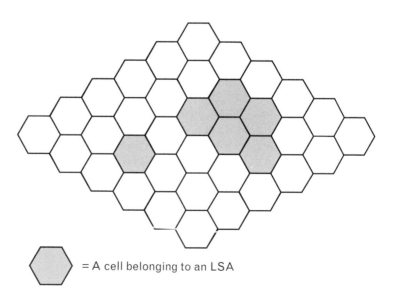

= A cell belonging to an LSA

Possible applications for SoLSA include private cells for a company. A company may have private cells to which only the LSA members (company employees) have access. The operator can offer lower tariffs for the calls made from these cells. The operator may also want to encourage the use of mobile phones at home by defining several LSAs which cover a town area by area, and defining lower tariffs (at least during off-peak hours) for the "home users" that are located in their own home area (LSA). The SoLSA concept is defined in [32].

12.11   Smart Antennas

Smart antenna techniques are already used in many wireless systems, but UMTS is the first system where they are considered already in the system specification phase. Smart antennas are especially attractive in WCDMA networks, as they could be used to reduce the intracell interference levels considerably. Interference is one of the most important and difficult issues in the WCDMA air interface, and any improvement in the interference level management will bring increased capacity.

A smart antenna is a vague concept and can include various different antenna structure implementations. In this section we briefly discuss the most important of them. Generally, a smart antenna is an antenna structure consisting of more than one physical antenna element, and a signal processing unit that controls these elements and combines or distributes the signals among these elements. Note that the antenna elements are not smart as such, but the smartness of the device lies in the controlling signal processing unit.

Earlier smart antennas were considered to exist only in the network (i.e., in base stations). Complex multiantenna element structures were considered too large and too expensive for small handsets. However, this is now changing with new antennas technologies. In the future we will see mobile communication devices with several small antennas. For example, it will be possible to put several small antennas into a laptop lid to provide it with smart antenna capability.

Why should we use smart antennas and what makes them so desirable? The traditional omnidirectional antennas radiate evenly and uniformly in all directions. This is nice, but the user who receives the signal can only be in one place at a time; thus, most of the energy the base station emits is wasted, and in CDMA systems, it becomes interference to other users and base stations. Smart antennas employ various techniques to achieve directional transmissions. The base station transmits a narrow beam only in the direction of the UE, and nothing in other directions. The UE may also employ smart antenna techniques. But smart antennas are not only about directional transmissions; they can also be employed to provide various forms of diversity to the signal. To be precise, directional beams or sectors are also one form of diversity; that is spatial diversity. In short, smart antennas increase the system capacity because they improve the crucial signal-to-noise ratio (SNR) by reducing the noise N, and possibly also increasing the signal strength S.

The simplest smart antenna type to implement is a structure that simply provides antenna diversity (see Section 3.4.4). The basic idea behind this scheme is that several antennas are used (either in the transmitter or in the receiver or in both) in the hope that signal fading in these new radio paths are not correlated. The receiver can choose the signal with the best quality, or combine several component signals using weighting or some other algorithm. The optimum combining algorithm depends on many factors.

The next smart antenna type considered is the switched-beam antenna. This antenna type forms several narrow beams that are fixed. They cannot be steered to follow the user. Instead, the receiver unit selects the best-received beam and uses it for both reception and transmission. If the user moves, then it may become necessary to perform a handover to another beam. This type of antenna is depicted in Figure 12.24. In this example, the signal from User A would be handled by Beam 2.

Adaptive antennas are still more intelligent and, thus, more complex to implement. This type of antenna modifies the antenna pattern dynamically in such a way that the main lobe tracks the user, and nulls in the antenna pattern are directed towards identified interferers. Because the antenna tracks the user, there is no need for handovers between sectors. Also, the interference suppression is quite efficient if the number of dominant interferers is small. And this is just the case in a typical WCDMA cell. In UTRAN, the users' connections have different data rates. The biggest interference will be

FIGURE 12.24
Switched-beam antenna.

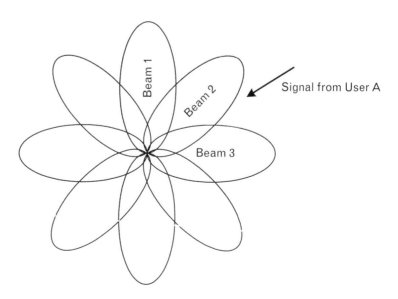

caused by high-data-rate users, and there can be only a few of them in one cell. Note that adaptive antennas are not so suitable for cells in which there are numerous low-bit-rate users making voice calls, and no high-bit-rate-users at all (i.e., in a very homogenous user environment). A cell with an adaptive antenna is shown in Figure 12.25.

Adaptive antenna beamforming is depicted in Figure 12.26. The same source signal is fed through several phase shifting networks. These shifts are calculated so that once the signals are sent via an antenna array, there will be constructive interference in the direction of the UE in question. In switched-beams antennas, beamforming is simpler because beams are fixed; thus, there is no need for dynamic beamforming weights.

Note that adaptive antennas enhance not only downlink performance, but uplink performance as well. A WCDMA cell employs orthogonal spreading codes in the downlink, but the uplink is unsynchronized; thus, this is not possible. Thus, in principle the signals in the uplink will cause more interference than their orthogonal counterparts in the downlink. Therefore, the use of adaptive antennas on the receivers in base stations can be regarded as even more important than their use in the downlink's narrow transmit beams. The interference levels are reduced in both links, uplink and downlink, if adaptive antenna techniques are used.

Only dedicated channels can be sent using directed beams (an exception to this rule is given in the next paragraph). Broadcast channels are common to everybody in a cell; thus, they are not directed to any particular user. Notice that it is the common broadcast channels that define the range of a cell. A mobile station cannot synchronize to a cell if it cannot experience acceptable reception of broadcast channels. Thus, using adaptive antennas in dedicated channels does not directly increase the range of a cell. However, it

FIGURE 12.25
Adaptive antenna.

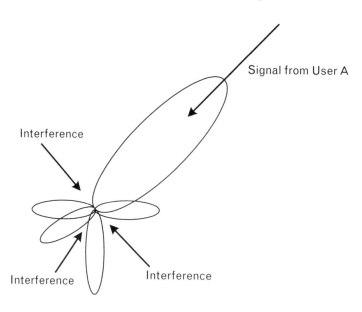

FIGURE 12.26
Beamforming in adaptive antennas.

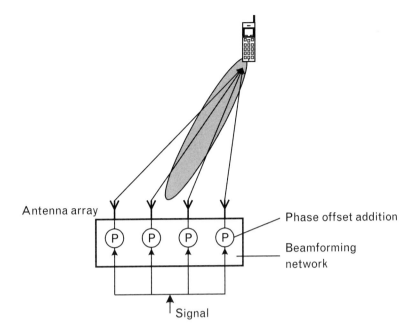

will reduce the overall interference level in a cell, and that makes it possible to increase the power levels of broadcast channels; thus, the radius can be increased. This is clarified in Figure 12.27.

However, in practice the directed beams will exceed the cell boundaries set by the common control channels. Also, for HO purposes the mobile station cannot just measure the omnidirectional antenna transmission and estimate the directed beam characteristics from it. Therefore, with every

FIGURE 12.27
Channel radius and smart antennas.

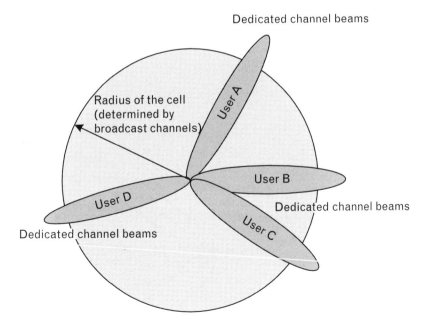

directed antenna beam, a special secondary common control physical chan-nel (SCCPCH) is sent. This can be used for channel estimation, but it can-not be used for synchronization; that is, a mobile station cannot synchronize to a cell via its SCCPCH, but only by using its primary CCPCH. Therefore, it does no harm if the SCCPCH transmission exceeds the cell boundary on top of the beam.

In a way this, concept is similar to sectored cells, except that here every user has his own sector, and these sectors can track the user and follow his movements. In WCDMA different beams use different scrambling codes. Otherwise, they are sent in the same frequency and timeslot. Thus, different beams do interfere with each other if they happen to overlap. Still, this tech-nique reduces intercell interference because without adaptive antennas all downlink transmissions would interfere with each other. With adaptive antennas only the overlapping beams interfere. When it comes to intercell interference, the change is not only a positive one because a narrow beam can reach quite far: further than with "traditional" antennas. Thus, it could introduce intercell interference to users even deep within other cells. Other base stations will not be bothered though (not directly that is), as their uplink receivers are on a different frequency in the FDD mode.

It is also possible that the uplink transmission consists of several multi-path components. In that case, the component signals have to be tracked independently, and then combined in a RAKE combiner (see Figure 12.28). Because all the components originate from the same UE, they all use the same spreading code. The interference in each multipath is estimated (from the CPICH), and the resulting estimations are used as weights when compo-nents are combined in the RAKE.

As a summary, the benefits of adaptive antennas include the following:

• Reduced cochannel interference;

• Increased range;

• Increased capacity.

There are also problems with this technology, such as the following:

FIGURE 12.28
*Multipath in uplink and
adaptive antennas.*

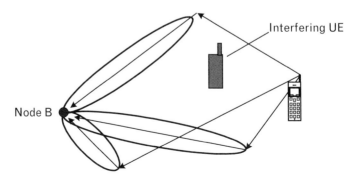

Interfering UE

Node B

• Different fading characteristics in uplink and downlink;

• Difficulty with calibrating adaptive antenna arrays.

In UTRAN-FDD the uplink and downlink use different frequencies; thus, their fading characteristics are different. Therefore, the direction of arrival (DOA) estimate gained from uplink reception in base station may not be optimal for downlink beamforming.

Adaptive antennas are currently a hot research topic, and new papers are published frequently. For example, see [33–36].

MIMO

Though they are an application for smart antennas, Multiple-Input-Multiple-Output (MIMO) antenna structures are something quite different compared to other smart antennas. They are thus discussed separately in this section.

Traditionally a good radio environment has been regarded as one with no obstacles between the transmitter and the receiver and with as few multipaths as possible between them. A line-of-sight situation between the transmitter and the receiver was always seen as preferable. However, the whole working principle of MIMO is based on multipath. One can only employ this technique in a rich multipath environment.

It is true that ordinary CDMA systems can also benefit from multipath. As shown earlier, a RAKE receiver consists of fingers; each finger can receive one multipath signal, all of which are then put together in a combiner. It is important to remember that in this case all the multipath signals are copies of the same signal. In MIMO, however, individual multipaths are employed to carry different data signals. Thus, in an ideal very rich multipath environment with several strong multipaths and M transmit and M receive antennas, it is theoretically possible to increase the channel throughput M fold. In MIMO, we assume the presence of a multipath environment and provide individual signals for all of the paths. The MIMO principle is depicted in Figure 12.29.

Each multipath can recycle the orthogonal code space, so the same spreading code can be used again in all multipaths. The signal processing unit in the receiver must be very clever so that it can separate individual component signals and track them in each receiver element. Optimally, the multipath environment should be such that the individual multipaths arrive at the receiver from different directions. Then each receiver element could act as a directional receiver and track only its own signal.

MIMO is a Release 6 work item within 3GPP. This means that the first 3GPP MIMO systems will be deployed well after 2005. Within 3GPP, MIMO is considered as an enhancement to HSDPA, so it would only be used in the downlink. There is no need for it in the uplink because the very

FIGURE 12.29
MIMO antenna system.

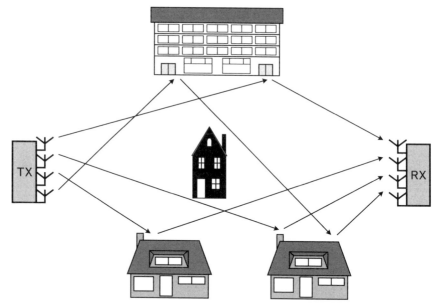

high data rates of MIMO are unlikely to be needed in the uplink. A cell's uplink throughput can be handled and increased using existing conventional methods. In 3GPP there will be a maximum of four antennas in the transmitter and receiver. This means that the throughput can be theoretically increased four fold. However, the MIMO technical report [37] gives 21.6 Mbps as the highest data rate one UE can expect to get from a MIMO capable HSDPA cell.

As pointed out earlier, the problem with MIMO is that it only works well in a rich multipath environment. This is not available everywhere, and as peculiar as it sounds, there may be problems with MIMO if the UE is in a very good radio environment. On the other hand, if radio conditions are really good, then the throughput can be increased by using other radio techniques, such as higher-order modulation (16QAM, or even 64QAM in the future), and reduced channel coding rates. Thus, MIMO can be seen as a method that increases the HSDPA throughput for mobiles further away from the base stations. If the UE is close to a base station, it does not need MIMO, as it can get high throughput by other means. 3GPP MIMO technical report is [37].

REFERENCES

[1]   Russian Federation Ministry of Defense, Coordination Scientific Information Center, GLONASS: http://www.rssi.ru/SFCSIC/english.html.

[2]   3GPP TS 22.071, v 5.1.1, Location Services (LCS); Service Description, Stage 1, 2002.

[3]   3GPP TS 23.171, v 3.7.0, Functional Stage 2 Description of Location Services in UMTS (Release 1999), 2002.

[4]   3GPP TS 23.271, v 5.2.0, Functional Stage 2 Description of LCS, 2002.

[5]   3GPP TS 25.305, v 5.4.0, Stage 2 Functional Specification of UE Positioning in UTRAN, 2002.

[6]   GSM 02.71 Digital Cellular Telecommunication System (Phase 2+); Location Services (LCS); Service Description, Stage 1.

[7]   GSM 03.71 Digital Cellular Telecommunication System (Phase 2+); Location Services (LCS); Functional Description, Stage 2.

[8]   3GPP TR 25.858, v 5.0.0, High Speed Downlink Packet Access; Physical Layer Aspects, 2002.

[9]   3GPP TS 25.308, v 5.2.0, High Speed Downlink Packet Access (HSDPA); Overall description; Stage 2, 2002.

[10] 3GPP TS 22.146, v 5.2.0, Multimedia Broadcast/Multicast Service; Stage 1, 2002.

[11] 3GPP TR 23.846, v 1.0.0, Multimedia Broadcast/Multicast Service; Architecture and Functional Description, 2002.

[12] 3GPP TS 23.140, v.5.2.0, Multimedia Messaging Service (MMS); Functional Description, Stage 2, 2002.

[13] IETF RFC 821 "Simple Mail Transfer Protocol," J. Postel, August 1982.

[14] IETF RFC 1957 "Some Observations on Implementations of the Post Office Protocol (POP3)," R. Nelson, June 1996.

[15] IETF RFC 1730 "Internet Message Access Protocol—Version 4," M. Crispin, December 1994.

[16] Short Message Peer-to-Peer Protocol Specification, SMPP Developers' Forum.

[17] 3GPP TS 22.140, v 5.1.0, Service Aspects; Stage 1; Multimedia Messaging Service, 2002.

[18] 3GPP TS 23.116, v 4.2.0, Super-Charger Technical Realisation; Stage 2, 2001.

[19] 3GPP TR 23.912, v 4.1.0, Technical Report on Super-Charger, 2001.

[20] 3GPP TR 23.908, v 4.0.0, Technical Report on Pre-Paging, 2001.

[21] 3GPP TS 29.120, v 4.0.0, Mobile Application Part (MAP) specification for GLR, 2000.

[22] 3GPP TS 23.002, v 5.6.0, Network Architecture, 2002.

[23] 3GPP TS 23.119, v 4.0.0, Gateway Location Register (GLR)—Stage 2, 2001.

[24] 3GPP TS 22.079, v 4.0.0, Support of Optimal Routeing (SOR); Service Definition—Stage 1, 2001.

[25] 3GPP TS 23.079, v 5.0.0, Support of Optimal Routeing (SOR); Technical Realisation, 2002.

[26] Aftelak, S., "New Speech Related Features in GSM," in GSM—Evolution Towards 3rd Generation Systems, Z. Zvonar, P. Jung, and K. Kammerlander (eds.), Norwell, MA: Kluwer Academic Publishers, 1999, pp. 17–43.

[27] 3GPP TS 26.071, v 4.0.0, Mandatory Speech Codec Speech Processing Functions; AMR Speech Codec; General Description, 2001.

[28] 3GPP TS 26.171, v 5.0.0, Speech Codec Speech Processing Functions; AMR Wideband Speech Codec; General Description, 2001.

[29] 3GPP TS 28.062, v 5.0.0, Inband Tandem Free Operation (TFO) of Speech Codecs; Service Description; Stage 3, 2002.

[30]   3GPP TS 23.153, v 5.0.0, Out of Band Transcoder Control—Stage 2, 2002.

[31]   3GPP TR 23.911, v 4.0.0, Technical Report on out-of-Band Transcoder Control, 2001.

[32]   3GPP TS 23.073, v 4.0.0, Support of Localised Service Area (SoLSA); Stage 2, 2001.

[33]   Mogensen, P. E., et al., "Antenna Arrays and Space Division Multiple Access," in *GSM—Evolution Towards 3rd Generation Systems*, Z. Zvonar, P. Jung, K. Kammerlander (eds.), Norwell, MA: Kluwer Academic Publishers, 1999, pp. 117–151.

[34]   Holma, H., and A. Toskala, *WCDMA for UMTS: Radio Access for Third Generation Mobile Communications*, New York: John Wiley & Sons, 2000, pp. 270–275.

[35]   Correia, L. M., *Wireless Flexible Personalised Communications*, Chichester, U.K.: John Wiley & Sons, 2001, pp. 194–222.

[36]   Boukalov, A., "System Aspects of Smart Antenna Technology," Presentation 6.5.1999, Radio Communication Systems Department/School of Electrical Engineering and Information Technology (EIT), Royal Institute of Technology (KTH), Stockholm, Sweden, accessed December 7, 2002, at http://www.s3.kth.se/radio/seminars/sa.pdf.

[37]   3GPP TR 25.876, v 1.1.0, Multiple-Input Multiple Output Antenna Processing for HSDPA, 2002.

CHAPTER 13

3G Services

A modern telecommunication network such as UMTS can provide a wide variety of services. The service concepts and definitions of UMTS are for the most part copied from the GSM world. But whereas in GSM the service parameters are often fixed, in UMTS they can be dynamically renegotiated whenever required.

13.1   Service Categories

The services provided by UMTS can be divided into four main classes:

1.   Teleservices;

2.   Bearer services;

3.   Supplementary services;

4.   Service capabilities (i.e., support for value-added services).

Each is discussed, one-by-one, in the following sections.

13.2   Teleservices

A teleservice is a type of telecommunication service that provides the complete end-to-end capability for communication between mobile users in accordance with standardized protocols. The user has no direct responsibility for the end-point applications in a teleservice. Contrast getting e-mail over a GSM handset attached to a laptop (a bearer service) with simply talking on a GSM phone (a teleservice). Teleservices make use of the whole OSI model protocol stack (except for a layered protocol between them), and they also include the terminal equipment functions. The difference between teleservices and bearer services is depicted in Figure 13.1. Teleservices utilize the bearer services provided by the lower layers (except for circuit-switched speech services).

Teleservices and bearer services must be decoupled. It is not good to map bearer services and teleservices to each other, as changes to one component require changes to the other.

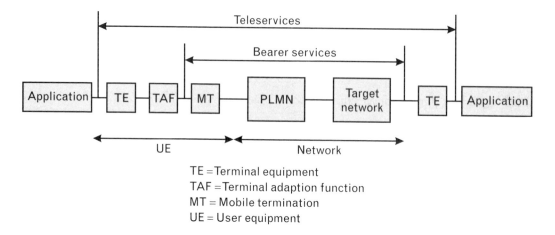

TE = Terminal equipment
TAF = Terminal adaption function
MT = Mobile termination
UE = User equipment

FIGURE 13.1   *Basic telecommunication services.*

Some teleservices need to be standardized so that they can interwork with corresponding teleservices provided by other networks. Others are used within only one network so that no internetwork standardization is needed. The following is a list of individual teleservices:

• Telephony;

• Emergency calls;

• Short message-mobile terminated/point-to-point;

• Short message mobile originated/point-to-point;

• Cell broadcast service;

• Alternate speech and facsimile group 3;

• Automatic facsimile group 3;

• Voice group call service;

• Voice broadcast service;

• Internet access.

This list is not exhaustive. New teleservices will certainly be developed in the coming years. On the other hand, not all of these services will be supported in phase 1 of UMTS. The GSM networks in 2000 typically supported only telephony (speech), emergency calls, and short message service (SMS).

13.3 Bearer Services

Bearer services are basic telecommunication services that offer the capability of the pure transmission of signals between access points. These services can be either circuit-switched or packet-switched. Bearer services concern only the three lowest layers of the OSI model. They are end-to-end transport services in which the user is responsible for the end-point entities.

A bearer service is defined using a set of characteristics that make it different from all the other bearer services. These service characteristics define such things as the traffic type, the traffic characteristics, and the supported bit rates. What makes UMTS so different from the other mobile telecommunication systems is that it allows for the negotiation of these parameters between the application and the network. Bearer services were fixed in 2G mobile networks. The selected bearer service was allocated when the connection was set up, and it then remained unchanged as long as the connection existed. There is a negotiation routine in UMTS in which the application requests a certain bearer service, and the network checks the available resources and then grants the requested service or suggests a lower level of service. The application at the user side either accepts or rejects the network's suggestion. It is also possible to renegotiate the properties of a bearer service during an active connection in UMTS. This property makes the UMTS bearer service much more flexible and allows the network resources to be much better utilized.

The network's transfer capabilities are characterized using the following variables:

- Connection-oriented/connectionless;

- Traffic type;

  - Constant bit rate (CBR);
  - Dynamically variable bit rate (VBR);
  - Real-time dynamically variable bit rate with a minimum guaranteed bit rate;

- Traffic characteristics;

  - Point-to-point (uni-/bidirectional, symmetric/asymmetric);
  - Point-to-multipoint (multicast/broadcast).

The information quality is characterized according to the following parameters:

- Maximum transfer delay;

• Delay variation;

• Bit error ratio (BER);

• Data rate.

All these parameters have an effect on the quality of the bearer service. The maximum transfer delay must clearly be quite small for certain applications, while for other applications (e.g., retrieval-type applications), it really doesn't matter much.

Delay variation (jitter) may be a problem, especially in packet-switched mode. This means that successive packets arrive at different time intervals. This can easily be fixed at the receiving end by using buffering, but on the other hand, buffering increases the overall delay. If the application has several separate media components (multimedia), then another type of problem may appear: skew. Skew represents the relative timing differences of the different media components. Consider a movie in which the picture and voice components should be synchronized to match each other. Multiplexing the components together in the transmitting entity is the most common way to prevent skew. If the synchronization is correct at the transmitting entity, it remains so during the transport because the components are transmitted together. Delay variation is discussed more in [1].

BER is the ratio between the incorrectly recovered bits and total number of transferred information bits. Some applications, like voice, are quite tolerant of high BER values. Typical data applications, however, often require low BERs. A poor BER value can be improved using error correction methods. There are two main methods: forward error correction (FEC) and automatic repeat request (ARQ). In the FEC method, the transmitting side adds redundancy to the data, so that the receiving side can reconstruct the original data even if a certain number of errors have been introduced to the data while in the transmission media. The amount of redundancy determines the amount of errors the receiver will be able to fix. In the ARQ method, the receiver uses error detection schemes to recognize corrupted data blocks, not correct them, and then requests the transmitting side to send these data blocks again. Both methods have advantages and disadvantages. FEC does not require a reverse channel and can provide constant delay, but the redundancy means a reduced information throughput. ARQ can provide high throughput while the channel quality is good, and in any case it can guarantee very low BER values. However, occasional retransmitted data means delay variations.

The data rate expresses the amount of transmitted data over a period of time. This is the total gross data rate, including all the control and management bits. Different applications have highly varying requirements for data rates. High-quality video requires hundreds of Kbps, while non-real-time applications can operate on very low data-rate connections. The maximum data rates UMTS can provide are 384 Kbps outdoors and 2 Mbps indoors in the first phase of UMTS. The latter rate may also be possible outdoors if the UE is very close to a base station and stationary.

Notice that these parameters usually have a great effect on each other. A certain parameter value can be improved easily while it may worsen the values of others at the same time. For example, the maximum transfer delay can be reduced if the BER values are not so important for an application.

13.4 Supplementary Services

A supplementary service (SS) complements and enhances bearer services and teleservices. They cannot exist without these basic services; that is, there is no such thing as a stand-alone supplementary service. An SS usually resides in the switch.

One SS may supplement several basic telecommunication services. Also, one basic telecommunication service may simultaneously use several SSs. The latter case requires that the interactions between active SSs be carefully specified.

An SS can be either general or prearranged. A general service is available to all subscribers without prior arrangements being made with the service provider. A prearranged service requires a subscription for the service with the service provider. Typically, a set of SSs is marketed as a service package, which can be subscribed to as a complete package.

Table 13.1 presents the currently defined SSs for 3G. Each entry also contains a reference to the corresponding technical specification (TS). Note that in telecommunication jargon, these SSs are most often referred to via rather complex acronyms, which are also given in this table.

A good starting point for SS studies is the general SS specification [2].

13.5 Service Capabilities

The three previously discussed service classes (bearer services, teleservices, and supplementary services; see Section 13.1) are all standardized, and their functionality is strictly specified. This means that no matter which operator provides them, they are always the same from the subscriber's perspective. This is, of course, a good property, but on the other hand it makes it difficult for an operator to provide unique services that can be used to differentiate the services offered by competing operators. A telecommunication service built with the above-listed standard service components will probably be very much like some other existing service. Operators have longed for a set

TABLE 13.1   SUPPLEMENTARY SERVICES

SUPPLEMENTARY SERVICE (OR SS GROUP + INDIVIDUAL SS)	ACRONYM	TS NUMBER
Call Deflection SS	CD	22.072
Number Identification SS		22.081
Calling Line Identification Presentation	CLIP	
Calling Line Identification Restriction	CLIR	
Connected Line Identification Presentation	CoLP	
Connected Line Identification Restriction	CoLR	
Call Offering SS		22.082
Call Forwarding Unconditional	CFU	
Call Forwarding on Mobile Subscriber Busy	CFB	
Call Forwarding on No Reply	CFNRy	
Call Forwarding on Mobile Subscriber Not Reachable	CFNRc	
Call Completion SS		22.083
Call Waiting	CW	
Call Hold	HOLD	
Multiparty SS		22.084
Multiparty Service	MPTY	
Comm. of Interest SS		22.085
Closed User Group	CUG	
User-to-User SS		22.087
User-to-user Signaling	UUS	
Charging SS		22.086
Advice of Charge Information	AoCI	
Advice of Charge Charging	AoCC	
Call Restriction SS		22.088
Barring of All Outgoing Calls	BAOC	
Barring of Outgoing International Calls	BOIC	
Barring of Outgoing International Calls Except Those Directed to the Home PLMN Country	BOIC-exHC	
Barring of All Incoming Calls	BAIC	
Barring of Incoming Calls when Roaming Outside the Home PLMN Country	BIC-Roam	
Enhanced Multilevel Precedence and Preemption	eMLPP	22.067
Call Transfer SS		22.091
Explicit Call Transfer	ECT	
Completion of Calls to Busy Subscribers	CCBS	22.093
Name Identification SS		22.096
Calling Name Presentation	CNAP	
Multicall	MC	22.135

of tools with which to build unique services, a means to distinguish themselves from the competition, and perhaps reduce churn.

Service capabilities are a set of building blocks that can be used to implement value-added services. As the value-added services themselves are not standardized, but only the building blocks, it is possible to implement them in a way that produces unique services. Unique services are more likely to attract and hold subscribers than constant price wars among operators with identical services.

Service capabilities are accessible to applications via a standardized application interface. They are provided by various toolkits and mechanisms, such as the SIM Application Toolkit (SAT) [3], the Mobile station Execution Environment (MExE) [4], the Customized Application for Mobile Network Enhanced Logic (CAMEL) [5], and the Intelligent Network (IN).

There are two types of service capability features:

1.   Framework service capability features;

2.   Nonframework service capability features.

Framework features are common utility features that are used by the nonframework features. They provide such things as authentication, authorization, registration, and notification services. Nonframework features are used by the applications as building blocks for value-added services. If possible, these features should be as generic as possible (i.e., not network specific), so that the applications using the features are easily portable.

The following is a list of nonframework features:

• Session control;

• Security/privacy;

• Address translation;

• Location;

• User status;

• Terminal capabilities;

• Messaging;

• Data download;

• User profile management;

• Charging.

13.6   QoS Classes

What makes UMTS so different from GSM and other 2G systems is the UE's ability to negotiate the QoS parameters for a radio bearer (RB). The negotiation procedure is always initiated by the application in the UE. It sends a request to the network defining the resources it needs. The network checks whether it can provide the requested resources. It can either grant the requested resources, offer a diminished set of resources, or reject the request altogether. The UE can then either accept or reject the modified offer. It is also possible to renegotiate these parameters while the connection is active if the requirements of the application change (UE-initiated renegotiation) or if the network resource status changes (NW-initiated renegotiation).

The parameters of an RB define the QoS offered for an application. In UMTS the QoS requirements can be divided into four classes:

1. Conversational real-time services;

2. Interactive services;

3. Streaming services;

4. Background services.

Table 13.2 depicts the different QoS classes and typical applications for each class.

13.6.1   Conversational Real-Time Services

The traffic in this conversational class is bidirectional and more or less symmetric. Examples of applications belonging to this class include voice, videophones, and interactive games.

The conversational real-time service class is technically the most challenging class. Because the services in this class are conversational, only a very short delay is acceptable (typically a few hundred milliseconds), and the delay variation must be negligible and hold relatively constant. The short delay requirement means that traditional retransmission protocols (ARQ)

TABLE 13.2   QoS CLASSES AND TYPICAL APPLICATIONS

QoS Class	Error-Tolerant Applications	Error-Intolerant Applications
Conversational	Voice, video	Interactive games, Telnet
Interactive	Voice messaging	Web browsing, ATM, e-mail server access
Streaming	Audio, video	FTP data transfer, still image
Background	Fax	E-mail (server to server)

cannot be easily used to fix transmission errors. Instead forward-error-correction (FEC) methods must be used. FEC increases the amount of raw data transmitted; it adds redundancy to the transmitted data, which the receiving end uses to remove transmission errors. The delay variation must be kept small during the entire transfer. The small delay requirement also means that buffers cannot be used on the receiving end to smooth the variations in delay.

If the transmitted data are either audio or video information, then some errors are acceptable on the receiving end, as these do not present a notable loss of the QoS in real-time services. A human user cannot sense small errors in voice or video information, but people are very sensitive to excessive delays in speech services. The transfer of other types of data makes the situation worse, as any errors in the received data will probably cause problems for the application using the data.

13.6.2 Interactive Services

The interactive class includes services in which a user requests data from a remote server, and the response contains the requested data. Examples of these services are Web browsing, e-shopping, and database inquiries.

The difference between conversational and interactive classes is that the data traffic in the conversational class is symmetric or nearly symmetric, whereas in the interactive class, the data traffic is highly asymmetric: one direction is used for payload data traffic, and the other mostly for control commands (requests and data acknowledgments). Moreover, the timing requirements are not quite so strict with interactive services as they are for conversational services. Conversational services allow for a maximum of a few hundred milliseconds of delay, but interactive services may in some cases tolerate a few seconds of delay. The upper limit for the delay is found to be around four seconds. Anything more than four seconds will probably start to annoy the user, even though the applications that use the data are not compromised. Interactive services do not tolerate any more transmission errors than conversational services. However, this goal is easier to achieve with interactive services, as the looser delay requirements make it possible to use more efficient error protection and correction methods in data transmission.

Delay variation is not really a problem with interactive services. The data are typically presented to the user only after all of the data has been received. If an application is sensitive to delay variations, this can be achieved with buffering on the receiving end as long as the maximum delay remains below the specified threshold.

The boundaries between the interactive services and the conversational services may sometimes be blurred. At one extreme, some applications may require a considerable amount of uplink data transfer at times. Because we can generally assume that the main data transfers in interactive services occur in the downlink, the kinds of services that seem more symmetric than most may actually belong to the conversational class at times. On the other hand, an interactive service with very few uplink transfer actions may actually be close to a streaming class service.

13.6.3 Streaming Services

This class of services, the streaming services, typically includes video and audio applications directed to a human user. The differences between the conversational and streaming classes can be demonstrated by considering the difference between talking on the phone (conversational) and listening to music on a CD (streaming). What makes streaming class different from the interactive class is that the data transfer in the streaming class is almost totally one-way and continuous: highly asymmetric. The receiving end does not have to receive the whole file before it can start presenting it to the user. There are some strict delay variation requirements for the data, which are presented to the user. If the service is a multimedia service, then there must not be any skew between the multimedia components. However, the requirements for maximum delay are rather slack, a good guideline value could be as long as 10 seconds.

The only data traffic in the opposite direction (usually in the uplink) consists of a few control signals, like starting and stopping the data stream. The lack of interaction between the communicating entities makes it possible to allow for long transfer delays. This again makes it easier to provide the application with a data stream with small delay variations between data entities. This can be handled with a reception buffer. The incoming data packets are buffered, and the packets are released from the other end of the buffer at a constant rate. This method can fix the delay variations up to the size of the buffer. It is not possible to fix a 6-second delay variation between successive packets with a buffer with room for only 5 seconds' worth of packets at a time.

The easiest way to prevent skew is to multiplex the multimedia components into one stream in the transmitting entity. This ensures that there will not be any delay variation between the components of the multimedia presentation. The transmission media can introduce delay variation between individual packets, but this can be fixed in the reception buffer as explained earlier.

If the streaming application is an audio or video application, then some errors may be permitted in the final data presented to the user. However, with other kinds of data, such as still images or telemetry, few errors are allowed.

What makes this class so attractive to the 3G system is that it is possible to provide the streaming class services via packet-switched networks. The small delay variations can be compensated for with reception buffers. Many 3G applications, such as video-on-demand and audio-on-demand, will be streaming services.

13.6.4   Background Services

The background class consists of those services that do not have precise delay requirements at all. The applications generally expect to receive the data within a defined time, or the time limit is quite high. However, it may use timers to make sure that the data transfer has not stalled altogether. Typical applications using background services are fax and SMS.

Because there are no delay requirements, the data transfer can be handled as a background activity. The data has to be error free, but this is especially easy to achieve in this case. Because there are no time constraints, the application can use retransmission protocols to ensure error-free data. However, the error detection algorithm must also be efficient. A prerequisite for an efficient error correction scheme is that an error is, in fact, detected first. The retransmission protocol will not ask for packet retransmission if it does not know that the packet was erroneous in the first place.

Delay variation is not considered with background services. The data are presented to the user only after the whole file has been received correctly.

The data transfers in the background class services are very asymmetric. The payload data is transferred in one direction only; the other direction may contain only a small amount of control signaling. The bandwidth requirement is not large in either direction, as the background character of the service makes it possible to transfer the files over a very slow connection.

13.6.5   QoS Service Classes and 3G Radio Interface

As one can see from the previous sections, most service classes have very asymmetric data transfer requirements. Only the conversational class of applications have symmetric or nearly symmetric bandwidth requirements. The other three classes (interactive, streaming, and background classes) need very asymmetric connections. When this is mapped into the radio interface, the three asymmetric service classes use mostly downlink resources, and very little uplink bandwidth. This asymmetric nature of the connections will be a concern for the UTRAN FDD mode, which has symmetric and fixed bandwidth allocations for both uplink and downlink. Asymmetric connections as such are easy to implement in the UTRAN FDD mode air interface; the network can assign highly varying spreading factors for the uplink and downlink if this is required. The danger is that UTRAN networks using FDD mode will run out of downlink capacity, while still having plenty of unused uplink bandwidth.

It is worth remembering that network planning studies have shown that a typical FDD cell capacity is uplink limited (i.e., a typical FDD cell can have slightly more downlink traffic than uplink traffic). This is because the capacity in a typical WCDMA system is not limited by the number of spreading codes, but by the interference (both inter- and intracell interference) caused by other users and base stations. Interference in the uplink is generally more severe because the uplink transmissions are not synchronous; thus, the codes are not orthogonal. However, the capacity difference between uplink and downlink in an FDD cell is not very significant. So it does not provide a robust solution to this asymmetric traffic problem.

One solution to the potential capacity problem is the TDD mode, in which the uplink and downlink capacity can be allocated dynamically. Each 10-ms radio frame in the TDD mode is divided into 15 timeslots, and these timeslots can be allocated dynamically as uplink and downlink slots, as long as there is at least one slot in each direction. This mode makes it possible to allocate, for example, 4 slots to the uplink and 11 to the downlink, thus providing much more capacity in the downlink direction. However, the TDD mode has severe interference problems in large- and medium-sized cells because of the intercell interference. Generally, TDD is best used indoors. Fortunately, the highly asymmetric applications are probably more attractive to slow moving indoor users; surfing the Internet while driving a car is, certainly, a "killer" application. Service classes in 3G are discussed in [6] and [7].

REFERENCES

[1]   Roberts, J., U. Mocci, and J. Virtamo, "*Broadband Network Teletraffic,*" COST 242 report, Berlin: Springer-Verlag, 1996, pp. 63–83.

[2]   3GPP TS 22.004, v 4.1.0, General on Supplementary Services, 2002.

[3]   3GPP TS 22.038, v 5.2.0, USIM/SIM Application Toolkit (USAT/SAT); Service Description, Stage 1, 2002.

[4]   3GPP TS 22.057, v 5.3.1, Mobile Execution Environment (MExE); Service Description, Stage 1, 2002.

[5]   3GPP TS 22.078, v 5.6.0, Customised Applications for Mobile Network Enhanced Logic (CAMEL); Service Description, Stage 1, 2002.

[6]   3GPP TS 22.105, v 5.1.0, Service Aspects; Services and Service Capabilities, 2002.

[7]   Holma, H., and A. Toskala (eds.), *WCDMA for UMTS: Radio Access For Third Generation Mobile Communications,* New York: Wiley, 2000, pp. 9–23.

CHAPTER 14

3G Applications

14.1 Justification for 3G

Why do we need 3G systems? Can they provide services that are impossible to deliver via 2G systems? In this section we will question the justification for 3G.

One much-quoted justification of 3G is the lack of available channels in current 2G systems; especially in urban areas, the capacity limit of the existing 2G systems may be reached in the near future. During peak hours the operators may soon be unable to provide service for everybody in traffic hot spots. But the lack of capacity in 2G systems alone doesn't make the building of 3G systems necessary. Current 2G usage is still mostly voice, and this traffic could be handled quite easily by just adding extra frequency allocations to 2G networks (e.g., from the UMTS spectrum). One GSM frequency carrier takes 200-kHz bandwidth and can accommodate eight traffic channels. Thus, one WCDMA 5-MHz frequency carrier could accommodate 25 GSM frequency carriers, translating into 200 GSM traffic channels. The true and practical number of traffic channels is lower, as control channels use part of the capacity, and also in TDMA systems the same frequency cannot be reused in nearby base stations. Still, a 5-MHz frequency slice would offer lots of new capacity for a GSM system.

The higher raw maximum data rate is not a good reason either. It is true that theoretically UMTS can provide approximately 2 Mbps in optimal conditions. But in practice 384 Kbps will be the maximum rate a 3G user can expect to receive in the first phase of UMTS. A 2.5G user can get the same rate (at least in theory) from an EDGE+GRPS network, although an approximately 200-Kbps maximum date rate may be closer to the truth. And note that our 2.5G user only needed one 200-kHz frequency band for this. The 3G connection used 5 MHz. Moreover, 200 Kbps is enough for most applications planned for 3G at the moment.

But, we can see that the character of mobile communications will evolve considerably in the coming years, and data in their many forms will become more and more important. Voice will remain an important component in telecommunications, but quite often it will be combined with other types of information to form multimedia applications. The current 2G networks were designed to transfer voice traffic only; real-time multimedia can be transferred via the GSM phase2+ air interface only with great difficulty.

GSM was designed at the end of the 1980s. The needs and expectations of the telecommunication world then were totally different from those of today. Although the GSM system has evolved over the years very success-fully to meet the new incremental demands, it has some problems that make it difficult to use for the emerging 3G applications. Probably the biggest problem is the relatively inflexible air interface. An application that gener-ates bursty traffic with several multimedia components cannot be effectively handled in GSM. The UTRAN can, however, allocate resources dynami-cally according to the instantaneous needs of the application. The resource usage can be readjusted for each radio frame (10 ms). This is not possible in GSM, although GPRS improves the capabilities of GSM a bit and EDGE offers even more flexibility.

It is the great flexibility of the UTRAN air interface that makes the dif-ference. The interface can handle highly variable bit rates. The same con-nection can transfer services with different quality requirements. Speech, video, and other forms of data can be multiplexed into multimedia services, and this mix can be transferred over the UTRAN. The offered bandwidth can be dynamically allocated as needed. Unused resources can be promptly deallocated such that the result is a highly spectrum-efficient system. Thus, even though GSM can provide roughly the same services as UMTS, UMTS can provide them more efficiently and economically.

However, GSM technology is surely not at the end of its life. The planned enhancements to GSM, such as HSCSD, GRPS, and EDGE, will improve its capabilities considerably, and the result, 2.5G, will be quite capa-ble of handling many of the same applications that 3G is designed to handle. GSM operators could also integrate some other radio access technologies into their networks to boost data rates; for example, WLAN systems can pro-vide bit rates up to several tens of Mbps in hot spots, which is much more than the standard UMTS can provide. So, given the competition from the 2.5G GSM networks, 3G systems will not be given an easy ride.

Some say that 2.5G is good enough in an era of content-limited net-works; true 3G content is too far off to justify 3G networks yet. Others say we have to deploy 2.5G in order to build up a nonvoice business that can support 3G investments. And still others say that since the network must be deployed before the new content can fill it, we need to jump to 3G as soon as possible to make sure the new content comes soon enough to attract the new revenues. Many suggest that 2.5G and 3G are the same in their pre-sumed ability to attract new revenues from existing subscribers. 2.5/3G is about abandoning the coverage-limited subscription-driven market in favor of a content-driven market. The more saturated a market, the more impor-tant 3G.

14.2   Path into the Market

There are many possible paths to deploy 3G systems. Some of them are discussed in this section.

In the first scenario, we look at the new 3G operator who already has some 2G network infrastructure. In this case, especially if there is no competition from aggressive pure 3G-only operators, the operator does not have to rush to develop and provide new 3G services. It gets revenue from the existing 2G network and can build the 3G network relatively slowly, first in hot spots, such as city centers, airports, and main roads. The services it will provide are initially, for the most part, the same services the operator is providing via its 2G network, possibly with slightly enhanced capabilities. This type of operator can afford to wait until 3G technologies and applications are ready; there is no need to deploy prototypes. 3G deployment can also be used to ease the congestion in the operator's 2G network. This kind of slow 3G deployment requires that users buy dual-mode mobiles.

In the second scenario, we look at the case where 3G is used to provide new and different services to customers right from the beginning. The 3G operator does not have a 2G network; it is a so called green-field operator. It does not get revenue from existing 2G networks, and, in fact, in some markets it has to pay heavy licence-fee installments. Therefore, new UMTS services will be provided as early as possible to get at least some positive cash flow. A green-field operator has to lure customers from the existing 2G and 2.5G networks. Thus, the new services have to be such that it is not possible to provide them via 2G networks with the same QoS. The operator has to find a "killer" service, or preferably a suite of killer services, which makes the use of this network especially attractive. Failing this, the pricing model of green-field operator services must be aggressive. The 3G network will be built as fast as possible given the commercial realities.

The life of a new 3G green-field operator will be difficult. It has to build a very expensive network without any revenue flow for a very long time. Moreover, some countries have granted the 3G licenses via bidding contests and the resulting enormous price tags are an additional burden for these operators. Luring customers to a new 3G network will be difficult, as existing operators can provide much wider service coverage with their 2G networks than a new operator can possibly provide with its 3G technology. This problem could probably be solved by regulatory means. Existing operators could open their 2G networks to the new operators in this kind of case. The 3G operator could become an MVNO in 2G. However, this generates new problems, such as security and pricing. Who sets the price tag for the resources used by the new operator? The free-market approach may not provide a solution, as there is very little competition in this market. There are probably only two to four existing GSM operators per country, and they are all probably very reluctant to offer their network capacity to their future competitors. On the other hand, prices set by telecommunication regulating agencies may not be a better solution. The wrong price tag may distort the market situation.

On the other hand, to reduce costs many 3G operators may end up building common radio access networks. There can be several levels in this sharing, as the operators may share the base station site, base station equipment, transmission medium, or the whole radio access network. However, note that in some markets, the licence regulations may prohibit sharing. If there is a common radio access network, then obviously these operators cannot use coverage as a competition tool.

14.3   Applications As Competition Tools

Applications and services will be an important component of competition between the operators. Pricing, coverage, and other issues are, of course, also important, but in the end it is the applications and services the consumers are looking for that count. Operators may not produce all these services by themselves (history suggests the new applications will almost certainly not come from the wireless sector), but they will probably provide services produced by third parties, and possibly handling the billing on behalf of these new third-party content providers. For these external content providers, 3G will open totally new opportunities. It has been very difficult to design new services for the existing 2G networks because of the inflexibility of the air interface and lower data rates. Note that the external wireless application content providers only provide the wireless application. The network operator is responsible for the technical issues related to the data-transmission needs of the application. The network operator allocates and manages the required bandwidth according to requested QoS parameters. A new mechanism is needed for the bandwidth negotiation between the operator and the third-party content provider. Only the operator knows what kind of bandwidth is available in each cell. On the other hand, only the content provider knows the bandwidth demand of a particular service and what kind of QoS can be provided with the given bandwidth. The bandwidth supply and demand must be made to meet each other with the help of some kind of brokering service.

Billing is also an important task for the operator. A successful service has to make money. An unprofitable service cannot survive. So far, the problem with many third-party content providers has been that there is no sensible method to invoice customers. A reasonable, below-the-pain-threshold charge is often so small that the charges from the invoicing will exceed it. Many low-value applications and services would make a nice profit, if there were a method to charge the customer a penny for each usage. One million users would generate one million pennies, and that is a lot of money. This kind of small-payment scenario is called a micropayment. Operators have to provide means to support micropayments for their content providers because otherwise these content providers cannot survive, and without content providers, a 3G operator will fail too. It will become like a toll road without traffic. Micropayments are discussed further in Section 14.4.6.

Nobody knows for sure what kind of applications will succeed in 3G. Lots of guesses and studies looking into the future have been made; a guess is called an analysis if it has been made by a research group. Quite often these predictions fail (in [1], Webb et al. suggest reasons why predictions fail). A successful analysis of an issue requires the study of its history, but in mobile telecommunications, the history is a very short one and cannot adequately provide hints on what is going to happen in the future. For example, a few years ago SMS was seen as a very clumsy way to send what was perceived as too-short text notes to other people. The service was seen as a silly and a hopeless case. Now tens of billions of SMS messages are sent every month, and a large part of other mobile applications are based on SMS technology (e.g., voice mail notifications and weather alerts).

One way to succeed in the 3G applications market is to start early and learn the applications business in 2.5G GSM (GSM enhanced with GPRS and EDGE). We can also examine highly content-driven systems like the i-mode offering in Japan, which is carried to subscribers on the relatively modest PDC system. This is a suggestion—not proof—that it is the content that matters, not the network technology. These 2.5G systems can provide higher data rates than the basic 2G systems, providing new opportunities for application designers. Indeed, it is quite attractive to start by designing applications to 2.5G GSM, as the core network is the same as in the UTRAN case; thus, the same applications can later be provided via the UTRA network. The problem is, as explained earlier, how to arrange the billing of customers.

It will be difficult for the wireless sector to invent and develop killer applications. Most development projects will result in failure. But it is worth trying, as the real money in the 3G telecommunication business will be in the applications. It has been estimated that the amount of money in the mobile services is already around 10 times the amount in the cellular infrastructure business [2].

14.4 Application Technologies

Application technologies are not and will not be specified by the 3GPP. There are, however, several emerging technologies that will be used to provide common platforms for mobile-telecommunication application development. They are typically promoted by organizations formed by companies in the telecommunications industry.

14.4.1   Wireless Application Protocol

The wireless application protocol (WAP) is an HTTP language derivative especially designed for the wireless environment. It will probably be the most widely used protocol in the first phase of UMTS, as it is already used in GSM networks. This head start will gain more momentum once GPRS networks are launched with their faster data capabilities. In plain GSM of today, the WAP applications clearly suffer from the slow data rates. Note that both WAP 2.0 and i-mode will use the same XML language, extensible HTML (XHTML). WAP is promoted by the WAP Forum (www.wapforum.org).

14.4.2   Java

Java is already known from the Internet. It is, in principle, a platform-independent protocol. The user terminals in 3G will contain a wide variety of different operating systems, which will be a problem for application software developers if traditional languages are used. Java provides the programmer with a way to get rid of platform dependency. Java applications can be retrieved over the mobile network and executed in the UE. However, the operating system of the UE must include the Java Virtual Machine to be able to execute Java code. Java is developed by Sun Microsystems, which owns the trademark.

14.4.3   BREW

Binary Runtime Environment for Wireless (BREW) is a platform-independent application execution environment for wireless devices. BREW is developed by Qualcomm. The BREW platform is based on C/C++, which is one of the most popular programming languages, so it is easy for application developers to adopt BREW. The idea behind BREW is very similar to Java, except that BREW is specifically developed for the wireless world, and that there is no need for platform-specific solutions in BREW, such as Java Virtual Machine. BREW is requires little memory on the phone, which makes it more easily ported to low-end mass-market phones. The BREW platform is free for both handset manufacturers and application developers. The business model is based on the licensing fees from network operators.

14.4.4 Bluetooth

Bluetooth is a protocol for short-range wireless links. It is not designed only for mobile telecommunication applications; a Bluetooth link can connect several electrical appliances together. A typical Bluetooth usage environment will be the peripheral devices of a PC. Most PC users surely want to get rid of the jungle of wires around their PCs, and Bluetooth will provide an affordable method to do this. In 3G Bluetooth links can be used to connect the UE to various devices, such as headsets, printers, and control devices. Bluetooth is promoted especially by Ericsson; indeed Bluetooth is a trademark owned by Ericsson.

14.4.5 I-mode

I-mode is an NTT DoCoMo proprietary data service technology used in 2G PDC (see Section 1.1.2) networks in Japan. I-mode is very much like what WAP + GPRS will be in the future. It is based on a packet data network (PDC-P), where the usage is charged based on the amount of data retrieved, and not on the amount of time spent in the network. I-mode is good for accessing Internet Web pages, as the applicable pages can be implemented using a language based on the standard HTML (compact HTML). I-mode has been a phenomenal success in Japan, and proves that there is a real market for mobile Internet services and mobile e-commerce. In June 2002 i-mode had 33 million subscribers, and the number is still growing rapidly. Operators all over the world are following the i-mode service very closely, as it provides them with examples of what kinds of applications could succeed in their own future 3G networks. However, any conclusions should be drawn with caution because in some aspects Japan is quite different from other market areas. The reasons i-mode is popular in Japan may not be the same everywhere. It is good to remember that i-mode data transmission speed is only 9.6 Kbps, so one does not need super-fast data connections for successful mobile services. Good services need good content.

14.4.6 Electronic Payment

Electronic money is an important enabler for e-commerce. A service or application has to make money. Even the most technically advanced, exciting, and addicting service will fail if it cannot generate revenue. So far the big problem with paying for services in mobile networks has been that there is no convenient method to make very small payments. Most services will be such that a user will not consume it unless it costs a very small sum of money. The business model of such a content provider is based on a large number of customers. The cost of one service usage must be so small that the user does not regard it as a real cost. If a morning cartoon delivered to your handset costs 1 penny, then you do not think twice about subscribing to the service ... if you happen to like cartoons.

What we need here are micropayments. These are very small payments, even fractions of the smallest local monetary unit. Currently, e-commerce is typically paid for by giving one's credit-card number to the content provider. This is not the way to handle micropayments, as the cost of billing is much higher than the cost of the purchased service or goods. Moreover, this is a rather nonsecure way to pay for things.

Because there is a lot of money in the payment business (pun intended), there are also lots of interested parties developing payment solutions. This section presents a few competing initiatives that aim to develop an electronic-payment solution for mobile commerce.

Radicchio [3] was the first to be launched, by Sonera SmartTrust, Gemplus, and EDS. The basic idea behind this scheme is that the SIM card inside a mobile terminal also functions as a credit card and the credit-card reader is the mobile terminal itself. A mobile terminal could include two SIM cards, one issued by the mobile phone operator to be used for traditional telecommunication service usage, and another issued by a bank or credit-card company to be used for e-commerce. The security of transactions would be guaranteed with the use of Public Key Infrastructure (PKI).

Ericsson, Motorola, and Nokia started a joint effort to develop a common framework for mobile e-business. Siemens has also joined this initiative, which is known as MeT (Mobile Electronic Transactions) [4]. The aim here is to develop an industry standard for mobile e-commerce that would make it easy and secure to use these services. Also, this scheme is based on a "removable security element"; that is, a SIM card or a SIM-like card to provide tamper-resistant security.

The Mobey Forum [5] is a consortium of several large banks (founding members are ABN AMRO Bank, Banco Santander Central Hispano, BNP Paribas, Deutsche Bank, HSBC Holdings, Nordea, and UBS; several other banks will have joined the consortium after this writing) and Nokia, Ericsson, and Siemens. The Mobey Forum concept includes a dual-SIM mobile in a similar way as the other initiatives. Of course, in Mobey Forum's model the secure payment card would be issued by a bank.

The youngest of these initiatives is the Mobile Payment Forum [6]. Its membership list includes credit-card companies and large operators among others. They intend to "complement the work that is already being done by other industry initiatives and organisations such as Mobey and MeT." Most probably they also want to make sure that the interests of credit-card companies and mobile operators are not forgotten.

Note that in addition to the previously discussed alliances dealing with security, there are other ones, such as the WAP Forum [7], the PKI Forum [8], and PayCircle [9]. The above-mentioned initiatives are not all necessarily competing ones; the ideas they promote can be combined. And the same company can be a member of several initiatives. It would be in the best interests of 3G m-commerce if the 3G community could agree on only one payment scheme. This is not a difficult task technologywise, but various commercial interests and patents will make it one. The members of these initiatives can be classified into five interest groups: cellular infrastructure manufacturers, banks, operators, credit-card companies, and payment technology developers.

Moreover, some operators would certainly like to promote a model in which they are the billing party for all purchases made via their network. Whatever services and goods are purchased, these are shown in the monthly mobile phone bill. This scheme would have certain advantages for the operator. First of all, the operator already has an efficient billing system that could be easily used for other services than pure telecommunication services. The investments required would be modest. Secondly, the operator could certainly charge a nice premium by doing the billing on behalf of other content providers. Thirdly, if operators handle the billing, they can accumulate important information about their users, and build an accurate user profile. This can be exploited in targeted advertising as explained in Section 14.7.

But the drawbacks in this are also substantial. For the 3G business to really take off, we need lots of content providers. If all 3G business is controlled and run by operators' portals, then all third-party service and content providers have to make contracts with all the operators they want to do business with. For a content provider aiming to offer its services globally, this would mean hundreds of contracts. It is not feasible for all content providers of the Internet to register with all operators and arrange their relationships via direct contracts. Most probably operators would not even want to offer contracts to all content providers of the Internet, as there will be millions of them. Thus, this kind of payment system would result in a closed club of content providers, the membership of which would be limited to those who can be accessed via the operator's portal. The problem with this is that most content providers in the Internet would be left out of this scheme. They would adopt an alternative payment scheme, and that would then become the mainstream payment method, leaving 3G operators to play with a secondary solution.

Prepaid customers are also problematic for operators. The identity of these customers is not known. It is difficult to build a user profile for prepaid customers, especially if the customer is only an occasional user or keeps on swapping between networks, depending on the cheapest air-time offer. It is impossible to provide efficient targeted advertising for this kind of customer. In addition, nothing is known about the credit history of a prepaid user, so he can only be sold services up to the credit in the prepaid card.

It is understandable that operators do not want to become just "bit-pipe" providers, but that they also want to participate in the value-added service business. However, this should not result in selecting their own "proprietary" payment solutions, as in the long run it would bring them more harm than good. A thriving 3G business would bring lots of traffic to 3G networks. This is not to say that operators could not be content providers as well. In fact, because they own the network, they have an advantage over other content providers in that it enables them to provide certain services better than others, for example certain location-based services.

A secure electronic method to identify someone allows all kinds of new applications. A secure electronic identification module could also function as an electronic identification document (i.e., an electronic passport). It could be used to identify the owner of the mobile terminal to a content provider if required. But it could also be used instead of paper identification documents in the nondigital environment. This kind of usage would mean that the identification module must be absolutely secure and tamperproof. It is a serious crime if credit-card numbers are stolen, as then someone's money can be taken. But cracking someone's identity module means stealing his electronic identity, and this is potentially a far more serious situation.

Note that these e-payment schemes make it possible to use the electronic payment module to also pay for other non-e-commerce purchases. It should be possible to use this module in the same way we use our credit cards today, except that this module is more secure and capable than a credit card. Furthermore, for example, your gas bill could be sent to your mobile device in electronic format, and you could pay it instantly (e.g., by keying in your payment-card PIN number).

These schemes would make it possible to prove the identity of the mobile terminal user, but it must also be possible for the user to choose not to reveal his identity. It is not necessary for every supermarket cashier to know the identity of a customer, if the customer can prove with his electronic payment module he has the necessary funds to pay for his purchases. There is more about mobile e-commerce in Section 14.7.

14.4.7   IPv6

At the moment, the Internet does not provide very good tools for the management of mobile devices. The currently deployed Internet Protocol version 4 (IPv4) has shortcomings for mobility [10]. It cannot easily do the following:

- Provide forwarding addresses to mobile devices reattaching to the network;

- Provide good authentication facilities, which are required to inform the routing infrastructure about the new location of the mobile;

- Enable mobiles to determine whether the new network to which they have reattached is the same as the old network;

- Enable mobiles to inform their communication partners about the changed location.

These problems can all be fixed with the new version called IPv6.

Moreover, the Internet is about to run out of network addresses. IPv4 uses 32 bits of address space, but this space is allocated rather poorly, and some countries already have real difficulties with their small allocations. IPv6 uses 132 bits of address space, which provides an extremely large number of unique addresses. These will be badly needed as shortly all mobile-telecommunication devices, such as mobile phones, will need their own IP addresses. And it is not only the traditional telecommunication devices that are after these addresses. Practically every electrical appliance could be equipped with a communication chip that could connect the appliance to the Internet. The Internet-capable fridge is not yet in shops, but who knows what kind of applications we will see in 10 years? From the point of view of the mobile communication industry, the lack of free addresses can also be seen as a good thing because it forces the Internet to adopt the new improved version of the protocol, which also addresses the problems with mobility. Without this element, the incentive to implement the changes in the Internet would be much smaller.

The goal of this development will be the all-IP network, where IP protocols are used in every node from the UE to the Internet server. A Nokia White Paper [11] separates three stages in the evolution toward the all-IP network (see Figure 14.1):

1.   GPRS core network connected to the Internet;

2.   3G RAN + core network connected to the Internet;

3.   All-IP network from the UE connected to the Internet proper.

The all-IP network means that all user data, including voice, will be sent in IP data packets over the radio interface.

A nice feature in IPv6 is that it can coexist with IPv4. The service provider can upgrade its infrastructure in several phases to comply with IPv6. End-to-end IPv6 can be provided through IPv4 networks because IPv6 packets can be encapsulated into the payload of IPv4 packets.

An IPv6 data packet will have a sizable header, which is not a problem in fixed networks, but the bandwidth in the radio interface is a scarce resource; thus, sending IPv6 packets as such over the radio interface is not a very attractive solution. It can be circumvented by means of header compression. In the Uu (radio) interface protocol stack, the header compression

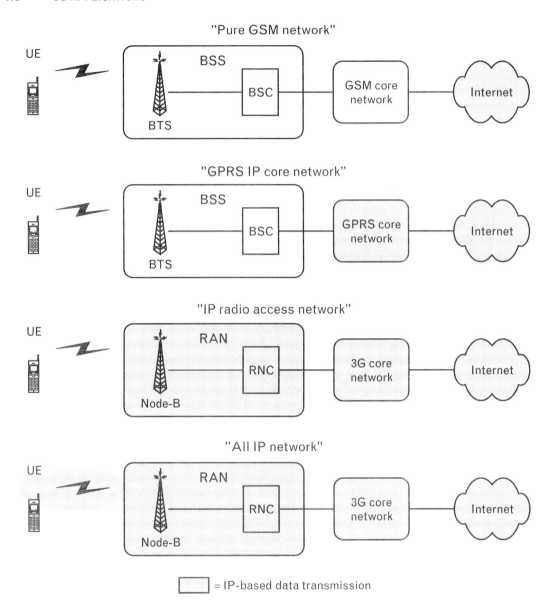

FIGURE 14.1   *Evolution toward the all-IP network.*

is a function of the packet data convergence protocol (PDCP) (see Section 7.10), which is positioned just below the data protocols (like IPv6). The Internet Protocol header compression is defined in [12].

IPv6 is promoted by the IPv6 Forum, a consortium of network vendors and content providers (www.ipv6forum.com). For more information about IPv6 and mobile Internet, consult [10, 11] and [13, 14]. The architecture of an all-IP network is discussed in [15].

14.5 Multimedia

14.5.1 Application Types

A multimedia service is a composite service that consists of several media components, such as speech, video, still images, and music. New components may be added to the mix during the connection and old ones removed.

Many of the multimedia applications to be used in 3G are already in use as single-media applications in 2G. For example, news may be delivered to terminals as SMS messages in 2G, but in 3G the same news service may include voice news accompanied with video clips or still images from the most interesting pieces of news.

Multimedia applications can be interactive or distributional. Interactive multimedia applications include some feedback from the user. The nature of this feedback determines whether the interactive application is a conversational, messaging, or retrieval service. Distributional applications do not require any feedback from the user, but they can be controlled by the user; for example, the user may have subscribed to certain distributional applications and receives only those services.

14.5.2 Technical Problems

14.5.2.1 Error Accumulation in Compressed Video

A multimedia presentation requires a large amount of data to be transmitted over the network. In principle, a displayed video clip has a data requirement:

$$\text{Data\_rate} = \text{Number\_of\_pixels} \times \text{Number\_of\_color\_info\_bits} \times \text{Refresh\_rate} \tag{14.1}$$

However, a typical video picture contains plenty of redundant information, both spatial and temporal. Spatial redundancy means that a video picture probably contains lots of adjacent similar pixels. Temporal redundancy instead means that a certain pixel remains unmodified as time passes.

Both kinds of redundancy can be exploited by a suitable coding scheme to reduce considerably the number of bits that have to be transferred. If a video picture contains a blue background, it is not necessary to send this color information separately for each pixel in the picture. And if some part of the picture remains static over several video frames (which is quite common in video clips), it is not necessary to repeat the same information all over again in each frame for these pixels.

The problem with highly optimized (compressed) video coding algorithms is that they are very sensitive to transmission errors in the compressed data. Each bit in the compressed data is probably significant, and an error can accumulate in both a spatial and a temporal sense. For example, if only the changes to the previous picture are coded and transmitted, then an error in the presentation will remain until the pixel(s) in question are completely drawn again. See [16] for the principles of MPEG video encoding.

A solution to this problem is to send the full picture information periodically in every *n*th frame (intraframes, or I-frames), and otherwise send only the changes to the previous frame (P-frames). The I-frame does not use any information from the previous frames, so it removes the accumulated errors. The faults in the video picture generated by the errors in P-frames can be seen only until the next I-frame. Of course, an error in the I-frame can be seen all the time until the next I-frame. See Figure 14.2.

Sometimes B-frames are also used. These are bidirectional prediction frames, which can take their motion compensation from either the previous or the next I- or P-frame, or from both of them. The B-frames have the lowest bandwidth requirement of all frame types.

14.5.2.2   Multimedia Synchronization

Two sorts of synchronization need to be handled in a typical multimedia presentation, intra- and intermedia synchronization. A multimedia presentation contains several (at least two) media components. If a component is tim-dependent, like a video or audio component, then it has to be synchronized to remove jitter (see Figure 14.3). The timing difference between successive data components has to be constant. This is known as intramedia synchronization.

Jitter is easy to remove if one component does not have to be synchronized with other components. The receiving entity can have a large enough reception buffer, which stores the incoming data packets and then releases

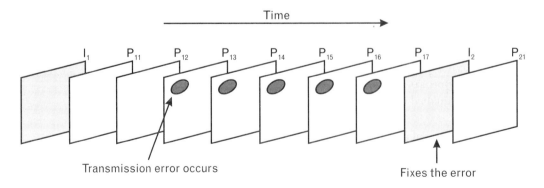

FIGURE 14.2   *I- and P-frames in video coding.*

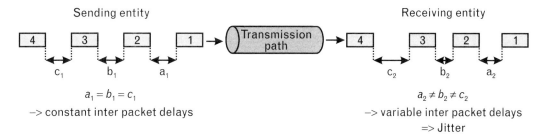

FIGURE 14.3 *Jitter.*

them at a constant rate (i.e., a "leaky bucket") (see Figure 14.4). The ability to fix jitter depends on the size of the buffer. The bigger the buffer, the larger the jitter that can be corrected, but also the longer the resulting delay. If the buffer becomes empty, then a new data packet cannot be released, and the interpacket delay grows longer than usual. If the buffer overflows, then packets are lost.

Jitter can be prevented in circuit-switched transmission networks by using synchronous transmission modes. However, in packet-switched networks packet-transmission delays and packet-switching delays may not be constant; thus, jitter is easily generated.

Intermedia synchronization involves maintaining the relative timing dependencies between different media components transmitted via parallel channels. The speech and sound of a multimedia presentation should match the actions in the video component. The timing difference between different media components is called skew (see Figure 14.5). Skew is caused by the different relative transmission delays of the parallel channels.

There are two basic methods to remove skew: production-level and presentation-level synchronization. In production-level synchronization, the different media components are interleaved and multiplexed after they have been produced and synchronized. The multiplexed compound data stream is then transmitted via a single traffic channel. Because only one channel and data stream is used, skew is impossible. Jitter is still possible in

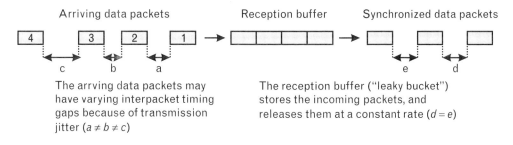

FIGURE 14.4 *Leaky bucket.*

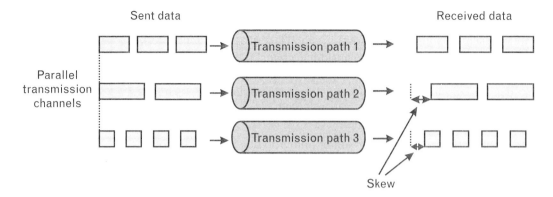

FIGURE 14.5   *Skew.*

the channel, but this can be handled with the methods described earlier. ITU H.320/H.324 and MPEG-based techniques all use this method.

In presentation-level synchronization, the media components are synchronized only in the receiving entity just before presentation. This requires some additional information to be transferred with the data packets so that this task can be carried out. It can be information about the generated delays during the transmission, or it can be timestamps. Moreover, this method requires more synchronization functionality from the underlying system because it has to manage the synchronization of the component traffic channels. Multimedia synchronization is discussed in [17].

14.5.2.3   Interworking

Most multimedia applications in 3G will not be internal only to that system. Fixed networks will be an important platform for multimedia development; they will typically have more transmission bandwidth, and fixed network users will also have better presentation facilities.

Therefore, many multimedia applications will be used over several different telecommunication networks. This may cause problems because interworking between various networks can generate additional delays in transmissions. Interworking includes the conversion from the protocols used in one system to the protocols used in another system.

14.6   Traffic Characteristics of 3G Applications

The new 3G applications will have diverse requirements for data traffic. These services can be roughly classified as real-time and non-real-time services.

Interactive applications require real-time services. Examples of these include speech and conversational video. They must have a short transmission delay, preferably less than 200 to 300 ms; otherwise, the delays will annoy the participants. The requirement for a short delay has a direct impact on performance. The longer the acceptable delay, the longer the interleaving can be, and a more efficient retransmission scheme can be used as well. Thus, conversational services will require more capacity than non-real-time services because of the lower delay requirements. Table 14.1 presents some delay values and their effects on voice communication. Notice that the indicated delay values are end-to-end delays. The air interface in UMTS is not the only instance that generates delays. There are also processing delays in terminal equipment, switching delays in the core network, and transmission delays in the backbone network.

With non-real-time services, like e-mail, file transfer, or Web browsing, the acceptable delays can be much longer. If the maximum allowable delay is only around 1 second, more efficient interleaving and coding techniques can be used, resulting in lower bandwidth requirements. In addition, there is enough time to use retransmission protocols, which lowers the bandwidth requirement even further.

Some typical transmission bandwidth requirements are given in Table 14.2.

To handle these different service requirements, UMTS will provide different bearers for each demand. Note that one multimedia application can (and most probably will) use several bearers, one bearer for each separate multimedia component. This will enhance the service quality and reduce the overall required bandwidth.

Note that although UMTS is a packet-based network, it can provide guaranteed throughput (with guaranteed delays), which is something the current Internet, for example, cannot provide. An excellent theoretical study on broadband network dimensioning is [20]. That book helps one to understand the problems met while designing a modern telecommunications network.

TABLE 14.1   END-TO-END DELAY EFFECTS ON VOICE COMMUNICATION

DELAY	EFFECTS ON VOICE COMMUNICATION
50 ms	No audible delay
100 ms	No audible delay if echo cancellation is provided and the link is of good quality
150 ms	Starts to have an effect on conversational voice communication
250 ms	Significant disturbance, speaking style must be adapted
400 ms	Upper limit to conversational audio set in [19]
600+ ms	No communication possible

From: [18, 19]

TABLE 14.2   TRANSMISSION BANDWIDTHS

APPLICATION	REQUIRED BANDWIDTH
ASCII PC-screen	9.6–14.4 Kbps
Voice	4–25 Kbps
HIFI-audio	32–128 Kbps
Video, VCR-quality	1.5 Mbps
Video, TV-quality, MPEG-2 compression	4 Mbps
Video, HDTV-quality, MPEG-3 compression	20–30 Mbps
Videophone, MPEG-4 compression	32–384 Kbps
Web-browsing	28+ Kbps

14.7 M-commerce

E-commerce, or electronic commerce or e-shopping, is a term that refers to commercial activity in the Internet. Mobile electronic commerce (m-commerce) is the word we use when the Internet's services are consumed or goods are purchased via a wireless link.

E-commerce will keep expanding very quickly in the future. It is also estimated that wireless access to the Internet will overtake fixed access by the middle of this decade. Therefore, it is likely that m-commerce will have a rosy future.

Because of the bright prospects for m-commerce, there are plenty of eager players attempting to take a slice of the business. The mechanism for the management of m-commerce is not yet clear. In current e-commerce schemes, payments are usually handled by giving the credit-card details to the e-trader via the Internet. This is quite unsuitable for small payments as it is insecure; thus, safer and easier methods are needed for m-commerce in 3G systems. These are discussed in Section 14.4.6.

Mobile phone operators would like to get a part of the huge revenue stream that will be generated by m-commerce. They already do have a billing relationship with their customers, and the same mechanism could be used to handle the billing of m-commerce purchases. The cost of items bought or the service consumed could be added to the monthly bill. The customers have already been credit checked once they have subscribed to the network. If they are prepaid customers, then they may not have been subjected to a credit check, but in this case they have paid for their network usage in advance, and this credit could also be used to finance their m-commerce transactions. However, this kind of scenario would certainly encounter opposition from the banking business. It would require the operators to expand their business from the pure telecommunications transmission business to retail and banking. Their operating licenses probably would not cover this kind of activity.

Banking companies have long experience with processing bills and payments. They would be more than happy also to process the bills generated by m-commerce. The mechanism for this could be a credit card embedded into a mobile terminal. It could be an enhanced SIM card, but the problem here is that a combined SIM card (containing both a mobile subscription from an operator and a credit facility from a bank or credit-card company) would confuse the situation in the customer's mind. It would also require extensive cooperation between the operator and the bank. The customer would negotiate two contracts at the same time: a mobile contract and a banking contract. A better solution is to have two SIM-cards in each m-commerce capable terminal, one issued by the operator (a traditional SIM) and another by the bank (a credit/payment card). This method would also provide the banks a way to promote their own brands.

The banking companies' biggest fear with this kind of arrangement is security. This is not because of any shortcomings in the cryptographic algorithms in UTRAN or any other technical issue. 3G will be a very secure environment, and SIM cards will have a better protection than the present-day credit cards already have. The most nonsecure component will be, as always, the human consumer. Currently mobile phones are not regarded as very valuable devices. They are left lying on tables, loaned to other people, and charged unattended. It does not help that in many countries mobile phone purchases are subsidized; thus, their purchase price is artificially low. Therefore, mobile phones are regarded as cheap commodities. However, people do not generally leave their credit cards on pub tables and are even less likely to leave their cash. The habit of carelessly handling mobile phones will be difficult to change, but it has to happen if these devices start to also contain electronic money. Yes, the credit SIM-cards will probably be protected with PIN numbers, but many people will choose quite unimaginative numbers like 1111, which will be quite easy to break. One attempt to solve this problem has been to add a fingerprint recognition module to a handset. This checks the fingerprints of the phone's user to prevent unauthorized usage. The drawback here is the increased cost because of the new hardware.

M-commerce will partly consist of the same services as fixed line e-commerce. One can buy goods and consume services. Everything available via the fixed Internet will also be available via the wireless Internet. However, the mobile nature of wireless terminals opens up other possibilities. The location-based mobile applications suggest totally new applications areas. Various information services, advertisements, and offers are much more valuable if they can be "precision-guided" to users whose locations are known. A list of restaurants in the town and their special offers of the day is a nice service when offered to the handset of a hungry businessman. But it is even better if this list contains only those restaurants located within half a kilometer of the location of the handset.

Another useful property of mobile phone networks, which can be exploited in m-commerce, is the possibility to relay information in real-time. Real-time information services as such are nothing new. TV and radio can relay real-time information very effectively, but these are broadcast services. One cannot subscribe to a service that can specify the reception of certain news clips at certain times via these services. A highly selective service is possible with m-commerce. Selected stock trading news can be sent immediately to the subscriber, and to the extent the handsets are always on, the information can actually be delivered to the subscriber in real-time.

These two examples (location-based advertising, and real-time customized news) are examples of information services. M-commerce also includes the trading of goods. In principle, one could purchase almost anything using m-commerce. However, the best chance of success will again be with services that exploit the special characteristics of the system, like mobility or the knowledge of the location of the UE. Buying jeans or foodstuffs using a mobile handset is probably not the easiest exercise around. However, paying for the purchase of jeans or foodstuff in a shop is another matter (see Section 14.4.6). Another example is the parking charge. Paying for parking is probably something that will be done mostly through mobile handsets by 2010. There are already functional m-commerce parking systems. In one scheme the mobile handset sends an SMS message indicating the start of the parking period, and later another one indicating that the parking period has stopped. Another solution could be to use the Bluetooth link in the handset to conduct transactions with the parking meter. The problem here is that each parking meter would require some rather expensive hardware and logic. What is good for the consumer is that he or she only pays for the actual parking time. There is also no need to worry about parking time running out and, thus, no need to rush out to the car to add extra coins to a parking meter.

M-commerce sceptics claim that people simply do not want to spend any more than they already do today for their communication services, and this is not enough to support 3G. However, there are clear signs that the spending will increase if there are new services and content to attract new revenue from existing subscribers. Today young people in Scandinavia and Japan already spend most of their disposable funds on mobile communication, even if this forces them to reduce spending on other things like movies and clothing. Moreover, once m-commerce systems are in place and functioning, people will move part of their shopping spending from the "real world" to the Internet. 3G is about new things, not just more communications services. For example, a customer may prefer to buy the latest hit single using m-commerce and have it delivered instantly to his mobile music device, rather than going to a traditional record store.

Once the micropayment scheme has been implemented, content providers can start offering new kinds of low-value services. In these services the revenue stream consists of very small trickles, but if there are enough of them, the stream can become very mighty indeed.

The very best applications for m-commerce are probably yet to be discovered. M-commerce will fundamentally change many things in our society. Some things will be done in a totally different way, and there will also be new kinds of applications that do not have equivalents in today's world. Start thinking now if you want to make money out of this.

14.8   Examples of 3G Applications

Everybody involved in 3G businesses should understand that it is not the technology people are buying and consuming, but the services and content. There has already been one notable failure in 2G, when WAP was marketed as a technology, but nobody told the customers what they could do with WAP. GPRS has also had a slow adoption process because there have not been enough GPRS-specific services on offer. This shows that customers are not eager to adopt new technologies, but they are looking for new services. I-mode is actually based on a relatively slow transmission technology, but provides great services.

14.8.1   Voice

Voice is and remains the most important type of application in mobile telecommunications. However, it will increasingly be combined with other forms of communication to form multimedia communication. Even the pure voice service can provide new possibilities for applications. It is already possible to set up multipoint conference calls, but this has not been widely exploited. And voice mail will be an attractive alternative for text-based mail systems, such as e-mail or SMS. It is also known that the increased usage of other applications also increases voice usage. For example, if you send a electronic picture postcard (MMS) from the Eiffel Tower to your friend in Helsinki, you'll probably receive a phone call from your envious friend quite soon.

ETSI has also specified the advanced speech call items (ASCI) features for GSM. These features are services that are accessible only to a closed user group. A GSM network enhanced with ASCI services is comparable to a Trans-European Trunked Radio (TETRA) network, and can be seen as a direct competitor to it. This is a pan-European standard used for public authority networks that are secure and closed to other users. They are networks with private access only.

The most important features of ASCI include the following:

- *Voice broadcast service (VBS):* the capacity for a single mobile to talk to a group of mobiles;

- *Voice group call service (VGCS):* the capacity for a group of mobiles to talk to each other;

- *Enhanced multilevel priority and preemption (EMLPP):* The fact that urgent calls can preempt less urgent calls.

ASCI features were developed after an initiative from the international railway body, the UIC (Union Internationale des Chemins de Fer). They needed a new private mobile radio system that would be compatible throughout Europe. ASCI can also be seen as a way for GSM operators to compete with TETRA networks. They can now argue that they can provide the same services as TETRA networks at much lower costs. ASCI is discussed more in [21]. All 2G cellular technologies (GSM, IS-136, IS-95) have recently added functions that mimic PMR (Private Mobile Radio) services, such as TETRA and iDEN. It is generally easier for a PMR technology to mimic a cellular feature than the other way around.

In 3GPP networks voice will be transferred in IP packets once the network becomes and all-IP network. This scheme is known as Voice over IP (VoIP). Actually VoIP is not an improvement as such for voice transfer. It is a mechanism that forces voice services into a packet-switched network not optimized to handle them; this had to be accomplished so that packet-switched IP could be used everywhere. We need to remember that circuit-switched systems were originally developed for voice transfer, and decades of optimization have made them very good at this task. Packet-switched systems will have problems with voice; this can be studied in [22].

14.8.2 Messaging

Messaging services will be an important application segment. The success of SMS messaging shows that there is a marketplace for services like these. SMS messages are a convenient way to send notes to other people. They do not interrupt the other person's tasks like phone calls do. They are eventually delivered even if the other person is not available because the phone is turned off or the subscriber is outside the network's coverage area. They do not require any interaction by the receiving person. SMS is efficient in social terms: Some people just cannot end a phone call without first telling their own life stories and those of all their relatives. Subscribers like the fixed-charge aspect of SMS and the precision and apparent permanence of text; it is easier to read a new address than to remember it during a phone call.

The basic text-based SMS will also be available in 3G, but the faster data rates of the new system make it possible to send much more than plain text in these messages. There is a new concept developed based on the notion of an enhanced SMS concept. This is called the multimedia messaging service (MMS).

This concept translates into a non-real-time messaging service that can deliver several multimedia components, such as text, (still) images, voice, and video. An MMS message can contain more than one component; these components are then combined in the user interface to produce a multimedia presentation. A simple MMS application could be an electronic picture postcard. Other MMS examples include electronic newspapers, news, traffic information, maps and driving instructions, music on demand, advertisements, and on-line shopping. MMS is discussed at length in Section 12.4.

E-mail is probably a very safe bet when predicting 3G applications. E-mail is widely used in the Internet and increasingly in mobile terminals. The barrier to its wider use has been the clumsy input facility provided by the standard handset. It is a difficult exercise for most subscribers to write an even modest e-mail message using the typical keyboard of a mobile phone, although the T9[1] system by Tegic [23] helps here a bit.

The GSM specification also includes a provision for SMS cell broadcast (SMS-CB) service. This is a service where SMS messages are broadcast to all mobile stations in an area (one cell or a set of cells). An enhanced version of this service, called the cell broadcast service (CBS) will be used in UMTS networks. CBS can deliver much longer text messages than the GSM SMS-CB service. A CBS message has a maximum size of 1,395 characters. These messages can be received by all terminals capable of receiving CBS that are in idle mode or in the CELL_PCH or URA_PCH connected states. Cell broadcast messages are assigned a message class type, which can be used by the UE to filter and receive only those messages that, moreover, are of interest to the user. The categories to be subscribed to could include, for example, news, traffic information, and weather forecasts.

A future enhancement of CBS, Multimedia Broadcast/Multicast Service (MBMS) is discussed in Section 12.4.

14.8.3 Internet Access

Internet access is an almost mandatory application for 3G mobile terminals. Over the last decade the Internet has grown to be a very important communication medium, and it continues to grow rapidly. The UMTS Forum estimates in [24] that the Internet will have 500 million users by 2005. Access to

---

1   T9 is a predictive text-input system. The system has a vocabulary of commonly used words and it tries to guess the word the user is currently typing in. Many handset manufacturers already include T9 as an option.

a communication medium as important as the Internet must be included in a 3G application portfolio.

Fortunately, this access will be relatively easy to implement in a 3G terminal. The 3GPP is specifying an all-IP network, which means that Internet protocols could be used all the way down to the terminal level. A mobile terminal would be an Internet node, just like any PC, with its own IP address number (see Section 14.4.7).

14.8.4   Location-Based Applications

The UTRAN system will contain several methods to determine the UE's location. This makes it possible to provide a totally new class of applications to mobile phone users: applications that use the knowledge of the mobile user's location.

The location data itself can be useful information for the user. The mobile terminal can inform a lost user about his or her location. In practice, the plain coordinates as such do not help much; they must be accompanied by some additional information, like the local map or instructions about how to get to a desired location from the current position. It is easy for the operator to provide the user with a local digital map, even if the user is lost, because his location is known to the operator. The map could even be sponsored by a local business, which could advertise itself in the map.

Location-specific billing is an application that could be used by a network operator. It could introduce a scheme where the usage of the mobile phone is cheaper from certain locations, for example, from the home or an office. For this kind of application, the location estimate does not have to be very accurate; a cell-coverage-based location method is adequate.

If the location of a mobile terminal is known, then it is possible to send it location-specific information, like advertisements, special offers, or traffic information. A mobile user could send a query requesting information about nearby bookshops, and the network could provide this information right away from its yellow pages directory as it knows the location of the requesting terminal. The service could possibly be free, if the user allows the network to send relevant advertisements with the information (e.g., Books Heaven is 200m to the north and today they offer the collected works of Nietzsche for only £9.99). However, sending advertisements blindly to all subscribers passing the shop-front is not a very clever idea. Most users would regard this as junk mail, and if they receive 10 advertisements on their handset after passing 10 shops on High street, they will probably change their operator. Moreover, sending advertisements cost real money for the shopkeeper.

Tracking is already used in the trucking business and with some parcel services. At the moment it is mostly done via satellites, but once UMTS networks gain wider coverage, tracking could be done in some applications also via UMTS. One new application could be the tracking of stolen cars. If a car includes a 3G chip (as most new cars will probably do in a few years' time), it can be tracked by the UMTS network as long as it stays within the coverage area.

Note that this coin has two sides. Your movements can also be tracked, even without your knowledge. If the UE has a GPS positioning capability, the operator could even determine the speed of your car. This would be very convenient for law enforcement agencies as there would be no need for special speed radars any more and they could send an instant speeding ticket via your 3G device. Are you afraid of Big Brother? You should be because he is getting even bigger.

The supporting technologies required for location services are discussed in Section 12.1. See also Section 14.9.3 for special terminals that exploit the location services.

14.8.5 Games

Games will be another major application segment in 3G. Most people do not admit that they like playing computer games, but despite this, the games are still selling extremely well. New service environments are typically justified by sensible applications, such as banking, share trading, and ticket purchases. In practice, most usage and revenue come from entertainment. There is no reason to prevent mobile entertainment from reaching a strong position in 3G. The success of entertainment and especially pornography in the terrestrial Internet predicts that this will happen.

Currently, the games some mobile terminals contain are pretty simple. The small displays and limited input devices severely restrict the applications. With bigger displays, more powerful processors, and (possibly) special 3G game terminals, most of these limiting factors will disappear. Compared to desktop PCs, mobile terminals will always lag behind in screen resolution, processor speed, and sound quality. Thus, the wireless industry has to find applications that will exploit the special characteristics of the mobile communication environment.

Networked games are played against other players over the network. These games are obviously in the interest of operators, as they eat up airtime. An operator could also set up automated machine players with varying skill levels, so that there would always be an opponent available. A simple example of networked games could be card games. These would be quite suitable for the 3G systems, as the game would not need a circuit-switched connection, but each played card could be sent via a packet-switched channel. Also the required graphics in the terminal could be fairly simple. The terminal manufacturers could also provide Bluetooth links for "local" multiplayer games. This solution would not consume any airtime; thus, operators would not support such games.

Game applications should be downloadable from the operator or from a gaming content provider. This way the operator can get additional revenue from gaming through commissions, and customers can try new games if they have grown tired of their old games. In multilevel games, the first levels could be free as that way the game becomes more widespread. Once the player is hooked, he is quite willing to pay for additional levels. The games should preferably be built on some common platform, such as WGE [25] or Java, so that they would be easily portable to various mobile terminals.

Note that good games in mobile terminals do not necessarily have to be very complex. Probably the two most widely known games in recent years have been Tetris and Tamagotchi. Both are very simple games, and at the same time, very addictive. They prove that it is the good idea that makes a game enjoyable, and not necessarily its visual appearance or complex logic.

The importance of games in mobile terminals should not be underestimated. A large segment of customers will choose their mobile terminals based on the recreational services they can provide. Young people especially will be attracted to models that can provide the best and trendiest leisure applications. Their purchase decisions are not necessarily based on the standby times or the weight of the handset or any other strictly technical issue.

14.8.6   Advertising

Generally, it is not a very good idea to send an advertisement to all subscribers in the network. Most subscribers are probably not at all interested in a certain advertisement and will regard it as junk mail. This will only irritate the customers, and they may change their operator because of this practice. Moreover, sending large amounts of advertisements will use a lot of capacity; thus, it is expensive. The operator should therefore build an accurate user profile database and send the advertisement to only those customers who are most likely to be interested in it. For example, if a customer has bought several books using his mobile device, it is probable that he is interested in receiving information on book sales or new books on the same subject.

Of course, the user must be given the last word in advertising. It must be possible for a user to reject all advertising. On the other hand, the operator can make the reception of advertisements more attractive, for example, by lowering the subscription charges. Note that the operator can put a nice price tag on this kind of targeted advertising. For example, an ink-jet cartridge shop would pay a lot for an advertising campaign that is directed to 3G customers that bought an ink-jet printer about six months earlier.

14.8.7   Betting and Gambling

This is yet-another interesting 3G application type. Especially in East Asia, betting is very popular, and many customers would certainly subscribe to

3G just because of this application. A betting application combined with a 3G payment card provides completely new possibilities. A user could follow a horse race from the stand and bet on the winning horse without leaving his seat. The bet is instantly deducted from his payment card and possible prizes could similarly be paid back in real-time. The user can also watch the race from in his armchair at home and make bets as if he were actually at the horse race.

Various game shows on TV could include viewer betting. However, there could be problems even in the most capable 3G network if 10 million viewers send in their bets at the same time.

Prepaid customers may not be allowed to enjoy this service. This is not because of missing credit history. If there is credit left in the prepaid account, then it could be used for gambling in principle. However, in most countries betting and gambling are forbidden for minors; thus, the identity of the customer must be known. This is not the case with normal prepaid customers.

14.8.8 Dating Applications

These are already very popular in Japan. Many people prefer to get to know other people without revealing their own identity first. The technical implementation of a dating application may vary. It can be a simple bulletin board with dating adverts combined with an anonymous e-mail server, or it could be a lonely-hearts mobile chat room. Users can also set up their own profiles (or the profile for the company they seek), and wait until the matchmaker application finds a suitable victim. Dating ads may also include still images and audio clips.

14.8.9 Adult Entertainment

And the last in this list, but certainly not the least profitable, is the adult entertainment sector. This is the most profitable entertainment business in general, and it will remain so in 3G. The premiums will be very high. It will be interesting to see how operators deal with adult entertainment. It is a lucrative market, and they would certainly want to take their share of the profit somehow. On the other hand, in some countries there may be regulations preventing operators from providing this kind of service, or it is simply not socially or politically acceptable for an operator to do so. In any case, it will be very difficult to censor adult entertainment services because, in most countries, these will be legal or only modestly regulated. UMTS is a global system; thus, these services can be accessed anywhere. Monitoring access to the fixed Internet is an almost impossible task, and here we have an Internet with mobile users, who can be anonymous if they have prepaid subscriptions.

These kinds of applications need big color displays and relatively high data-transmission capability for downloading still images and video clips. Payment for services could be handled instantly with the embedded payment card.

14.9   Terminals

Most 2G terminals today were designed to be used for voice communications. A typical handset still resembles the good old telephone receiver. Some say the lingering familiarity makes it an "invisible technology" and assures its success. To the extent 3G terminals depart from the familiar realm, they become "visible" and risky. The good old telephone receiver has been enhanced with a small display and a numeric keypad, but the basic idea is the same. The microphone can be placed close to the mouth, while the speaker is by the ear at the same time. One can easily see that this kind of handset is optimized for voice. It is very difficult to do anything else with it. Even sending a short SMS message requires a considerable amount of finger acrobatics, and it takes time.

In 3G, speech communication will still be the dominant component, but there will be many other ways to communicate besides voice. Therefore, the old type of voice handset cannot be used satisfactorily in all communication situations found in the next generation. This means that the mobile terminal market will most probably diversify. There will be a wide variety of different types of terminals, as one size does not fit all. It is important that services that rely on visual presentations, for example, have adequate displays. Other devices may need a proper keyboard, and some others hi-fi loudspeakers.

New services have to be easy to use. From this we can further assert that user devices must be easy to use. The user device must be very easy to configure; preferably any configuration necessary will be done automatically without user input. And the user interface must be intuitive. It is impossible to overstate how difficult these matters are to deal with and how important they are.

There have already been some attempts in 2G to develop different kinds of terminals for users. The most well-known and successful of these is probably the Nokia Communicator. This is a rather large GSM handset. In its "closed" position, it functions as a standard GSM handset; while "opened," it provides a bigger display and a small qwerty keyboard. This kind of handset is better suited for textual communication like sending and receiving e-mails. However, the slow data rates of standard GSM are an obstacle for anything else, such as Web browsing, not to mention any kind of multimedia communication. Of course, this is not the fault of the handset design, but merely of network's data-transmission capability. Nokia Communicator's success probably comes from letting the user make it an "invisible technology" when using it as a phone, or a "visible technology" when configured as a PC; it does not try to be both at the same time.

Another trend in mobile terminals will be "mass customization" of the terminal. The standard mass-produced phone will not appeal to all customers, and they will probably want to customize their phones. Currently, it is already possible to download new ring tones and change the colorful covers on some handsets. Further customization items could include downloadable applications that run on WAP or Java platforms. Each customer could build a personalized application mix for his or her mobile terminal. Furthermore, it should be possible to customize the individual applications.

As said, we will need a wide variety of different user device types. This is true, but it might be useful to think about this from another point of view. It can also be said that many existing devices, currently without communication capability, will include communication chips in the future. This list includes vending machines, household appliances, game modules, cameras, stereos, cars, and so forth. And there will also be new types of devices we do not know about yet.

14.9.1 Voice Terminals

Voice terminals will remain the most important terminal segment in 3G, at least for the foreseeable future. Interactive discussions are an efficient and natural way to communicate. However, this does not mean that in 10 years' time the voice terminals will still be old telephone receiver derivatives. When the technology improves, the handsets can be smaller and lighter. If a terminal is a pure voice terminal, then there is no need for a display, and if the input can be done via voice recognition, there will be no need for a keypad either. For example, the earpiece and the microphone could be two separate devices; the earpiece could be embedded in spectacle frames, and the microphone and the central unit could be carried as a modern wristwatch. The interdevice communication could be handled via a wireless link (e.g., Bluetooth).

It is also possible to exploit voice transfer in more specialized terminals. A portable MP3 player is used for playing MP3 compressed music. This kind of device could also include a 3G communication chip, which would enable it to download MP3 music from the Internet to a local storage device via a wireless link. A typical MP3 compressed piece of music takes about 4 MB, an amount of data that can be transferred over a UMTS air interface in a reasonable time.

14.9.2   Multimedia Terminals

Multimedia communicators can be used when the communication includes components other than voice. This kind of device has to contain a display for presenting received video or still images. If the subscriber wants to send either of these, there also has to be a small camera (video camera for video transmissions). Textual data can be supplied via voice recognition or a handwriting-recognition device. There will probably not be a keyboard for text input as this would have to be quite large to be used easily. Moreover, a keyboard is a slower way to input text than either voice or handwriting recognition, provided that these new technologies really do work properly. The size of this kind of device has to be bigger than the voice-only terminal. However, it should be light enough and small enough to be carried in a pocket. If the device contains a camera, then it should be possible to place the device on a table so that it remains stable. It is probably quite difficult to include a camera in a handheld device because the task of focusing the camera on the speaker is not the easiest we can imagine, especially if the camera and speaker move all the time. On the other hand it could be possible to implant a 3G communication chip into a camcorder or a digital camera. We are also disassociating the camera twice: once by removing it from the more familiar camera-type device, and again by stuffing the camera into a telephone. Double disassociations seldom work in initial market introductions: If they are viable, they may work in a second introduction. The SIMON failed in the early 1990s, but the wireless PALM seems to have succeeded, even though both are roughly the same device. Neither looks like the Nokia Communicator, which tries to avoid the double disassociation.

14.9.3   Navigation Devices

The 3G systems will include a location service, so one natural terminal type for 3G is a navigation device. There already exists at least one attempt to provide something like this, namely the Esc! phone from Benefon. It includes a big display, and it can download maps from the network. The location of the phone can be seen from the displayed map. In the future we will also see GPS devices that include 2G or 3G communication chips.

Navigational devices can take many forms. A hiker in the forest would probably want to have a small lightweight navigation device with only the basic voice call functionality. When outside the 3G coverage, the location data could be provided by the GPS system. Hikers often have specialized needs for their map applications, which are served well with modern GPS terminals.

A truck may also include a tracking device that monitors the movements of the vehicle and provides a navigation aid for the driver. Here the weight or the power consumption of the device is not an issue, but the device has to be easy to use. It should not disturb the driver in his main task, which is driving. All such functions, which can be automated, should be implemented. Any necessary input to the device should be done in a way that does not require unnecessary work, for example, via speech recognition.

Most passenger cars will probably include some kind of navigation device in a few years' time. Similarly here, the weight, the size, or the power consumption of the device is not an issue. The same device probably scans for all traffic announcements from the radio waves, as well as from 3G cell broadcast service (CBS). If traffic jams are detected, the device proposes an alternative route. All input should be possible via speech recognition, and possibly also the output via speech generation. It will be possible to make hands-free voice calls via this device.

The existence of special navigation terminals does not exclude the use of navigation applications in standard mainstream terminals. But for certain special needs, a purpose-built navigation terminal is probably the right solution.

Privacy issues must be solved before 3G navigation devices enter widespread use. Who has the right to access the location data and when?

14.9.4 Game Devices

Mobile communication will be used more and more for recreational purposes. At the moment, some mobile phones already include simple games, which can be played while the phone is in the idle mode. Some games also provide a multiplayer option, where communication with other players with similar terminals is handled via a wireless link (e.g., infrared). However, these games are merely an add-on that the terminal manufacturers use to lure customers to buy their handsets. The games do not generate any new transactions while they are played; that is, the operator or any other third party does not get any new revenue from them.

There is, however, quite a lot of development activity in the games business regarding mobile telecommunications. It is not yet clear what kind of games will be successful, but in any case the old voice-optimized handset is not the best device on which to play these games. A special gaming device will have to include a big, fast color display and special input devices, such as a joystick. Also, proper loudspeakers are needed, possibly of hi-fi quality. Depending on the games played, a Bluetooth device for multiplayer games may be included. There is more about gaming applications in Section 14.8.5.

14.9.5 Machine-to-Machine Devices

Once 3G matures, it is quite probable that the number of nonhuman 3G users will vastly exceed the number of human 3G users. With some applications, the terminals do not have to contain all the possible input and output devices. Many electric appliances will include embedded mobile phone chips, and then only part of the functionality of a standard mobile terminal is required.

In some cases the terminals do not require any kind of input device, as they are designed to relay only downlink information. Typical applications using this kind of device could include information services that broadcast information to a group of people. For example, bulletin boards may display real-time traffic information, such as waiting times for the next service in bus stops and train stops. Billboards may broadcast advertisements in city centers, and so on. This kind of device must have big displays, but no input devices.

In some other applications, the terminal will handle only uplink traffic; that is, it transmits, but probably does not receive anything except control commands for the machine in question. A mobile phone chip inside a vending machine can inform the warehouse when supplies are running low or if there is a malfunction in the cooling or money changing mechanisms. It is also possible for the machine to accept electronic payments made with a mobile phone; thus, it has to have the means to relay the payment data forward to a collection point.

Laptop and notebook computers may include 3G communication chips as standard. Here the input and output devices are already provided by the parent device, so the actual 3G device does not have to include these.

What is common to all these machine devices is that they are probably much cheaper to produce than devices for human usage. The device does not need a display, and probably does not need a battery as the parent device will have a power supply anyway. The data rates generated can be predicted accurately; thus, the chip can be optimized for this amount of traffic. User interface hardware or software is not needed.

References

[1]   Webb, W., *The Future of Wireless Communications*, Norwood, MA: Artech House, 2001.

[2]   Fernandez, B. A., "The Future of Mobile Telephony," paper presented at Mobile Telephony and Telecommunications conference, Madrid, Spain, May 22–23, 2000. Available at http://www.cordis.lu/ist/ka4/mobile/index.htm.

[3]   http://www.radicchio.org.

[4]   http://www.mobiletransaction.org.

[5]   http://www.mobey.org.

[6]   http://www.mobilepaymentforum.org.

[7]   http://www.wapforum.com.

[8]   http://www.pkiforum.com.

[9]   http://www.paymentgroup.org.

[10] Latid, L., "IPv6—The New-Generation Internet," *Ericsson Review No. 1,* 2000, at http://www.ericsson.com/review.

[11] "IP-Radio Access Network," Nokia White Paper, February 2000, at http://www.nokia.com/press/background/pdf/IP-RAN.pdf.

[12] IETF RFC 2507: "IP Header Compression," M. Degermark, B. Nordgren, S. Pink, February 1999.

[13] Eriksson, G., et al., "The Challenges of Voiceover-IP-over-Wireless," *Ericsson Review No. 1,* 2000, at http://www.ericsson.com/ review/2000_01/article96.shtml.

[14] Andersson, C., and P. Svensson, "Mobile Internet—An industry-wide paradigm shift," *Ericsson Review No. 4,* 1999, at http://www.ericsson.com/review/ 1999 04/article92.shtml.

[15] 3GPP TR 23.922, v 1.0.0, Architecture for an All IP Network, 1999.

[16] Roberts, J., U. Mocci, and J. Virtamo, *"Broadband Network Teletraffic,"* COST 242 report, Berlin: Springer-Verlag, 1996, pp. 18–46.

[17] 3G TR 22.960, v 3.0.1, Mobile Multimedia Services Including Mobile Intranet and Internet Services, 1999.

[18] Kyas, O. *ATM Networks,* 2nd ed. London: International Thomson Computer Press, 1997, p. 13.

[19] 3GPP TS 22.105, v 5.1.0, Service Aspects; Services and Service Capabilities, 2002.

[20] Roberts, Mocci, and Virtamo, *"Broadband Network Teletraffic,"* COST 242 report, Berlin: Springer-Verlag, 1996.

[21] Webb, W., "Advanced Speech Call Items," in *GSM-Evolution Towards 3rd Generation Systems,* Z. Zvonar, P. Jung, and K. Kammerlander (eds.), Norwell, MA: Kluwer Academic Publishers, 1999, pp. 45–64.

[22] Eriksson, G., et al. "The Challenges of Voice-over-IP-over-Wireless," *Ericsson Review No. 1,* 2000, at http://www.ericsson.com/review.

[23] http://www.tegic.com.

[24] "The Future Mobile Market," UMTS Forum Report No. 8, March 1999, accessed December 7, 2002 at http://www.umts-forum.org/reports.

[25] Wireless Game Engine (WGE), at http://www.9dots.net.

CHAPTER 15

The Future

The first commercial UMTS networks in Europe were launched in 2002. In Japan NTT DoCoMo launched its own WCDMA network somewhat earlier at the end of 2001. NTT DoCoMo's system is called Freedom of Multimedia Access (FOMA). However, this was not an IMT-2000 compatible network, but a proprietary standard that was based on an old interim UMTS specification release. Thus, even though this system is based on the same principles as UMTS, it is not compatible with it, and UMTS devices cannot roam into FOMA, or vice versa. In Europe, the first UMTS trial network was launched on the Isle of Man in 2001.

This is the beginning. At first, most of these networks will be local, providing only urban coverage, while 2G GSM provides wider-area coverage. Note, however, that green-field operators (i.e., operators without an existing 2G network) must build out their networks to cover wide areas as soon as possible, unless they are given airtime from existing 2G networks; see the mobile virtual network operator (MVNO) concept in Section 10.4.1.

15.1   New Spectrum

Even before the first commercial 3G networks were launched, it was already known that the original spectrum allocation for terrestrial UMTS systems will not be sufficient in the long run. The industry's own figures for successful business plans show that the current spectrum will not be enough. In Europe the current terrestrial spectrum allocation is 2×60 MHz in paired bands and 35 MHz in unpaired bands. Depending on the country, this is typically divided between four and six operators. Even the luckiest operators have only 2×15 MHz of paired spectrum, which gives them three frequency carriers to use.

As radio spectrum is a scarce resource, it is difficult to find new spectrum that could be allocated for UMTS worldwide. Worldwide spectrum-allocation recommendations are made by the ITU in World Radio Conference (WRC) meetings. The ITU has already started to look for additional spectrum for IMT-2000. Note that the original IMT-2000 spectrum recommendation was released as long ago as 1992. It included the following spectrum bands:

- 1,885–2,025 MHz;

- 2,110–2,200 MHz.

The new bands accepted at WRC-2000 in Istanbul include the following:

- 806–960 MHz;

- 1,710–1,885 MHz;

- 2,500–2,690 MHz.

The old allocation was 230 MHz altogether, and the new decision adds 519 MHz to it. However, these numbers are misleading in that the real available number will be much less. Different bands may be available in different regions, but no region may claim all of them. Still, this sizable addition is a remarkable change in regulatory attitudes represented by the tiny spectrum allocations in the precellular era, offering significant reserve capacity for the coming years. It should be noted, however, that this was merely an agreement to draft a recommendation about the subject. The ITU's recommendations are not binding. Regional and national organizations can overrule the ITU if they so wish. Therefore, it remains to be seen how much of this band will actually be licensed to 3G operators. In any case, this process will take several years. Also note from Figure 15.1 that the new bands overlap existing 2G spectrum.

The spectrum-allocation process is a very slow one because these recommendations are global; thus, there are plenty of competing opinions and interests. Also, because a spectrum band is often already used by someone else, the allocation process includes moving existing users to some other spectrum allocation. This takes time and may represent an expense for the old spectrum users who may force it onto the new users.

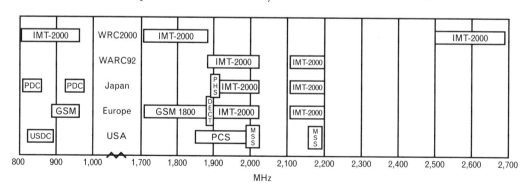

FIGURE 15.1   *New IMT-2000 spectrum from WRC2000.*

15.2   Satellites

Satellites offer one way to provide mobile-telecommunication services. Satellite-based mobile-communication services are referred to by many different names, such as Mobile Satellite Services (MSS), Satellite Personal Communications Networks (S-PCN), and Global Mobile Personal Communications by Satellite (GMPCS). All these concepts mean more or less the same thing: the provisioning of personal communication services via satellites to users with handsets comparable in size to 2G (terrestrial) mobile handsets. It can be questioned whether MSS systems should be discussed in this chapter. Certainly they have great potential, but it may be that this potential will never be fully realized.

Using satellites for telecommunication purposes is not revolutionary. They have been used for various communication purposes for decades, even for providing mobile-telecommunication services. Fixed-satellite services are involved in relaying terrestrial telecommunication transmissions. When Pete Smith from Glasgow calls his Aunt Agatha in Brisbane, Australia, the call is probably relayed via a satellite link. Satellite broadcasting services are used to broadcast television channels over large areas. The recipient needs a satellite dish, which can be either an individual TV owner or a satellite-TV broadcasting distribution company. Satellite positioning is also widely deployed in a variety of applications, and its advantages can be enjoyed by just about anyone who can afford relatively low-cost and attractive, yet functional, devices.

MSS is a mobile-telecommunication service in which the mobiles are linked directly to the satellite (Figure 15.2). This kind of system has many advantages, as it can provide global (or wide-area) coverage with cost independence. The list of disadvantages is also long and includes limited system capacity, limited indoor service, high investment requirements, relatively long signal-propagation delays, and relatively low data-transmission capacity.

15.2.1   The Market for MSS Networks

Today's problems with MSS are partly due to the unforeseen success of terrestrial cellular systems. The design work for the satellite systems currently being deployed was started during the first half of the 1990s. Back then, nobody could have predicted the success of mobile cellular systems (now there is GSM coverage even in Greenland). The coverage areas of these networks were limited, and often consisted only of towns and major roads. Thus, there seemed to be a lucrative market for satellite services. They could have been used to complement the terrestrial cellular services, and also provide coverage in the countryside, on the sea and big lakes, in countries without adequate telecommunication systems, and so on. An important

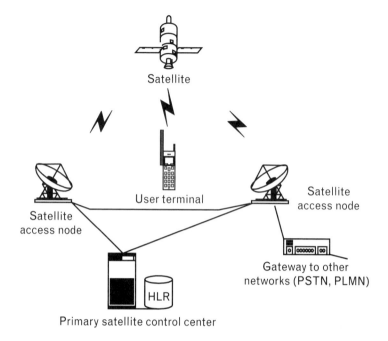

customer segment was intercontinental business users. 2G cellular systems were not global, but regional; thus, a business user travelling on another continent probably could not use his or her domestic mobile phone because the local cellular standard was different.

But it did not work like that. People continued to buy mobile phones, and the mobile networks had to expand further and faster than had been anticipated. Currently it is already quite difficult to find a place in Western Europe that is outside cellular coverage. The same is also true (or will be quite soon) in other developed countries. If there is a place where there are people with communication needs, quite soon some network operator will build a cellular network there.

International business users, those subscribers with plenty of money and whose expenses are paid by their employers, already seem to be a lost opportunity for the MSS players. One can already buy dual- or triple-mode phones that work in both the United States and Europe. It is not even necessary to buy a multimode phone to be able to use the mobile phone services on another continent, as GSM systems provide SIM roaming. This means that a user can take his SIM card on a journey, rent a GSM mobile phone (probably using different frequency bands) in the destination country, and use his SIM card with it. Moreover, the incompatibility between European and Japanese standards will be gone if they both adopt the 3G-UTRAN network. However, it is still true that there are so many different mobile-phone standards in the world, building a multimode phone capable of handling all of them is not feasible.

In the third world, there would surely be plenty of potential users for satellite services; the problem here is obviously the expense of satellite services. Subsidies are needed, either by the local governments or by international aid organizations. Unfortunately, MSS companies are commercial businesses, and they'll have to yield a profit or go bust. Thus, they cannot be the subsidizing entity. A possible scenario would be to install a WLL system in a village, which is then connected to a satellite. Another, cheaper, solution is a fixed phone booth that is connected directly to a satellite.

Because of these limitations, MSS companies are sure to have done some innovative thinking and redesigning of their business models over the last few years. Still, the recent history of MMS consists almost entirely of bankruptcies. There is nothing wrong with the technology; it is working fine. There just does not seem to be market left for these systems.

Note that in any case, satellites will be used increasingly in telecommunications. Location services need satellite support, and GPS, GLONASS, and Galileo satellite location systems will be widely used. Satellite broadcast services keep on expanding. Broadband satellite networks are also in the pipeline, although their fate could be linked to the fate of MSS networks. It may well be that terrestrial broadband provisioning expands faster than satellite broadband planners had anticipated, and there is no market left for them either.

15.2.2 Satellite Orbits

Satellite systems can be classified based on height of their orbits. There are four types:

1. LEO (low Earth orbit);
2. MEO (medium Earth orbit);
3. GEO (geostationary orbit);
4. HEO (highly elliptical orbit).

MSS networks typically use LEO or MEO orbits. Regional systems favor GEO satellites.

LEO systems have the lowest orbit of all. They are located between a height of 200 km and 1,400 km, most often between 700 and 1,000 km. There are two no-go areas for satellites, namely the inner and outer Van Allen belts. The Van Allen belts are two rings around the Earth, containing high-energy charged particles. The inner and stronger belt is at 300 km and the outer belt is at 16,000 km.[1] LEOs and MEOs have circular orbits. The advantageous properties of LEO systems include the following:

1   Van Allen radiation belts were discovered in 1958 by U.S. physicist James van Allen.

• Low propagation delay;

• Low transmitter power requirement;

• Higher elevation angles;

• Higher operating reliability due to redundancy.

The problems include the following:

• Satellites move fast in relation to the Earth, so frequent handovers (HOs) are needed.

• A great number of satellites are needed to cover the Earth.

LEO satellites are close to the Earth; thus, the signal delays are quite small and transmitter powers can be smaller than with other orbital systems. It is even possible to have limited service indoors with LEOs. However, the low orbit also means that a great number of satellites is needed (e.g., Globalstar with 48; and Iridium with 66). LEO satellites move very fast in relation to the Earth's surface, which means that HOs are needed repeatedly and frequently to keep the connection active, even when the user remains stationary. For example, an Iridium cell moves so fast that an HO is needed about once every minute. The number of satellites also means that the control and management of the system is a sizable task. On the other hand, because there are so many satellites, losing a few of them does not significantly affect the usability of the whole system from the subscriber's perspective. Globalstar and Iridium are both LEO MSS networks.

MEO satellites typically orbit at around 10,000 km. The properties of MEO systems are much like LEOs. The differences include slightly longer propagation delays because of the higher orbits. The higher orbits also means greater path losses. The signals cannot penetrate buildings; thus, MEO systems are strictly confined to outdoor uses. The attractive properties of MEOs include the smaller number of satellites required. Because the orbit is much higher than with LEOs, only a fraction of the satellites required in LEO systems are needed. ICO plans to use only 10 satellites in two orbits (+2 spares) to cover the entire Earth. The lifetime of MEO satellites is also typically longer than in LEO systems (10–15 years versus 5–8 years for the LEO birds). ICO is a MEO system.

GEO systems orbit the Earth at exactly 35,786 km. This distance is special because the satellites remain in a fixed position relative to the surface of the Earth. The satellite's velocity is the same as the rotation speed of the Earth at 35,786 km. This makes it possible to provide regional services via a single GEO satellite. Just "park" the satellite over the service area and it stays there. Similarly, GEO is suitable for broadcast services. Because these satellites remain "stationary," it is easy to direct the reception antenna toward the satellite. The high orbit also means that only three to four satellites are needed to cover the entire globe (or close to it).

GEO systems, however, have several problems that make them unsuitable for mobile-communication services:

- Signal-propagation delays are very long (the round-trip delay is 250 ms, which makes traditional retransmission protocols unusable).

- Mobile terminals need high transmitter powers and large antennas.

- They cannot provide service for areas near the Earth poles because GEO satellites are located above the equator.

- Getting satellites into geostationary orbit is more expensive than getting them into LEO or MEO orbits.

The Inmarsat satellites are in geostationary orbit. They can provide telecommunication services via portable terminals, but because of their size, these can hardly be called mobile phones. The smallest Inmarsat terminals are about the size of a large laptop computer. Inmarsat is, nevertheless, an important player in satellite communications, as it started to provide these services back in 1979, and for two decades it was the only service to be able to do so on a global scale.

Planned regional GEOs include ACeS (Southeast Asia) and Thuraya (Middle East and Africa).

HEO satellites use an elliptical orbit, and their distance from Earth varies between 1,000 km and 39,400 km. The nearest point to Earth in this orbit is called the perigee and the farthest point the apogee. The idea behind HEOs is to use the communication services of a satellite when it is near its apogee. In this position, its orbit speed is relatively low; thus, it remains near the apogee for a long time. HEOs must regularly pass through the outer Van Allen belt, and they receive radiation from it.

The different orbit types and the Van Allen belts are depicted in Figure 15.3. This picture is taken from [1]. The same source contains an excellent collection of up-to-date satellite information.

15.2.3 Examples of MSS Systems

This section explains the technical details of some currently operational MSS systems, or systems that are in advanced planning stages. This is a purely technical presentation and does not discuss the economical situation of each network, as this information tends to be quite dynamic in nature.

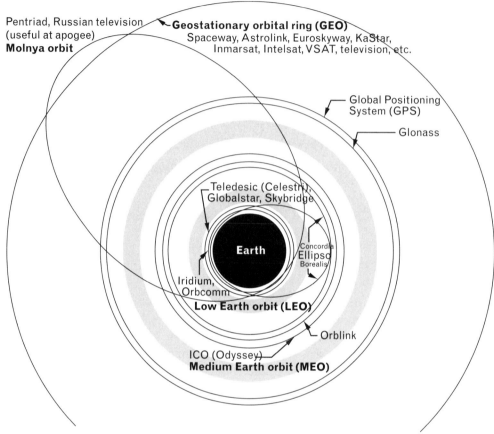

Orbital altitudes for satellite constellations

    peak radiation bands of the Van Allen belts (high-energy protons)
    orbits are not shown at actual inclination; this is a guide to altitude only

from Lloyd's satellite constellations http://www.ee.surrey.ac.uk/Personal/L.Wood/constellations/

FIGURE 15.3   *Satellite constellations.*

15.2.3.1 Iridium

Iridium was the first MSS system to be launched. Its biggest investor was Motorola. The system development started in 1990, the first satellites were launched in 1997, and at the end of 1998, commercial operation began. Thereafter, Iridium has had a rocky path, but it is still functional.

A fully functional Iridium system has 66 satellites. The original plan had 77 satellites (hence the name Iridium—the number of electrons in an Iridium atom), but this was reduced later to only 66 satellites (an atom with 66 electrons is called dysprosium—one can understand the decision to keep the old name). The satellites use a polar orbit at a height of 780 km. There are altogether 6 orbits with 11 satellites per orbit.

Each satellite provides 48 cells or spot beams. The cluster size is 12, so each frequency could be reused four times in each satellite. The bandwidth-allocation for the Iridium system is 5.15 MHz (1,621.35–1,626.5 MHz). As the channel bandwidth is 31.5 kHz and the spacing between channels is 41.67 kHz, this gives a total of 124 frequency carriers for the system.

Each Iridium carrier contains four duplex channels. That means that both the uplink and downlink are sent via the same carrier. A TDMA frame lasts 90 ms and starts with a paging and control slot, followed by four uplink and four downlink timeslots (Figure 15.4).

The satellite component of the Iridium system is rather complex. The signals can be relayed via intersatellite links. Each satellite can connect into the previous and the next satellite in the same orbit, and also to the nearest satellite in adjacent orbits. For example, in Figure 15.5, satellite 44 could connect into satellites 43, 45, 33, and 55. This arrangement reduces the amount of required gateway stations on Earth, as the signal can be relayed over long distances via satellites before it is sent back to Earth. Also, because of the relatively low orbit, it is impossible to cover large oceans without intersatellite links. However, intersatellite relaying is a technically complex task because the distances and relative directions of satellites on the interorbital links change constantly.

The system as designed could serve 1.4 million users. At the time of this writing, a large part of Iridium capacity is used by the U.S. Defence Department. See [2] for further information on Iridium.

15.2.3.2 Globalstar

Globalstar's commercial operation started at the end of 1999. Its owners include space and telecommunication companies such as Loral, Qualcomm, Hyundai, and Alcatel. The Globalstar system has 48 satellites in eight orbits. The orbit height is 1,414 km. The system coverage is not global, but excludes the polar areas (north from 74°N and south from 74°S). The coverage area of a satellite is divided into 16 cells, as shown in Figure 15.6. The air interface technology is based on the IS-95 system (it is CDMA); the

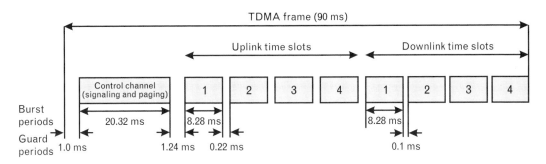

FIGURE 15.4   *Iridium TDMA frame.*

FIGURE 15.5
Iridium satellite relays.

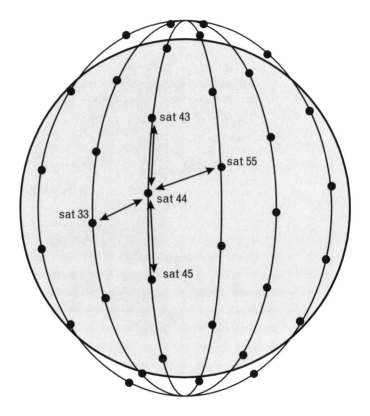

carrier bandwidth is 1.25 MHz. The system bandwidth allocation is 1,610 to 1,621.35 MHz in the uplink and 2,483.5 to 2,500 MHz in the downlink (yes, the downlink has more bandwidth). Globalstar does not use intersatellite links; thus, a large number of Earth stations is needed (100+) because a relaying Earth station must always be found inside the satellite coverage area; otherwise, calls cannot be connected. On the other hand, the satellites are relatively simple and inexpensive. The land segment is based on GSM technology.

The reader may have noticed that Iridium and Globalstar have applied different design principles in their networks. Iridium has a complex space segment, but a simple land segment; Globalstar has a simple space segment, but a complex land segment. Both approaches have their good and bad points. A complex space system has to use expensive satellites that can do intersatellite relaying. A complex satellite in space is also risky. If something goes wrong in that complex machine, it is rather difficult to send a telecom engineer to fix it. However, the system can do with a few Earth stations. This results in an inexpensive land segment. In the other approach the space segment is simple and bent-pipe type satellites can be employed. These are, in principle, just repeaters in the sky, which relay the signal from the handset to the land station, and vice versa. But then the land segment has to be rather complex. A large number of Earth stations are needed, sometimes in

FIGURE 15.6
Globalstar cell layout.

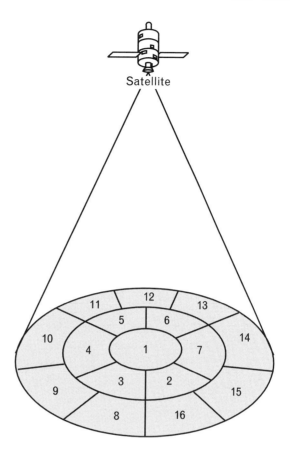

FIGURE 15.6
Globalstar cell layout.

difficult locations. It is expensive to manage this large land-based global tele-communications network.

Globalstar mobile handsets are all multisystem handsets, the terrestrial system's cellular technology being either GSM, AMPS, or CDMA. More about Globalstar may be found at [3].

15.2.3.3 ICO

Intermediate Circular Orbit (ICO) was founded in 1994. The biggest owners were Inmarsat and Hughes. This system uses MEO satellites; thus, the technology differs from LEO systems like Globalstar and Iridium. ICO needs only 10 satellites in two orbits to cover the whole globe. The orbit height is 10,355 km, and both orbits should include an additional spare satellite (the total installation being thus 10 + 2 or 12 satellites). At this height each satellite revolution around the globe takes about 6 hours.

Because of the higher orbit, ICO needs only 12 Earth stations, which are called satellite access nodes (SANs). There are no intersatellite connections, but calls are routed directly via SANs to the ground network. The air interface technology is based on GSM. There are 163 cells for each satellite.

The bandwidth allocation for ICO is 1,980 to 2,010 MHz for the uplink and 2,170 to 2,200 MHz for the downlink (this spectrum is actually allocated by the ITU for the satellite component of IMT-2000). The channel bandwidth is 25.4 kHz and there can be 750 carriers per satellite. Each carrier is divided into six timeslots. This gives a theoretical maximum of 4,500 channels per satellite.

As in the Globalstar system, ICO mobile handsets will be multisystem handsets. For more information about ICO, see [4].

15.2.3.4   Teledesic

Teledesic was founded in 1990 by Bill Gates and Craig McCaw. The Teledesic system is not a mobile communications system and will concentrate on broadband data traffic to fixed terminals. It aims to provide fast (up to 64 Mbps in the downlink and only 2 Mbps in the uplink) broadband data services to areas that are not served by terrestrial fibre-optic networks. It has registered the trademark Internet-in-the-Sky for this purpose.

The data-transmission capability is immense when compared with other LEO/MEO narrowband systems. These other systems typically reach only 9.6 Kbps. To be able to provide this kind of data rate, Teledesic needs a wide spectrum allocation. This is found in the Ka-band; the uplink uses a band between 28.6 and 29.1 GHz, and the downlink between 18.8 and 19.3 GHz. The problem with frequencies this high is the high signal attenuation caused by moisture or any kind of obstacle.

The Teledesic system consists of 288 small satellites. These are divided into 12 orbits, giving 24 satellites per orbit. This installation covers 95% of the globe. Each satellite has 576 cells, and each cell can provide up to 1,800 channels with a 16-Kbps data rate. The satellites have intersatellite connections, and they in fact form a packet-switched network in the sky. The network is capable of providing different kinds of QoS classes for users.

The radio interface between the user terminal and the satellite is different in uplink and downlink. In the uplink direction a technology called multifrequency time-division multiple access (MF-TDMA) is used. In the downlink a different technology, asynchronous TDMA (ATDMA), is in use.

In the uplink direction MF-TDMA employs multiple frequency channels in parallel (usually 1–128 frequencies) for data transmission. Each frequency channel can provide a 16-Kbps data rate. Each channel is exclusively allocated to only one user at a time.

In the downlink direction ATDMA does not allocate the channels for different users. In fact, there are no channels in the same meaning as in the uplink direction. One TDMA frame consists of 1,800 timeslots, each lasting 2.276 ms. A user terminal does not know which timeslots are allocated to it, so it has to receive all packets (timeslots) and check the address information in each of them. This consumes power, but it is not seen as a problem, as Teledesic terminals are of the fixed-site type; thus, they probably have an external power supply.

The layout of cells in the Teledesic system differs fundamentally from other satellite networks, as its cell footprints on Earth are stationary, even though the satellites are moving relative to Earth. This concept and the differences from other systems are explained further in Section 15.2.4.

Teledesic plans to start its operation in 2005. See also [5] for further information.

15.2.3.5 Orbcomm

Orbcomm is a LEO system; the orbital height is 775 km. There are 35 very small satellites in space (as of June 2000) and more will be launched in the future. Orbcomm does not use intersatellite links; thus, because of the low orbit, it needs lots of ground stations. The radio interface technology is CDMA.

Orbcomm is a narrowband messaging system aimed primarily at tracking applications for equipment and goods rather than at mobile human subscribers. The data rates are quite modest. As the applications suggest, the system is used to deliver short paging-type messages; it can do this in both the uplink and downlink directions. The system is mainly used for monitoring and tracking applications. The satellites receive small data packets from sensors in vehicles, containers, vessels, or remote fixed sites and relay them to Earth stations. The system can also be used to send short alphanumeric messages for personal messaging.

The system uses the 137–138-MHz and 400-MHz frequency bands for transmissions down to mobile or fixed data-communications devices; and the 148–150-MHz frequencies for transmissions up to the satellites.

Orbcomm's network control center (NCC) is located at Orbcomm's headquarters in Dulles, Virginia. More information can be found through Orbcomm's Web page [6].

15.2.3.6 Skybridge

Skybridge is a broadband satellite system designed for data transfer. The terminals are of the fixed type. They are composed of an outdoor unit (typically on the rooftop) and an indoor unit. This equipment can be connected to a wide variety of appliances: PC, telephone, TV, LAN, and so forth.

Two distinct families of terminals are being developed: a residential terminal, which can provide a bit rate of up to 20 Mbps on the forward link and 2 Mbps on the return link; and a professional/collective (shared by several simultaneous users) terminal, which can provide the maximum bit rate of 100 Mbps on the forward link and 10 Mbps on the return link.

The space component is composed of a constellation of 80 bent-pipe[2] LEO satellites plus spares. There are two identical subconstellations of 40 satellites each with 20 planes, each plane containing 4 satellites. All 80 satellites are in a circular orbit at an altitude of 1,469 km, with an inclination of 53 degrees. The satellites are simple; no intersatellite links exist. Each satellite creates 18 spot beams. These satellites are served by approximately 140 gateway stations.

Both the uplink and downlink traffic will use at least 1.05 GHz of spectrum in both directions. The uplink is allotted a spectrum between 12.75 and 14.5 GHz and the downlink between 10.7 and 12.75 GHz. The system is expected to be ready for service in 2003.

Skybridge is developed by Alcatel. The latest information about this system can be gathered from [7].

15.2.3.7 Thuraya

Thuraya is a regional MSS system. It is a UAE (Abu Dhabi)–owned mobile satellite system. It will provide telecommunication services mainly to the Middle East and Africa. The Thuraya mobile satellite system is a turnkey project being built by Hughes at a cost of US $1 billion.

The space segment comprises one operational satellite in GEO orbit (launched in late 2000) located 36,000 km above the equator with one spare satellite on the ground ready for deployment. The service life of Thuraya satellites will be 12 years.

Satellites will have 250 to 300 spot beams. They also have digital beam-forming capability, which will allow for reconfiguring beams in the coverage area, enlarging beams, and activating new beams. This also allows the system to maximize the coverage of hot spots, or those areas where extra capacity is required. The uplink frequencies are 1,626.5 to 1,660.5 MHz and the downlink 1,525 to 1,559 MHz. The radio interface technology is FDMA. The system is capable of handling as many as 13,750 simultaneous calls.

Thuraya's handheld mobile terminals are comparable to GSM handsets in terms of size, appearance, and provided services. The system is further discussed in [8].

15.2.4 Location in Satellite Systems

The concept of location can differ in satellite networks from how it is understood in terrestrial cellular networks. In a land-based network, the base stations, thus, the cells, are stationary. Therefore, it is suitable to use cells for location-determination purposes. A location area is formed by a cell

---

2 Bent-pipe type satellites are simple satellites. In short, they are repeaters, which retransmit incoming signals on a different frequency.

or group of cells and assigned a unique identity. The mobile station informs the network every time it camps onto a cell that belongs to a different location area from the previous cell. The location information of the mobile is stored in a register in the network, so that the network knows the whereabouts of the mobile and can route mobile-terminated call attempts (paging) to the appropriate cell(s).

Figure 15.7 depicts the example of a mobile station moving from cell 15 to cell 16; this does not trigger a location update because the cell stays within location area A. However, once the mobile moves further from cell 16 into cell 22, a location update is needed because cell 22 belongs to location area B.

In a typical satellite system (LEO, MEO, or HEO, but not GEO, with its stationary satellite footprint), the satellites are moving in relation to the Earth, possibly at a very high speed. This means that attaching the location concept to cells is not a good idea. Especially in LEO systems, the mobile station has to reselect a new cell every few minutes. If the location identity had been attached to cell(s), the resulting location-update signaling traffic would easily clog the network. Note that these location updates would be generated by every switched-on handset, so the amount of signaling traffic would be very high.

A better solution is to attach the location concept to geographical coordinates on Earth. There are several methods a mobile can use to calculate its location, some of which are discussed in Section 12.1. Note that the discussion in that section concentrates on location determination in terrestrial cellular networks. Satellite networks can also use other proprietary methods to solve this problem. Generally, the location determination in satellite networks for location-update purposes does not have to be very accurate. The satellite cells are quite large, and the network cannot address the paging message to a smaller area than a cell. Once the location is known, the mobile compares it with the stored location value, and if they differ more than some predefined tolerance, then a location update is triggered. These coordinates are stored into a network database, possibly with a location error estimation. This error estimation depends on some kind of specific location-determination method. While this method is used, a stationary mobile

FIGURE 15.7
Location areas in a terrestrial network.

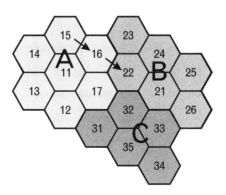

station should never generate location-update traffic to the network except if periodic location updates are required. When the network wants to initiate a connection, it checks which satellite cell(s) cover the stored location of the mobile and sends a paging message via these satellite beams. In Figure 15.8, the network would send the paging via cells 11, 16, and 17. The mobile station should update its location in the network registers if it notices that it has moved more than a predefined amount from the stored location.

It is also possible to use "stationary" cells in satellite systems. Teledesic, for example, plans to use such. The footprints of satellite beams do not move continuously along with the satellites. Instead they are directed at certain fixed areas on the ground. These areas are called supercells. While a satellite moves (and a Teledesic satellite moves fast relative to the Earth's surface), the satellite beam tracks the same supercell on the ground for a long time, until it moves to track the next supercell assigned to a particular satellite beam. This operation will not be done before a beam from another satellite has started to cover the supercell in question.

The Teledesic system intends to divide the Earth's surface into 20,000 supercells. Each supercell is further divided into nine smaller square cells. These nine cells are scanned sequentially by a satellite beam, but a satellite can scan 64 supercells simultaneously.

Ground-based cells require less administration than satellite-based cells. They will generate much less HO signaling traffic. As Teledesic terminals are mostly fixed and stationary, there should not be any HOs in that system.

15.2.5 Restricted Coverage

Satellites systems are global, or at least regional with very large regions. This means that satellite network operators must get operating licenses from all countries within the planned coverage area. Sometimes this is not as simple as it sounds. There may be various factors preventing a successful outcome. All countries may not want to grant operating licenses to satellite operators. A satellite operator may be a competitor to local telecommunication operators (in many countries, international telecommunication connections are handled by state-owned companies, and often they generate sizable profits). Also, it is possible that a satellite network is not allowed to provide its services to some countries for political reasons. Whatever the reason, the satellite

FIGURE 15.8
Location concept in satellite network.

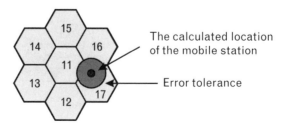

operator may have to implement a method to prevent the use of the network in defined areas, even if these areas are within the system coverage area.

A possible mechanism to handle this problem is to reject location-update requests if they are sent from barred areas. If the location information in the location-update-request message indicates that the mobile is now in a restricted area, a reject message is sent as a response. This denies the mobile handset from either receiving or initiating normal calls. Emergency calls may or may not be allowed. The problem with this scheme is that if the location-determination accuracy in the MS is poor, then such a mobile located near the restricted area may be denied service (see Figure 15.9).

15.2.6 Diversity

Satellite signals are sensitive to fading due to shadowing. Any obstacles in the transmission path will cause severe signal attenuation. Only a line-of-sight (LOS) connection is typically good enough for satellite connections. Shadowed or reflected signals are a problem, especially in urban areas. The probability of shadowing grows when the satellite elevation angle drops.

An efficient way to combat shadowing is to use satellite diversity (see Figure 15.10). This concept means that a logical connection to a mobile station is implemented using two (or more) physical connections. The same signal is duplicated in the network and relayed to the mobile station via more than one satellite. The mobile station receives these signals, and combines them into a composite signal, which most probably is of better quality than the component signals. The idea behind diversity is that if some obstacle causes shadowing to the mobile, another satellite in another direction can still serve the mobile station. The drawback is that one logical connection uses more physical resources than a connection without satellite diversity.

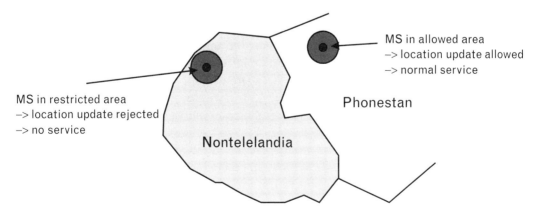

FIGURE 15.9   *Restricted area in satellite networks.*

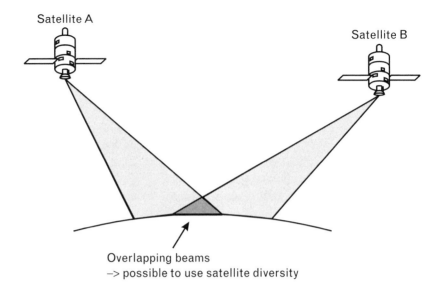

FIGURE 15.10
Satellite diversity.

Satellite A

Satellite B

Overlapping beams
-> possible to use satellite diversity

This method can only be used in systems that have several satellites visible at the same time. Of the systems presented earlier, at least Globalstar and ICO use satellite diversity.

15.2.7 Satellite Paging

Satellite networks can typically provide only outdoor coverage. This is because of the high level of signal attenuation caused by the long distance from the transmitter to the receiver. However, most of the time users will be indoors, and the inability to reach them there would severely hamper the usefulness of the system.

The most common solution is to include a high-power paging service in the satellite network. Once a satellite notices that it has not received a response to an ordinary paging message sent to start a mobile-terminated call procedure, it starts to send special high-power paging messages via a special channel to the mobile terminal in question. This signal should be able to reach a mobile even when it is located indoors. Once a user gets this notification, he or she knows that there is an incoming call waiting, and it can be received by going outdoors (or to a window that has a LOS to the satellite).

Notice that ordinary paging and high-power paging are two separate concepts. Ordinary paging messages are used to start an incoming call procedure. But this can only be done outdoors. High-power paging messages are an alert service, which can be used to move the user outdoors.

Another solution to this problem is to use small antenna repeaters to relay the signal indoors. A satellite antenna is placed on a rooftop or attached to a window, and then the signal is relayed indoors either via a wire or using a wireless connection. Note that this idea is used with many fixed-terminal satellite systems, such as Teledesic. A fixed outdoor antenna installation is not a problem with those systems. For mobile users, the extra antenna should be lightweight and easy to use and deploy, so that even a small antenna attached to a window with a suction cup and a wireless link to a mobile station would work.

15.2.8   IMT-2000 Satellite Component

Five proposals for the IMT-2000 satellite radio transmission technology (RTT) have so far been made to the ITU. It is quite uncertain whether there will ever be a satellite component in 3G standards, especially given the current problems if MSS systems. It may well be that if some of the earlier mentioned MSS systems prove to be commercially viable, it will then be given a IMT-2000 satellite system status. However, note that in IMT-2000 frequency spectrum there are allocations for MSS systems.

IMT-2000 satellite component proposals include the following:

- SAT-CDMA (by TTA Korea);

- SW-CDMA (Satellite W-CDMA, by the European Space Agency);

- SW-CTDMA (Satellite Wideband Hybrid CDMA/TDMA, by the European Space Agency);

- ICO RTT (Intermediate Circular Orbit by ICO Global Communications);

- Horizons (by Inmarsat);

More information about ITU IMT-2000 proposals can be gathered from [9].

15.3   3G Upgrades

As will be explained in the following chapter, the 3GPP standard does not form one fixed and immutable set of specifications, but it is a continuously evolving system. New upgrades are published in releases. The first deployed networks are Release 99– or Release 3–compatible systems. Thereafter we will see Release 4 systems, Release 5 systems, and so forth. The basic principle with new releases is that they should be backwards-compatible with earlier releases. That means that a user with a Release 3 UE will be able to use this UE also in a Release 5 network without problems, although he may not be able to take advantage of its new Release 5–specific features.

Table 15.1 contains the most important new features in Releases 5 and 6. These items are not all further explained here, as it would quite easily fill another book.

Some trends can be identified from this table. 3GPP is clearly aiming at increase the data rates and capacity of the air interface, especially in the downlink. If and when 3G takes off, the downlink as defined in Release 3 will become a bottleneck. On the other hand, this list includes several initiatives that try to make the life of an application developer easier. The goals

TABLE 15.1   NEW FEATURE IN RELEASES 5 AND 6

RELEASE 5	RELEASE 6
IP transport in the UTRAN	Multiple Input Multiple Output (MIMO) antennas
Evolution of the transport for CN	Feasibility study for the viable deployment of UTRA in additional and diverse spectrum arrangements
High Speed Downlink Packet Access (HSDPA)	RAN improvements
Improvements of radio interface	Security enhancements
RAN improvements	Multimedia Broadcast/Multicast Service (MBMS)
IP-based Multimedia Services (IMS)	Push Services
Extended streaming	WLAN/UMTS interworking
Global text telephony	MExE enhancements Release 6
OSA improvements	LCS enhancements 2
CAMEL phase 4	IMS phase 2
MExE enhancements	Generic user profile
Wideband Adaptive Multirate Codec (WB-AMR)	Digital rights management
(U)SIM toolkit enhancements	Speech-recognition and speech-enabled services
LCS enhancements	Feasibility study on priority service
Charging and OAM&P	Release 6 OSA enhancements
User equipment management	Emergency call enhancements Release 6
Intradomain connection of RAN nodes to multiple CN nodes (Iu-Flex)	Support for subscriber certificates
GERAN enhancements	UICC/(U)SIM enhancements and interworking
End-to-end QoS	Subscription management
Messaging enhancements	Trace management
Service change and UDI fallback (SCUDIF)	Presence
Network sharing (NetShare)	UE functionality split
	QoS improvements

here are open and common application platforms so that the same application could be run on different user devices without modifications.

15.4   Downlink Bottleneck

In the following, the downlink bottleneck problem and its possible solutions are addressed further. Many predicted 3G services are downlink-heavy, requiring considerably more downlink than uplink capacity. However, the mainstream 3G system, namely the 3GPP FDD mode, employs symmetric spectrum bands; that is, the amount of bandwidth allocated for both the uplink and the downlink carriers is identical. This also translates into roughly similar uplink and downlink maximum data rates. However, many predicted usage scenarios will require much higher capacity in the downlink. Internet browsing, news-on-demand, and mp3 music downloads are examples of applications where the downlink carries the majority of the traffic, and the uplink typically only carries control commands for the purpose of downlink traffic management. This will result in a situation where the downlink has used up all its capacity and becomes a bottleneck, whereas the uplink has plenty of capacity available. A solution whereby new symmetric spectrum bands are allocated for 3G is not acceptable, as this will result in under-utilized uplink carriers. Radio spectrum is a scarce and expensive resource not to be wasted.

15.4.1   TDD

The first and the most obvious solution is already specified by 3GPP. As explained earlier, the 3GPP concept encompasses two different air interface technologies: FDD and TDD modes. The mainstream mode is FDD. The first networks to be launched will use FDD technology, which, as described earlier, will employ symmetric spectrum bands and, thus, will suffer from the downlink bottleneck problem.

The TDD mode, however, is a time-division technology in which the same frequency carrier is used for both the uplink and the downlink traffic. There are 15 timeslots in each frame, and the number of timeslots allocated for uplink and downlink can be dynamically chosen, thus providing a means to adjust the capacity according to traffic distribution between uplink (UL) and downlink (DL).

The problem with this scheme is that the dynamic timeslot allocation is dynamic only in theory. In practice the timeslots cannot be freely allocated in each cell, as cells near to each other and employing different allocations will interfere with each other. This problem was explained earlier in Chapter 1. UEs, particularly in the boundary areas of such cells, may experience problems with overlapping uplink and downlink timeslots. The simplest solution to this problem is to use the same fixed uplink/downlink split in all cells, for example 10 timeslots down/5 timeslots up, as in Figure 15.11.

Probably a bigger obstacle for operators to adopt the TDD mode is the fact that it is technically so different from the FDD mode. So far all announced infrastructure contracts have been for the FDD mode technology. If a FDD operator wants to upgrade its network with TDD mode equipment, it will be a difficult exercise. It may be possible to integrate both FDD and TDD modes within one base station, but it is not a trivial upgrade. Moreover, customers would have to buy new expensive dual-mode handsets. The TDD mode is problematic in that being a time-division technology, it is rather difficult to build high-performance receivers that perform well when they are moving at high speed.

Note, however, that in China there is much interest in a locally developed 3G system, TD-SCDMA, that is based on the TDD technology (see Section 1.4.2). TD-SCDMA is quite similar to the wideband TDD mode, except that its carrier bandwidth (1.6 MHz) and chip rate (1.28 Mcps) are one third of the corresponding values of wideband TDD. It may well be that TD-SCDMA is adopted at least by some Chinese operators.

15.4.2   HSDPA

High Speed Downlink Packet Access (HSDPA) is a major work item for Release 5 in 3GPP. It was already given a thorough treatment in Chapter 12, so the technical details of HSDPA will not be further discussed here. HSDPA is designed to enhance the downlink data transmission capacity in 3GPP systems. It will increase the theoretical maximum data rate the system can provide, but what is even more important, it will increase the throughput in practical usage scenarios. The dynamic and adaptive nature of HSDPA channels mean that they will be able to transmit the highest possible amount of data frame after frame, even in poor and changing radio channel conditions. The HSDPA improvements are limited to the downlink; thus, it is very relevant to the discussion in this section.

HSDPA, as specified in Release 5, can provide theoretical maximum throughput of 14.4 Mbps, but this will increase later, for example with the adoption of MIMO antennas. MIMO is a Release 6 feature, and a MIMO-

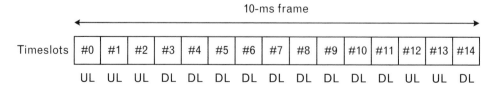

Figure 15.11   TDD frame with asymmetric DL/UL timeslot allocation.

enhanced HSDPA system (with four Tx/Rx antenna pairs) could possibly provide downlink rata dates in excess of 20 Mbps in an ideal environment.

It must be noted here that HSDPA employs a shared downlink channel for data transmission; thus, the theoretical data rates given earlier have to be shared between all active HSDPA users in the cell or sector. If HSDPA channels are deployed on the same carrier as the "normal" 3GPP channels, then a further problem may come from the lack of available spreading codes, as all these channels will share the same code space.

15.4.3 WLAN Interworking

Not all wireless data is transferred via cellular networks. In fact, the majority of wireless data is still in noncellular systems. WLAN devices have been around for many years, and their numbers have grown with increasingly popular laptop computers. So far WLAN systems have been used in offices, on campuses, and in other private environments. Only members of the private organization can use the system; thus, these systems are called *private WLAN* systems. Recently some operators have introduced *public WLAN* systems. These pilot networks typically cover airports, hotels, and conference centers. All users subscribing to the network have access to it. The charging is based either on a fixed monthly fee, the amount of data transferred, or some combination of these.

A public WLAN radio access network interworking with a cellular network offers an efficient way to enhance mobile cellular networks. It can be combined with an existing 3G or even a 2G network to provide extra capacity in traffic hot-spots.

There are several different WLAN technologies specified, and more are in the pipeline. The most important are the IEEE-defined 802.11 family of standards, and the ETSI-defined HiperLAN standards.

The 802.11 standards family [10] consists of the basic 802.11 standard, and a set of amendment standards, such as 802.11a and 802.11b. Each of these amendments also has to implement the 802.11 core standard. The most important standards at the moment include the following:

- 802.11a: 5-GHz band, max. 54 Mbps, to be widely used in North America, and also in Europe when the 5-GHz band becomes available;

- 802.11b: 2.4-GHz band, max. 11 Mbps, already widely used in Europe;

- 802.11g: 2.4-GHz band, max. 20+ Mbps, backwards compatible with 802.11b with 802.11a style modulation.

HiperLAN2 is a new WLAN technology [11, 12] that operates in the 5-GHz band and can provide raw data rates up to 54 Mbps. HiperLAN2 has not gained much support in the industry yet, but it is technically quite advanced and may attract customers once the spectrum situation in the 5-GHz band in Europe clarifies.

Two different methods have been proposed for integrating WLAN and cellular technologies: loose and tight interworking [13–15]. *Loose interworking* means that WLAN and cellular systems remain almost entirely apart. The WLAN access network is attached to the Internet backbone, and the cellular radio access network into the cellular core network. The access networks do not have anything in common, but the core networks are connected to each other (see Figure 15.12). Some extensions are proposed for the WLAN user device, so that certain essential tasks for a public access system can be handled, such as charging and authentication. Nokia has demonstrated a working solution in [16], in which the WLAN device contains a GSM SIM card. SIM-based charging and authentication control signaling is relayed to the GSM core network elements via the Internet. The loose interworking model has the advantage that very few changes to system specifications are needed, and in many cases standard WLAN and cellular equipment can be used. Again, new dual-mode handsets are needed, but these would be relatively simple in this case.

Tight interworking means that WLAN technology is employed as a new radio access technology within the 3G cellular system. Regardless of the access technology, there would only be one common cellular core network. At least two methods to integrate WLAN technology into 3G networks have been proposed. In [15] a WLAN access point (AP) would be attached to a radio network controller (RNC) via the Iub interface (see the lower part of Figure 15.13). The ETSI BRAN project [17] proposes a scheme in which the whole radio access network (including the base station controller) is WLAN specific, and this access network would attach into the 3G core network via Iu-interface (see the upper part of Figure 15.13).

At the time of this writing, the loose interworking model seems to be receiving the most support, probably because it requires very few changes to existing systems; thus, it is a cheaper option than tight interworking. 3GPP is also studying this option and will probably define a loose interworking model that is independent of any particular WLAN technology.

The cost of WLAN deployment depends primarily on the required coverage area. WLAN access points as such are mass-market devices and are very affordable, but their range is quite limited. A wide-coverage WLAN network would be quite expensive to build, not only because of the large number of access points needed, but also because of the cost of the broadband transmission media from the APs.

However, it can be argued that an omnipresent WLAN layout is not necessary. Users will only consume very high data-rate services in relatively

FIGURE 15.12
Loose interworking model.

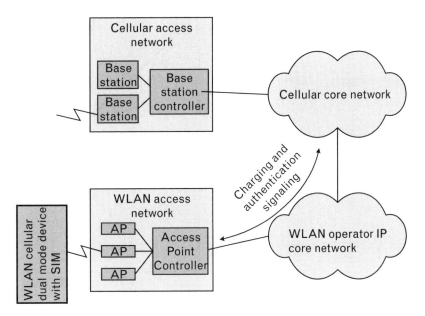

FIGURE 15.13
Tight interworking models.

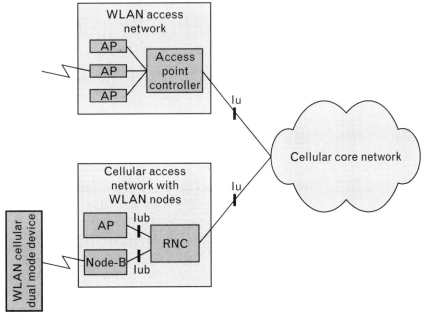

constrained spaces. These services require more attention from a user. The applications may be interactive requiring constant input and may need eye contact. It seems that the higher the application data rate, the lower the mobility of a user consuming it owing to the fact that high-bandwidth applications will require more attention from the user. This effect is demonstrated in Figure 15.14. A WLAN device can be moved around, but when it communicates, it is likely to stay motionless. Thus, it can be argued that

FIGURE 15.14
Mobility/data rate.

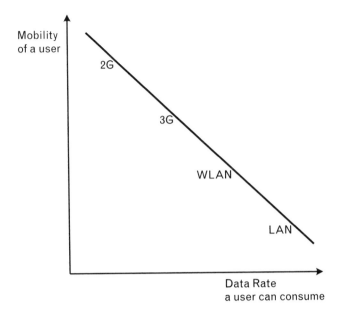

WLAN devices will only be used indoors, for example in offices, airport lounges, cafes, or at home. Low(er) speed data services can be handled by 3G, or even by 2.5G cellular networks. These are likely to be ubiquitous, or nearly so. Thus, using the WLAN interworking model to add downlink capacity to places where it is needed is likely to be quite an affordable option.

15.4.4 Variable Duplex Distance

In the first phase of 3GPP, the networks will use a fixed duplex frequency spacing. This means that the frequency separation between the uplink and the downlink carrier is kept the same. For the frequency band used in Europe, this separation will be 190 MHz. Once the UE has found a suitable cell and read the system configuration from the cell's downlink system broadcast, it will send its access burst on a frequency carrier that is exactly 190 MHz below the frequency of the downlink carrier it read the system information from.

Fixed duplex separation brings problems because there has to be one uplink carrier for each downlink carrier, even if there is much less data on these uplink carriers. If there were a method for the network to indicate to the UE that variable duplex frequency spacing was being employed, it would be possible to associate several downlink carriers with only one uplink carrier (see Figure 15.15). An operator could increase its capacity in traffic hotspots simply by increasing the number of downlink carriers, and associate those with existing uplink carriers, providing that there is enough capacity in those uplink channels.

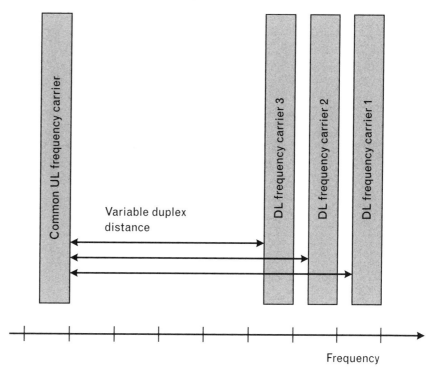

FIGURE 15.15
Variable duplex carriers.

This solution is quite attractive in that very few changes are required to the current specifications. It has been proposed in [18] that an upgraded variable duplex network should broadcast a new system information message SIB 5 bis containing the duplex distance for this carrier, and only those UEs that can read this message can actually access the network. The obvious problem with this solution is that old UEs without the variable duplex capability will not be able to access the network at all.

For the handset transmitter architecture, this solution is easy to implement, providing that the duplex distance remains large enough.

Note that the fixed duplex spacing in current 3GPP standards is in principle only applicable for common channels (i.e., in practice for the random access channel, RACH, that is used for sending the access request to the network). After receiving the access request from the UE, the network has the option to instruct the UE to switch to another uplink frequency once it starts using the dedicated channels that are used for user data transmission. But in fact, this new carrier assignment does not help with the initial problem, as each downlink carrier still needs to have a twinned frequency in the uplink, even if this is only used for sending the access request. Only the variable duplex distance scheme removes the need for twinned uplink carriers.

However, it must be noted that this scheme would not help at all with the existing spectrum allocations. Typically, each 3G operator has so far been granted only 2 or 3 paired carriers (i.e., 2 to 3 uplink and as many downlink carriers). An operator could not obtain any gain from variable duplex distance capable handsets in this case. But this scheme is an enabling technology. It would make it possible for operators to apply for asymmetric frequency spectrum in the future and use it effectively.

15.4.5 Hierarchical Cell Structures

This solution is in fact an old one, already employed by many 2G cellular networks.

This concept involves using overlapping cell structures especially in high traffic density areas. Each hierarchy level requires preferably its own frequency carrier in WCDMA. An operator with three frequency carriers will be able to use three hierarchical levels. Indeed, the UMTS Forum recommends that a 3G operator should be given at least 2 × 15 MHz paired bands (i.e., three duplex frequency carriers) plus 5-MHz unpaired spectrum [19]. Most 3G licences granted so far have been for 2 × 15– or 2 × 10–MHz spectrum allocations.

Macrocells are typically the highest hierarchy level, where a macrocell may be several kilometers in diameter. A microcell is several hundreds of meters in diameter, and a picocell a few tens of meters in diameter or smaller. The smaller the cell size, the higher the system capacity. Each cell can reuse the available code space. In principle, each cell in WCDMA can deliver about the same amount of data in the downlink direction regardless of its size. Increasing the number of cells by making them smaller increases the amount of data the network can transmit in the same area.

Hierarchical cell structures are a rather costly way to increase the downlink capacity. Traffic hotspots will require lots of new base stations. The problem is not only the cost of the actual new equipment, but also the difficulty of acquiring sites for the new base stations. In addition, the uplink spectrum would be underutilized. On the other hand, the users would probably like this solution, as they could continue using their old standard handsets.

The basic problem with this approach is that hierarchical cell structures do not increase the (quite limited) maximum data rate one cell can provide. A UE just cannot get 5 Mbps from a standard FDD mode base station, no matter what size the cell is.

15.4.6 Comparing the Schemes

All the proposed solutions have their advantages and disadvantages. In the end the choice is the mobile operators' who have to consider whether the solution solves the problem (downlink bandwidth), what it costs, and if the new system is easily manageable (i.e., how to do radio and capacity planning).

It should be understood that the pure maximum data rate of a system does not reveal everything about the practical achievable data rate. For example, new WLAN technologies can deliver very high maximum data rates, but these can only be achieved with relatively large data packets. However, typical user applications in the mobile world use short data packets. Thus, the practical throughput in WLAN systems can often be much smaller than the theoretical maximum.

Another important parameter to consider is the transmission delay. Even if the data throughput is high enough, the system may be unusable for an application if the delay is too long or there is too much jitter.

It is difficult to compare the costs of the proposed enhancements. The total cost of a new system consists of several components: R&D, network equipment costs, handset costs, deployment costs, and operating costs. Also the planned coverage area and the deployment schedule have an impact on the cost of the system. An ideal upgrade would be one that could coexist with the old system, and that could be deployed gradually to areas where extra capacity is needed.

The capability to do efficient radio and capacity planning is also important, especially when high-speed systems are considered. CDMA cells have a tendency to "breathe" when they are working close to their capacity limit. This is a phenomenon that must be avoided at any cost. Radio planning is thus especially important in CDMA, as capacity is most often interference limited, and not code limited. With a very high-speed system this is, however, a difficult task as the amount of data transmitted can vary a lot from frame to frame, and as a consequence, so can the interference level.

For WLANs, the radio-planning problem is a very different, though not less challenging, one. The new WLAN technologies typically employ OFDM modulation, so they do not have the CDMA-specific cell-breathing problem. However, WLANs operate in licence-exempt frequency bands, where anybody is free to setup his own system. Thus, it will be impossible to estimate accurately in advance the level of external interference. Interference may become a problem with closely located APs. For example airports are areas where all operators would like to set up their own WLAN systems. Operators should either agree on the frequencies that each one can use (the problem is that there are very few nonoverlapping frequencies), or even better, set up a joint WLAN access network. However, it is still possible that a neighboring private WLAN system could cause interference. Bluetooth devices could also become a major problem in the 2.4-GHz band. Some relief to this problem can be expected if the 802.11h standard is employed. This is a spectrum-managed extension to the 802.11a standard.

Spectrum availability is also an important factor when choosing the new technology. Much depends on whether there is new spectrum available for these upgrades, or if the operator has to cope with the existing spectrum.

And if there is spectrum available, is it paired or unpaired, and what does it cost? Again WLAN interworking is a special case, as the used spectrum is licence exempt. Thus, in this case the spectrum is available, but the operator does not have exclusive access to it.

Some of the proposed solutions are already specified and are ready to be employed when the need arises. The TDD mode has been a part of 3GPP standards from the beginning. Its problem may be that it is so different from the mainstream FDD mode that investments required to deploy it would be too large.

HSDPA is a 3GPP internal scheme to increase downlink throughput. The current 3GPP specifications do not properly support very high-speed data transfer. HSDPA introduces a new downlink channel that is optimized for this purpose. The system includes new features, such as new modulation schemes, shorter frames, and a more efficient re-transmission protocol. HSDPA can be implemented on existing carriers, but significantly higher data rates can only be achieved if new frequency carriers can be used for HSDPA channels. HSDPA is a Release 5 item, so the system will be available in time to solve the expected problem.

WLAN interworking is technically a quite different solution. WLAN capable devices could reach even higher maximum data rates than HSDPA devices, and WLAN would also improve the uplink throughput. The loose interworking model favored by some recent studies would provide an easy way to achieve higher data rates. It would also be possible to use standard WLAN equipment for WLAN-cellular interworking. The problem with this solution is that the character of WLANs makes it very difficult to do any radio planning. Because the loose interworking is, as the name says, loose, any intersystem procedures are difficult to implement. Crude roaming is possible, but, for example, service continuity could be a problem, and certainly the QoS of the active service will suffer when a device roams from one system to another. However, the loose interworking model is technically easy to implement, and in 2002 there are already several pilot systems running in Scandinavia and the United States.

Tight WLAN interworking models have not received much support because they would require new standards to be specified (which is always a slow process), whereas the loose interworking model allows quicker deployment of new systems. However, a tightly integrated WLAN-cellular system would have its benefits, and it is possible that these will be introduced later. Note that 3GPP has considered the use of WLAN-style OFDM wideband carriers to be used in HSDPA systems at some later stage. Thus, the tight interworking model could actually become HSDPA phase II.

Variable duplex distance frequency carriers could provide an easy way to increase downlink capacity, as very few changes to current specifications would be needed. The drawback is of course that new (unpaired) radio spectrum is needed. In addition, the new carriers could only be used by new handsets that have the variable-duplex capability. Fixed-duplex handsets could end-up transmitting their uplink using the incorrect frequency carrier. However, variable duplex distance is still in the feasibility-study phase in 3GPP at the time of this writing.

Hierarchical cell structures are an old but effective solution to capacity problems in 2G, and they will also be employed in 3G. This scheme is very useful, but it has its limitations. High downlink data rates will require large numbers of very small cells, and access points are always costly to set up, as are their back-haul transmission media. Moreover, the basic FDD mode cannot provide very high data rates, no matter how small the cells are.

The best solution is probably a combination of the previously described techniques. And the optimum combination depends on whom we ask. Hierarchical cell structures will be implemented by 3G operators anyway. In addition, they may want to add HSDPA channels to boost the capacity of existing carriers. Because HSDPA is tightly integrated into the 3G system, the services that are used over HSDPA can be managed and charged using standard 3G procedures. Variable duplex would also be a nice addition to the mix, if new spectrum areas were made available.

There are also non-3G operators in the mobile telecommunications business, for example 2G and Wireless Internet Service Provider (WISP) operators. They will want to provide the same (or better) services as 3G operators; thus, they may encounter similar problems with bandwidth. Obviously, these operators cannot adopt 3G-specific solutions, such as HSDPA (even though MIMO could also be adopted in 2G systems). However, WLAN may be quite an attractive alternative for them. It is a relatively inexpensive solution, especially if compared to the licence fees 3G operators have paid in the United Kingdom and Germany.

A 2G operator could upgrade its cellular network to 2.5G with GPRS and EDGE technologies, and then add WLAN coverage to traffic hot-spots (offices, airports, city center pedestrian areas, cafes, etc.) This mix would be very competitive with a 3G network.

For WISPs WLAN interworking provides a way into the lucrative mobile data market. WISPs could act as WLAN access network operators, selling air-time for actual cellular operators (even for several of them at the same time), or they could gain access into cellular networks by becoming virtual network operators and start selling their 4G services to users under their own brand. The cellular component in this case could be a 2.5G network; it does not necessarily have to be a 3G system.

After all, the downlink bottleneck is a problem that operators would be quite happy to encounter. If they have this problem, it means that they also have lots customers using high-bandwidth services. Symmetric uplink-downlink capacity is suitable only for traditional voice calls, but voice calls will not save 3G business models, as they will not deliver sufficient revenue to operators and service providers. If the 3G cellular business is to succeed, it will have to provide bandwidth-heavy services; thus, in the long run, solutions to the downlink bottleneck problem have to be found.

15.5   4G Vision

Foretelling the future is problematic. It is not yet clear exactly what features 3G will actually contain and what kinds of services it will provide. It is, therefore, rather difficult to predict what kind of system will constitute the 4G mobile-telecommunication system. The only safe description is that it is something that will come after 3G. Mobile telecommunication is an area that has developed enormously during the last few years, and the pace of development is not going to get any slower. In 1990 GSM was still under development and not a single GSM network was operational. There were some national (and even regional) 1G mobile phone networks, but they had only a few users, and the offered services typically included only speech with precious few supplementary services. The total subscriber base was probably only about two million worldwide. Not even the bravest telecommunication analyst would have dared to predict 500 million mobile phone users worldwide 10 years later, with penetration rates exceeding 70% in some markets. Our analyst wouldn't have been a telecommunication analyst for long, but locked behind closed doors and provided with some therapy.

Thus, trying to predict the telecommunication systems and services that will be in use in 2010 is outright foolish. Despite this, we will have a try in this section.

Note that the term 4G is not defined yet and nobody knows exactly what it is, so anyone can claim to know the truth about it. The term is used quite liberally and has therefore developed a sense of legitimacy. Marketing personnel love to talk and write about 4G. There are frequent press releases stating that this or that technology will implement 4G. It is good to be a bit suspicious about these.

There will not be a single network branded as 4G in the same way that there is 2G-GSM or 3G-UMTS. Instead, 4G will be a collection of networks and a wide variety of smart devices communicating with each other. Many of those component networks are already in existence in some form, but they keep evolving, and gradually they are given interworking functionality (see Figure 15.16).

The future 4G-compatible networks will certainly include the UMTS and the enhanced GSM deployments, but they may also include wireless LAN (WLAN) networks (e.g., the 802.11-family and HiperLAN2), satellite networks, broadcast networks such as digital audio broadcasting (DAB) or even digital video broadcasting (DVB), and many other types of networks we don't know about yet. The 4G technology will provide for a collection

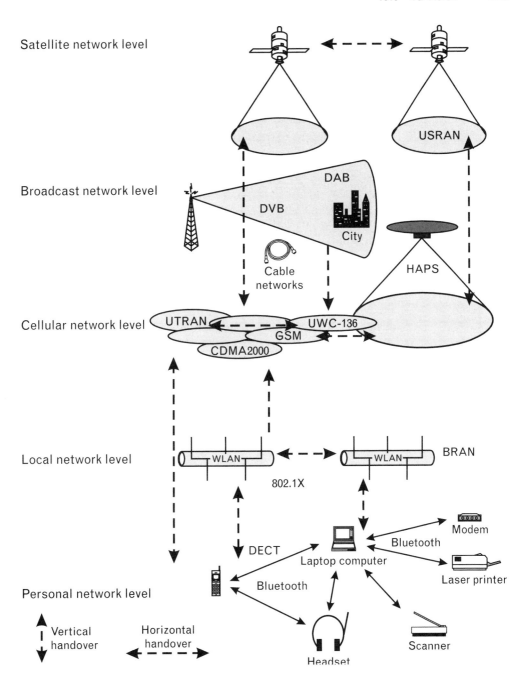

Satellite network level

Broadcast network level

Cellular network level

Local network level

Personal network level

FIGURE 15.16   *4G vision.*

of different kinds of multiple access networks in which a user can gain a portal to the Internet (or whatever the entity) by the most appropriate means. This will be a departure from a one-network-can-do-anything approach.

These networks must cooperate seamlessly. The user will not have to know which network provides the service. The smart device can analyze its environment and choose the best available service provider and technology (service discovery). The selection is made according to the user's preferences stored in a user profile. This evaluation is a continuous process, and if the smart device finds a better system, it can perform a seamless HO to the more appropriate radio interface. The evaluation is based not only on radio path characteristics, as in current networks, but on other parameters, such as QoS, cost, and security.

Both the network entities and the smart devices will be able to configure themselves so that they can adapt to new requirements. This includes software downloads if a device notices that there is a new system available that it cannot use with its current configuration. There will be a wide variety of smart-device classes, the smartest of which will be authentic multipurpose devices capable of altering their functionality according to current requirements. Most devices will, however, be cheaper application-specific appliances that can do only one thing, but do it well and reliably (multipurpose devices such as PCs tend to be quite unreliable).

There will be several "layers" of networks. At the lowest level, the network could consist of the users' smart devices communicating directly with each other ad hoc networks. A laptop computer can send data via a Bluetooth link to a printer and receive data from a fixed network modem via another Bluetooth link. A remote control terminal can send a command via an IR link to a central control unit at home, which then adjusts heating and lighting via wireless links. All these devices are personal; they must include suitable security applications so that nobody else can use them; that is, a neighbor shouldn't be able to adjust your home's heating and security systems. Only very limited mobility is allowed, as links are typically quite short-range. Data transmission speeds may be hundreds of Mbps for some applications.

The next level could consist of LANs, such as WLANs. These networks are typically shared, so there may be other people using them. Mobility is still quite local; a WLAN network may cover an office building or a house. WLANs may also be set up in airports, conference centers, and other traffic hot spots. Data transmission speeds are quite high, certainly tens of Mbps, maybe even more in future WLAN technologies.

The cellular level may consist of 2G and 3G technologies, such as GSM/GPRS and UMTS, or actually enhanced forms of such current systems. The network coverage is very wide, but the offered maximum data rate is not the same everywhere in the coverage area and depends on the UE's location. The transmission speeds may be several Mbps in traffic hot spots, but in the countryside far away from a base station only tens or hundreds of Kbps. Roaming is global.

High-altitude platform station (HAPS) systems can supplement terrestrial cellular systems. These are base stations that are put high above the ground into solar-powered air ships or into ultralight airplanes. The concept has many desirable properties, such as the fact that connections are typically line-of-sight (LOS), and that it is easy to retune and reconfigure the network if necessary. Extra capacity can be allocated to new areas almost instantly when the need arises. On the other hand, it may be difficult to keep the base stations stationary in relation to the ground in certain circumstances.

The highest level of networks (quite literally) are the satellite systems, which will provide true worldwide coverage. Data speeds are lower than those provided by the latest cellular systems. Although the coverage is global, most satellite technologies cannot provide indoor coverage (but they can provide indoor paging).

DAB and DVB networks also provide wide-coverage broadcast services. These services may include return channels that are arranged via other wireless networks, typically via cellular systems.

The user devices must be able to perform both horizontal and vertical HOs. Horizontal HOs will take place within one network layer; vertical HOs will be made between different layers. For example, when a user with a personal communicator arrives in an office building, the smart device makes an HO from an outside cellular network to an indoor WLAN. Technically, managing all the different vertical HOs will be challenging.

At the moment, a mobile user takes out a contract with an operator (or a service provider) who also handles the billing. In 4G there will be several network layers available, and it would be very impractical for the user to make separate contracts with all of his operators. Therefore, the concept of contracts and billing may have to be changed for 4G. A customer could take out a contract with a service broker that stores the user's profile in location server and handles the billing. Different service and content providers and network operators will provide their own services to service brokers, and then get their fees directly from the brokers. The user profile is used by the service broker to decide what kind of services and via what kinds of networks will be provided to the user. The profile is also used in the internetwork HO algorithm to decide the most suitable network for the user.

Core networks will be broadband, and the consensus at the moment seems to be that they will be based on IP technology.

How can this all be achieved? There won't be any 4G standardization programs running for a long time yet. All the 4G components, such as GSM, UMTS, WLAN, DAB, and the like, will evolve separately over the years, and it is their interworking that really needs research effort. UMTS and WLAN interworking studies are already ongoing. The development work will be market driven; only those features and functionalities that are thought to have commercial potential will be implemented.

There will be more than two billion mobile users worldwide in 2010, and about half of them will use 3G systems. This is only the number of users;

there will be even more wireless telecommunication devices around. Each user will probably own several mobiles, or whatever people will call these telecommunication modules in the future. Quite often, these modules will be embedded into other equipment, such as cars, laptop computers, and household appliances. Moreover, there will be communication chips in various kind of vending machines, traffic-control systems, payment terminals, and so on. These modules are starting to understand each other little by little, so the 4G rollout will be a gradual process.

The European Commission has set up a research program called Information Society Technologies (IST) [20], which, among its many projects, handles the issues that will need to be solved for 4G.

REFERENCES

[1]   Lloyd's Satellite Constellations: http://www.ee.surrey.ac.uk/Personal/L.Wood/constellations.

[2]   http://www.iridium.com.

[3]   http://www.globalstar.com.

[4]   http://www.ico.com.

[5]   http://www.teledesic.com.

[6]   http://www.orbcomm.com.

[7]   http://www.skybridgesatellite.com.

[8]   http://www.thuraya.com.

[9]   http://www.itu.int/imt/2_rad_devt/proposals/index.html.

[10]   IEEE 802.11 Web site: http://www.ieee802.org/11.

[11]   Khun-Jush, J., et al., "HiperLAN Type 2 for Broadband Wireless Communication," *Ericsson Review No. 2*, 2000, http://www.ericsson.com/review.

[12]   Johnsson, M., "HiperLAN/2—The Broadband Radio Transmission Technology Operating in the 5 GHz Frequency Band," HiperLAN2 Global Forum White Paper, 1999, http://www.hiperlan2.com/presdocs/site/whitepaper.pdf.

[13]   ETSI TR 101 957 Broadband Radio Access Networks (BRAN); Requirements and Architectures for Interworking between HiperLAN/2 and 3rd Generation Cellular Systems.

[14]   Telia, "Interworking: UMTS and WLAN," presented at the 3GPP Future Evolution Workshop, Helsinki, Finland, October 2001, available at ftp://ftp.3gpp.org/workshop/Future_evolution_0110_Helsinki/tdocs/Set_01/FEW-012.zip.

[15]   Nortel Networks, "Stand-Alone Data Cells for UMTS," presented at the 3GPP Future Evolution Workshop, Helsinki, Finland, October 2001, available at ftp:// ftp.3gpp.org/workshop/Future_evolution_0110_Helsinki/tdocs/set_03/Few-032.zip.

[16]   J. Ala-Laurila, J. Mikkonen, and J. Rinnemaa, "Wireless LAN Access Network Architecture for Mobile Operators." *IEEE Communications Magazine,* November 2001, pp. 82–89.

[17]   ETSI TR 101 683 Broadband Radio Access Networks (BRAN); HiperLAN Type 2; System Overview.

[18] 3GPP TR 25.889 Viable Deployment of UTRA in Additional and Diverse Spectrum Arrangements; feasibility study, 2002.

[19] UMTS Forum Report #5: "Minimum Spectrum Demand per Public Terrestrial UMTS Operator in the Initial Phase," available at http://www.umts-forum.org.

[20] http://www.cordis.lu/ist/ka4/mobile/index.htm.

CHAPTER 16

Specifications

All modern public telecommunication systems are built to conform to some kind of standard. These modern systems are multivendor systems. Standards define interfaces, reference points, and models that help us organize large and complicated systems into functioning networks. The general approach is to make sure equipment from different manufacturers can work together over open interfaces. It is especially important that the air interface is truly open, so that any mobile terminal works with any network in that system, regardless of the network equipment vendor. This requires that all equipment comply with the relevant specifications. For UMTS, these specifications are formulated by 3GPP, the Third Generation Partnership Project. UMTS is a very complex system and the specifications must be very detailed to ensure error-free interoperability.

The purpose of this chapter is to explain the specification development process within 3GPP. This is an important process, but one that is not well known. People tend to think that the UMTS standard consists of only one set of specifications, which, once composed, will be fixed and unchanged, except for some relatively minor error corrections. This is not the case, as the UMTS standards evolves continuously. The 3GPP specifications are issued in releases, and every new release is a complete set of specifications. But even within a release, the specifications can be updated with new versions of the release.

3GPP specifications only state how a certain feature should function. They do not stipulate how that feature or function should be implemented. As long as the external functionality of a feature as seen over an open interface or reference point conforms to the published recommendations, the implementation is deemed to "conform" to the standards. In addition to the normative parts we expect to see in standard recommendations, a specification may sometimes include an informative section or annex that helps to understand the meaning of the specification. This supplemental information may even include a sample implementation. However, the informative parts of a specification are just for information, and they do not set any requirements for an implementation.

Specifications may also include optional features, which may or may not be implemented in a product. Especially in the early phases of the 3GPP work, a convenient way of solving a disagreement over what features should be included in a specification was to include all competing proposals as optional features and leave the decisions to the manufacturers or the operators as to what feature to use. The problem with this approach is that it makes the products quite complex as the peer entity must be prepared to support all optional features the other entity may have implemented. There is a clear tendency, therefore, to reduce the number of optional features in later releases sharply.

16.1   Specification Process

First of all, it must be stressed that the work of 3GPP is very much dependent on the support of its member companies. The 3GPP itself has a relatively small number of support personnel, and they do not compose these specifications by themselves, but only provide administrative support for the process.

The main work force for 3GPP comes from the cellular telecommunication industry. These include infrastructure and handset vendors, operators, service providers, and technology developers. It is in the interests of these companies to provide the required man-power for the work. The sooner a specification is ready, the sooner a company can start selling its products. If a company has developed and patented certain technology, it is also important for it to participate in the standardization process and try to steer the work so that its technology is included in the specifications.

There are two types of 3GPP standardization meetings: Technical Specification Group (TSG) meetings, also known as plenary meetings; and technical Working Group (WG) meetings. Each group is given well-defined tasks and responsibilities as shown in Figure 16.1.

TSGs usually meet four times a year, except TSG GERAN, which meets five times each year. These meetings are attended by relatively senior experts from participating companies. TSGs steer the work of the WGs that serve and support them. TSGs may identify potential new features, initiate technical studies on the feasibility of these features, and based on the results, possibly launch the actual specification work in WGs. TSGs also accept or reject the specification drafts and change requests made by their WGs.

As we would expect in this kind of arrangement, WGs hold their meetings more often, typically four to eight times a year; which works out to about one or two meetings between each TSG plenary. These WG meetings are attended by technical experts who know the discussed features very well. WGs carry out technical studies, compose draft specifications, and write change requests to existing specifications. All these are submitted to the TSG for information or for acceptance. Note that a WG can change a draft specification by itself because it is only an internal working document at that stage, but once the draft has been formally accepted by the TSG, all further changes to it have to be accepted by the TSG itself.

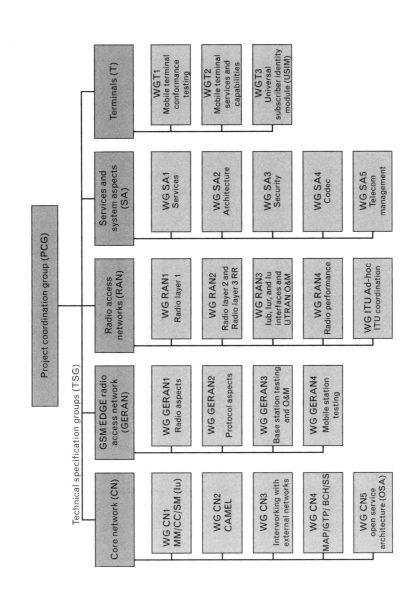

FIGURE 16.1 *3GPP organizational chart.*

Even though all TSGs are in principle equal, TSG SA is, in practice, "more equal" than others. TSG SA discusses the system aspects of 3GPP, and that makes its decisions rather more important than others'. For example, important new features are first considered by TSG SA because only that group can consider them from the whole system point of view. If TSG SA sees that a feature could be beneficial, then the work may start in other groups.

At some point the TSG states that the specification is frozen. This means that no new functionality will be added to it in the current release, only corrections are allowed. However, in the past 3GPP has frozen quite unfinished specifications, which have been lacking necessary technical details and parameters; thus, the term *corrections* actually meant adding contents to a frozen specification framework. Thus, "frozen" does not necessarily mean that the specification is ready for implementation. Sometimes it takes a long time for a WG to get all the technical details right.

Originally there were four TSGs; the fifth group, GSM EDGE Radio Access Network (GERAN) was set up in mid–2000 to continue the specification work of advanced GSM features. 3GPP inherited this task from ETSI SMG (Special Mobile Group). In addition to TSG and WG meetings, 3GPP can also arrange workshops and ad hoc meetings to discuss new and interesting issues or solve specific problems. Note that the specification work does not cease outside these meetings. Each group in 3GPP has its own e-mail discussion list, and a large part of the group's work is done via these lists. The working practices and methods, which are used in 3GPP, are defined in [1].

16.2   Releases

As discussed earlier, 3GPP specifications are issued in phases called *releases*. Originally these releases were named after the year they were supposed to be issued. However, it was quickly discovered that the 3GPP system was becoming more complex than predicted and that the standardization work progress was too slow. It was noticed for example that Release 99 networks could not have been launched until 2002 at the earliest. To prevent confusion, it was then decided to change the numbering so that Release 2000 became Release 4, Release 2001 was renamed as Release 5, and so forth. Release 99 is thus called Release 3 in some sources, but generally the older Release 99 term is still used instead of Release 3. This is also the case in this book. A new release will be issued roughly once a year or slightly less frequently. A release will not remain unchanged once it has been published as corrections and clarifications can be issued with new versions of the release.

Release 99 is the first release that is implemented in live networks. A Release 99 specification can be recognized from the version number 3.x. In simplified terms, this release includes a GSM core network combined with an all-new WCDMA-based radio access network. The services this release can provide are the same as available from GSM, albeit with better attributes in some cases. The handsets are more complex than in GSM, especially as in practice they have to be dual-mode UMTS-GSM handsets in Europe due to patchy 3G coverage. This release was functionally frozen in December 1999. However, it wasn't until 2002 that all the technical details were agreed upon and the ambiguities were removed from the specifications to everyone's satisfaction.

Release 4 is the first upgrade for 3GPP systems. Many operators can also implement a Release 4 network as their first 3GPP system. The following are the most important new features in Release 4:

- Transcoder-Free Operation (TrFO);

- Tandem Free Operation (TFO);

- Virtual Home Environment (VHE);

- Full support of Location Services (LCS);

- The new narrowband, or low chip rate, TDD mode (1.28 Mcps);

- UTRA repeater;

- Robust Header Compression (ROHC).

Release 5 is probably a "bigger" release than release 4. It contains several large and important enhancements for 3GPP systems. One could also quite justifiably say that Release 5 is the first true 3GPP system, as only Release 5's new features make it possible to provide genuine 3G-specific services with attributes that people are expecting from 3G. These features include the following:

- High Speed Downlink Packet Access (HSDPA)

- Wideband AMR (WB-AMR) codec

- IP-based multimedia services (IMS)

- Intra Domain Connection of RAN Nodes to Multiple CN Nodes (Iu-Flex)

- Reliable end-to-end QoS for packet-switched domain

Release 6 features were still under discussion at the time of this writing, but most probably it will contain at least the following new items:

- MIMO antennas

- Multimedia Broadcast/Multicast Service (MBMS), first phase

- WLAN-UMTS interworking

Future releases will then include new features, making the system more flexible, and more powerful. Note that each release has its own independent set of specifications. If a feature is used in several releases (as is the case usually), then each release has its own copy of the specification, and they may be changed independently of each other.

16.3   3GPP Specifications

16.3.1   Series Numbering

3GPP specifications and technical reports are divided into 15 series, which are numbered from 21 to 35. The following list describes the contents of each series. Note that some series may contain only a few specifications, whereas the largest one, the 25-series, includes several hundred specifications. These series apply to 3G only, or to GSM and 3G. There are also other series that apply to GSM only. Series 01 to 12 are pre–Release 4 GSM specifications. Series 41 to 52 are Release 4 and later GSM specifications.

- *21-series* Requirements specifications.

- *22-series* Specifications for services, service features, or platforms for services.

- *23-series* Technical realization series containing specifications describing interworking over several interfaces.

- *24-series* Specifications of signaling protocols between the MS/UE and the core network (i.e., Non–Access Stratum protocols).

- *25-series* UTRA aspects.

  - *25.100-series* The radio performance of UTRAN.
  - *25.200-series* UTRA physical layer 1.
  - *25.300-series* UTRA radio interface architecture, layer 2, and layer 3 aspects.
  - *25.400-series* The Iub, Iur and Iu interfaces within UTRAN.

- *26-series* Codecs (speech, video, etc.).

- *27-series* The functions necessary to support data applications.

- *28-series* Signaling protocols between radio subsystem and periphery of CN (not used in Release 99).

- *29-series* The protocols within the core network.

- *30-series* 3GPP program management.

- *31-series* The user identity module (UIM) and the interfaces between UIM and other entities.

- *32-series* Operation and maintenance aspects of the 3GPP Mobile System network (includes charging).

- *33-series* Security aspects.

- *34-series* Test specifications.

- *35-series* Encryption algorithms for confidentiality and authentication, and so forth.

These specifications are available at http://www.3gpp.org/. There are two kinds of specifications in this Web site: technical specifications (TS) and technical reports (TR). Technical Reports differ from technical specifications in that technical reports are 3GPP working documents, often describing some prestudy carried out before the actual technical specification could be written. These reports can be further divided into two groups. Those reports that are not intended to be transposed into publications are numbered as aa.8bb (e.g., TR 25.834 is a technical report on "UTRA TDD Low Chip Rate Option; Radio Protocol Aspects"). TRs intended for publication are numbered as aa.9bb (e.g., TR 25.950 is a technical report on "UTRA High Speed Downlink Packet Access"). Technical reports are useful reading for people who are studying 3G because they actually contain quite a lot of explanatory text, whereas technical specifications are quite often strictly normative, and difficult to understand. However, technical reports are just reports, and they should never be used as specifications when implementing something. For that purpose, always use the latest specification from the relevant release.

16.3.2   Version Numbering

As explained earlier, 3GPP specifications are evolving continuously, and new versions are released as a result of that. This section explains the version numbering scheme in 3GPP. We start by taking TS 25.331 v3.4.1 as an example. TS 25.331 is the Radio Resource Control (RRC) Protocol Specification. All 3GPP specifications are numbered as versions using an $x.y.z$ notation:

- x gives the release number. Release numbers 0, 1, and 2 indicate that this specification is a draft that is not yet approved by a Technical Specification Group (TSG). For example, $0.y.z$ indicates an early draft, and $2.y.z$ suggests something more mature. A release number of "3" indicates a Release 99 document, while release number 4 indicates a Release 4 specification, etc.

- y gives the technical version number. For the first version of a release, y is always 0. The version number is incremented by one each time a technical change is introduced to the specification. Each TSG meeting that approves one or more change requests will make such an increment. The new version will contain all approved change requests from a TSG meeting.

- z gives the editorial version number. An editorial change is one that does not affect the technical contents of the specification. These can include correction of typos or a clarification of the meaning of a specification. This field is reset back to zero every time the y field is changed.

Thus, version 3.4.1 indicates that this is a Release 3 specification, a technical version 4 (which is actually the fifth technical version, as the numbering starts from 0, and an editorial version 1 (i.e., the second editorial version after 0).

16.3.3 Backwards Compatibility

New releases and versions of the specifications should, in general, be backward compatible with older releases and versions. Backwards compatibility means that equipment that conforms to older versions of the specification can coexist in the system with equipment that conforms to later versions of the same specification. This is especially important for users who will buy expensive 3G devices. They have to be sure that their devices will work in all 3GPP networks, present or future, regardless of the particular release the network operator conforms to. Of course, we cannot expect a Release 99 handset to use all the services a Release 5 network provides, but because Release 5 is backwards compatible with Release 99, the handset can still access all its Release 99 services via the newer network.

REFERENCE

[1]   3GPP TS 21.900, v 4.0.0, Technical Specification Group Working Methods, 2001.

Cellular User Statistics

TABLE A.1   SUBSCRIBERS BY TECHNOLOGY (IN MILLIONS; OCTOBER 2002)

TECHNOLOGY	SUBSCRIBERS	PERCENTAGE (OF TOTAL DIGITAL)
GSM (including all variants)	763.7	71.5
WCDMA	0.142	0.0
CDMA	136.6	12.8
US-TDMA	107.6	10.1
PDC	59.5	5.6
Total Digital	1,067.6	100.0
Total Analog	32.4	

Source: EMC World Cellular Database.

TABLE A.2   MOBILE CELLULAR SUBSCRIBERS IN THE WORLD (IN MILLIONS)

1991	1992	1993	1994	1995	1996	1997	1998	1999	2000	2001 (est)	2002 (forecast)
16	23	34	56	91	144	215	319	491	741	1,030	1,390

Source: ITU.

TABLE A.3   MOBILE SUBSCRIBERS PER 100 INHABITANTS IN SELECTED COUNTRIES IN 2001

COUNTRY	%	COUNTRY	%	COUNTRY	%	COUNTRY	%	COUNTRY	%
Africa	2.95	Americas	26.35	Asia	9.33	Europe	43.77	Oceania	44.95
Botswana	16.65	Argentina	18.61	Bahrain	42.49	Austria	80.66	Australia	57.75
Congo	4.82	Brazil	16.73	Brunei	28.94	Belgium	74.72	New Zealand	62.13
Egypt	4.33	Canada	32.00	China	11.17	Croatia	37.70		
Ethiopia	0.04	Chile	34.02	Hong Kong	85.46	Cyprus	46.43		
Gabon	20.45	Jamaica	26.94	India	0.63	Czech Republic	65.88		
Kenya	1.60	Mexico	21.68	Indonesia	2.47	Denmark	73.67		

Table A.3 (Continued)

Country	%	Country	%	Country	%	Country	%
Libya	0.90	Panama	20.70	Iran	2.67	Estonia	45.54
Mauritius	25.00	Paraguay	20.40	Israel	80.82	Finland	77.84
Morocco	15.68	Puerto Rico	30.65	Japan	58.76	France	60.53
Mozam-bique	0.84	United States	44.42	Jordan	14.39	Germany	68.29
Namibia	5.59	Uruguay	15.47	Republic of Korea	60.84	Greece	75.14
Nigeria	0.28	Venezuela	26.35	Kuwait	24.82	Hungary	49.81
Seychelles	55.15			Lebanon	21.25	Iceland	82.02
South Af-rica	21.00			Macao	43.41	Ireland	72.94
Sudan	0.33			Malaysia	29.95	Italy	83.94
Tanzania	1.19			Pakistan	0.55	Luxembourg	96.73
Tunisia	4.01			Philip-pines	13.70	Netherlands	73.91
Uganda	1.43			Qatar	29.31	Norway	82.53
Zambia	0.92			Saudi Ara-bia	11.33	Poland	26.02
Zim-babwe	2.41			Singapore	72.41	Portugal	77.43
				Taiwan	96.55	Russia	3.79
				Thailand	11.87	Slovenia	75.98
				UAE	71.97	Spain	65.53
						Sweden	79.03
						Switzerland	72.38
						Turkey	30.18
						United Kingdom	78.28

Source: ITU.

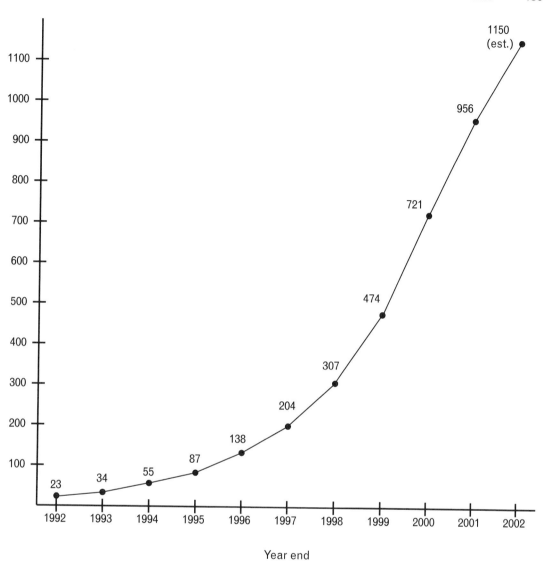

FIGURE A.1   *Subscriber growth worldwide (in millions). (Source: EMC World Cellular Database.)*

Appendix B

3GPP Specifications

Note that the following table lists 3GPP Release 5 specifications as of June 2002. Other Releases may have slightly different lists.

TABLE B.1   3GPP Specifications and Reports

Type	Number	Title
TS	21.103	3rd Generation Mobile System Release 5 Specifications
TS	21.111	USIM and IC Card Requirements
TR	21.801	Specification Drafting Rules
TR	21.877	Radio Optimization Impacts on the Packet Switched (PS) Domain Architecture
TR	21.900	Technical Specification Group Working Methods
TR	21.905	Vocabulary for 3GPP Specifications
TS	22.001	Principles of Circuit Telecommunication Services Supported by a Public Land Mobile Network (PLMN)
TS	22.002	Circuit Bearer Services (BS) Supported by a Public Land Mobile Network (PLMN)
TS	22.003	Circuit Teleservices Supported by a Public Land Mobile Network (PLMN)
TS	22.004	General on Supplementary Services
TS	22.011	Service Accessibility
TS	22.016	International Mobile Equipment Identities (IMEI)
TS	22.024	Description of Charge Advice Information (CAI)
TS	22.030	Man-Machine Interface (MMI) of the User Equipment (UE)
TS	22.032	Immediate Service Termination (IST); Service Description; Stage 1
TS	22.034	High Speed Circuit Switched Data (Hscsd); Stage 1
TS	22.038	USIM/SIM Application Toolkit (USAT/SAT); Service Description; Stage 1
TS	22.041	Operator Determined Call Barring
TS	22.042	Network Identity and Time Zone (NITZ) Service Description; Stage 1
TS	22.053	Tandem Free Operation (TFO); Service Description; Stage 1
TS	22.057	Mobile Execution Environment (MexE) Service Description; Stage 1
TS	22.060	General Packet Radio Service (GPRS); Service Description; Stage 1
TS	22.066	Support of Mobile Number Portability (MNP); Stage 1
TS	22.067	Enhanced Multi-Level Precedence and Pre-Emption Service (EMLPP); Stage 1
TS	22.071	Location Services (LCs); Stage 1

TABLE B.1   (CONTINUED)

TYPE	NUMBER	TITLE
TS	22.072	Call Deflection (CD); Stage 1
TS	22.076	Noise Suppression for the AMR Codec; Service Description; Stage 1
TS	22.078	Customized Applications for Mobile Network Enhanced Logic (CAMEL); Service Description; Stage 1
TS	22.079	Support of Optimal Routing; Stage 1
TS	22.081	Line Identification Supplementary Services; Stage 1
TS	22.082	Call Forwarding (CF) Supplementary Services; Stage 1
TS	22.083	Call Waiting (CW) and Call Hold (HOLD) Supplementary Services; Stage 1
TS	22.084	Multiparty (MPTY) Supplementary Service; Stage 1
TS	22.085	Closed User Group (CUG) Supplementary Services; Stage 1
TS	22.086	Advice of Charge (Aoc) Supplementary Services; Stage 1
TS	22.087	User-to-User Signaling (UUS); Stage 1
TS	22.088	Call Barring (CB) Supplementary Services; Stage 1
TS	22.090	Unstructured Supplementary Service Data (USSD); Stage 1
TS	22.091	Explicit Call Transfer (ECT) Supplementary Service; Stage 1
TS	22.093	Completion of Calls to Busy Subscriber (CCBS); Service Description, Stage 1
TS	22.094	Follow Me Service Description; Stage 1
TS	22.096	Name Identification Supplementary Services; Stage 1
TS	22.097	Multiple Subscriber Profile (MSP) Phase 1; Service Description; Stage 1
TS	22.101	Service Aspects; Service Principles
TS	22.105	Services & Service Capabilities
TS	22.112	USIM Toolkit Interpreter; Stage 1
TS	22.115	Service Aspects Charging and Billing
TR	22.121	Service Aspects; the Virtual Home Environment; Stage 1
TS	22.127	Service Requirement for the Open Services Access (OSA); Stage 1
TS	22.129	Handover Requirements Between UTRAN and GERAN or Other Radio Systems
TS	22.135	Multicall; Service Description; Stage 1
TS	22.140	Service Aspects; Stage 1; Multimedia Messaging Service
TS	22.226	Global Text Telephony; Stage 1: Service Description
TS	22.228	Service Requirements for the IP Multimedia Core Network Subsystem; Stage 1
TS	22.233	Transparent End-to-End Packet-Switched Streaming Service; Stage 1
TR	22.928	IP-Based Multimedia Services Examples
TR	22.941	IP Based Multimedia Framework; Stage 0

TABLE B.1   (CONTINUED)

TYPE	NUMBER	TITLE
TR	22.944	Service Requirements for UE Functionality Split
TR	22.976	Study on PS Domain Services and Capabilities
TS	23.002	Network Architecture
TS	23.003	Numbering, Addressing and Identification
TS	23.007	Restoration Procedures
TS	23.008	Organization of Subscriber Data
TS	23.009	Handover Procedures
TS	23.011	Technical Realization of Supplementary Services
TS	23.012	Location Management Procedures
TS	23.014	Support of Dual Tone Multi Frequency (DTMF) Signaling
TS	23.015	Technical Realization of Operator Determined Barring (ODB)
TS	23.016	Subscriber Data Management; Stage 2
TS	23.018	Basic Call Handling; Technical Realization
TS	23.034	High Speed Circuit Switched Data (HSCSD); Stage 2
TS	23.035	Immediate Service Termination (IST); Stage 2
TS	23.038	Alphabets and Language-Specific Information
TR	23.039	Interface Protocols for the Connection of Short Message Service Centers (SMSCS) to Short Message Entities (SMES)
TS	23.040	Technical Realization of Short Message Service (SMS)
TS	23.041	Technical Realization of Cell Broadcast Service (CBS)
TS	23.042	Compression Algorithm for SMS
TS	23.048	Security Mechanisms for the (U)SIM Application Toolkit; Stage 2
TS	23.053	Tandem Free Operation (TFO); Service Description; Stage 2
TS	23.057	Mobile Execution Environment (MexE); Functional Description; Stage 2
TS	23.060	General Packet Radio Service (GPRS) Service Description; Stage 2
TS	23.066	Support of GSM Mobile Number Portability (MNP) Stage 2
TS	23.067	Enhanced Multi-Level Precedence and Preemption Service (EMLPP); Stage 2
TS	23.072	Call Deflection Supplementary Service; Stage 2
TS	23.078	Customized Applications for Mobile Network Enhanced Logic (CAMEL); Stage 2
TS	23.079	Support of Optimal Routing (SOR); Technical Realization; Stage 2
TS	23.081	Line Identification Supplementary Services; Stage 2
TS	23.082	Call Forwarding (CF) Supplementary Services; Stage 2
TS	23.083	Call Waiting (CW) and Call Hold (HOLD) Supplementary Service; Stage 2

TABLE B.1 (CONTINUED)

TYPE	NUMBER	TITLE
TS	23.084	Multiparty (MPTY) Supplementary Service; Stage 2
TS	23.085	Closed User Group (CUG) Supplementary Service; Stage 2
TS	23.086	Advice of Charge (AOC) Supplementary Service; Stage 2
TS	23.087	User-to-User Signaling (UUS) Supplementary Service; Stage 2
TS	23.088	Call Barring (CB) Supplementary Service; Stage 2
TS	23.090	Unstructured Supplementary Service Data (USSD); Stage 2
TS	23.091	Explicit Call Transfer (ECT) Supplementary Service; Stage 2
TS	23.093	Technical Realization of Completion of Calls to Busy Subscriber (CCBS); Stage 2
TS	23.094	Follow Me Stage 2
TS	23.096	Name Identification Supplementary Service; Stage 2
TS	23.097	Multiple Subscriber Profile (MSP) Phase 1; Stage 2
TS	23.107	Quality of Service (QOS) Concept and Architecture
TS	23.108	Mobile Radio Interface Layer 3 Specification Core Network Protocols; Stage 2 (Structured Procedures)
TS	23.116	Super-Charger Technical Realization; Stage 2
TS	23.119	Gateway Location Register (GLR); Stage 2
TS	23.122	Non-Access-Stratum Functions Related to Mobile Station (MS) in Idle Mode
TS	23.127	Virtual Home Environment (VHE)/Open Service Access (OSA); Stage 2
TS	23.135	Multicall Supplementary Service; Stage 2
TS	23.140	Multimedia Messaging Service (MMS); Functional Description; Stage 2
TS	23.146	Technical Realization of Facsimile Group 3 Service—Non-Transparent
TS	23.153	Out-of-Band Transcoder Control; Stage 2
TS	23.172	Technical Realization of Circuit Switched (CS) Multimedia Service; UDI/RDI Fallback and Service Modification; Stage 2
TS	23.174	Push Service; Stage 2
TS	23.205	Bearer-Independent Circuit-Switched Core Network; Stage 2
TS	23.207	End-to-End Quality of Service Concept and Architecture
TS	23.218	IP Multimedia (IM) Session Handling; IM Call Model; Stage 2
TS	23.221	Architectural Requirements
TS	23.226	Global Text Telephony; Stage 2: Architecture
TS	23.227	Application and User Interaction in the UE; Principles and Specific Requirements
TS	23.228	Ip Multimedia Subsystem (IMS); Stage 2
TS	23.236	Intra-Domain Connection of Radio Access Network (RAN) Nodes to Multiple Core Network (CN) Nodes

TABLE B.1   (CONTINUED)

TYPE	NUMBER	TITLE
TS	23.240	3GPP Generic User Profile Requirements; Stage 2; Architecture
TS	23.271	Functional Stage 2 Description of Location Services
TS	23.278	Customized Applications for Mobile Network Enhanced Logic (CAMEL)—IP Multimedia System (IMS) Interworking; Stage 2
TR	23.815	Charging Implications of IMS Architecture
TR	23.871	Enhanced Support for User Privacy in Location Services (LCS)
TR	23.875	Support of Push Service
TR	23.908	Technical Report on Pre-Paging
TR	23.909	Technical Report on the Gateway Location Register
TR	23.910	Circuit Switched Data Bearer Services
TR	23.911	Technical Report on Out-of-Band Transcoder Control
TR	23.912	Technical Report on Super-Charger
TR	23.955	Virtual Home Environment (VHE) Concepts
TS	24.002	Gsm-Umts Public Land Mobile Network (PLMN) Access Reference Configuration
TS	24.007	Mobile Radio Interface Signaling Layer 3; General Aspects
TS	24.008	Mobile Radio Interface Layer 3 Specification; Core Network Protocols; Stage 3
TS	24.010	Mobile Radio Interface Layer 3—Supplementary Services Specification—General Aspects
TS	24.011	Point-to-Point (PP) Short Message Service (SMS) Support on Mobile Radio Interface
TS	24.022	Radio Link Protocol (RLP) for Circuit Switched Bearer and Teleservices
TS	24.030	Location Services (LCS); Supplementary Service Operations; Stage 3
TS	24.067	Enhanced Multi-Level Precedence and Pre-Emption Service (EMLPP); Stage 3
TS	24.072	Call Deflection Supplementary Service; Stage 3
TS	24.080	Mobile Radio Layer 3 Supplementary Service Specification; Formats and Coding
TS	24.081	Line Identification Supplementary Service; Stage 3
TS	24.082	Call Forwarding Supplementary Service; Stage 3
TS	24.083	Call Waiting (CW) and Call Hold (HOLD) Supplementary Service; Stage 3
TS	24.084	Multiparty (MPTY) Supplementary Service; Stage 3
TS	24.085	Closed User Group (CUG) Supplementary Service; Stage 3
TS	24.086	Advice of Charge (AOC) Supplementary Service; Stage 3
TS	24.087	User-to-User Signaling (UUS); Stage 3
TS	24.088	Call Barring (CB) Supplementary Service; Stage 3
TS	24.090	Unstructured Supplementary Service Data (USSD); Stage 3
TS	24.091	Explicit Call Transfer (ECT) Supplementary Service; Stage 3

TABLE B.1   (CONTINUED)

TYPE	NUMBER	TITLE
TS	24.093	Call Completion to Busy Subscriber (CCBS); Stage 3
TS	24.096	Name Identification Supplementary Service; Stage 3
TS	24.135	Multicall Supplementary Service; Stage 3
TS	24.228	Signaling Flows for the IP Multimedia Call Control Based on SIP and SDP; Stage 3
TS	24.229	IP Multimedia Call Control Protocol Based on SIP and SDP; Stage 3
TS	25.101	UE Radio Transmission and Reception (FDD)
TS	25.102	UTRA (UE) TDD; Radio Transmission and Reception
TS	25.104	UTRA (BS) FDD; Radio Transmission and Reception
TS	25.105	UTRA (BS) TDD: Radio Transmission and Reception
TS	25.106	UTRA Repeater; Radio Transmission and Reception
TS	25.113	Base Station and Repeater Electromagnetic Compatibility (EMC)
TS	25.123	Requirements for Support of Radio Resource Management (TDD)
TS	25.133	Requirements for Support of Radio Resource Management (FDD)
TS	25.141	Base Station Conformance Testing (FDD)
TS	25.142	Base Station Conformance Testing (TDD)
TS	25.143	UTRA Repeater; Conformance Testing
TS	25.201	Physical Layer—General Description
TS	25.211	Physical Channels and Mapping of Transport Channels onto Physical Channels (FDD)
TS	25.212	Multiplexing and Channel Coding (FDD)
TS	25.213	Spreading and Modulation (FDD)
TS	25.214	Physical Layer Procedures (FDD)
TS	25.215	Physical Layer; Measurements (FDD)
TS	25.221	Physical Channels and Mapping of Transport Channels Onto Physical Channels (TDD)
TS	25.222	Multiplexing and Channel Coding (TDD)
TS	25.223	Spreading and Modulation (TDD)
TS	25.224	Physical Layer Procedures (TDD)
TS	25.225	Physical Layer; Measurements (TDD)
TS	25.301	Radio Interface Protocol Architecture
TS	25.302	Services Provided by the Physical Layer
TS	25.303	Interlayer Procedures in Connected Mode
TS	25.304	UE Procedures in Idle Mode and Procedures for Cell Reselection in Connected Mode
TS	25.305	User Equipment (UE) Positioning in Universal Terrestrial Radio Access Network (UTRAN); Stage 2

TABLE B.1   (CONTINUED)

TYPE	NUMBER	TITLE
TS	25.306	UE Radio Access Capabilities Definition
TS	25.307	Requirements on UEs Supporting a Release-Independent Frequency Band
TS	25.308	UTRA High Speed Downlink Packet Access (HSPDA); Overall Description; Stage 2
TS	25.321	Medium Access Control (MAC) Protocol Specification
TS	25.322	Radio Link Control (RLC) Protocol Specification
TS	25.323	Packet Data Convergence Protocol (PDCP) Specification
TS	25.324	Broadcast/Multicast Control (BMC)
TS	25.331	Radio Resource Control (RRC) Protocol Specification
TS	25.401	Utran Overall Description
TS	25.402	Synchronization in UTRAN Stage 2
TS	25.410	UTRAN Iu Interface: General Aspects and Principles
TS	25.411	UTRAN Iu Interface Layer 1
TS	25.412	UTRAN Iu Interface Signaling Transport
TS	25.413	UTRAN Iu Interface RANAP Signaling
TS	25.414	UTRAN Iu Interface Data Transport & Transport Signaling
TS	25.415	UTRAN Iu Interface User Plane Protocols
TS	25.419	UTRAN Iu-BC Interface: Service Area Broadcast Protocol (SABP)
TS	25.420	UTRAN Iur Interface: General Aspects and Principles
TS	25.421	UTRAN Iur Interface Layer 1
TS	25.422	UTRAN Iur Interface Signaling Transport
TS	25.423	UTRAN Iur Interface RNSAP Signaling
TS	25.424	UTRAN Iur Interface Data Transport & Transport Signaling for CCH Data Streams
TS	25.425	UTRAN Iur Interface User Plane Protocols for CCH Data Streams
TS	25.426	UTRAN Iur and Iub Interface Data Transport & Transport Signaling for DCH Data Streams
TS	25.427	UTRAN Iur and Iub Interface User Plane Protocols for DCH Data Streams
TS	25.430	UTRAN Iub Interface: General Aspects and Principles
TS	25.431	UTRAN Iub Interface Layer 1
TS	25.432	UTRAN Iub Interface: Signaling Transport
TS	25.433	UTRAN Iub Interface NBAP Signaling
TS	25.434	UTRAN Iub Interface Data Transport & Transport Signaling for CCH Data Streams
TS	25.435	UTRAN Iub Interface User Plane Protocols for CCH Data Streams
TS	25.442	UTRAN Implementation-Specific O&M Transport

TABLE B.1   (CONTINUED)

TYPE	NUMBER	TITLE
TS	25.450	UTRAN Iupc Interface General Aspects and Principles
TS	25.451	UTRAN Iupc Interface Layer 1
TS	25.452	UTRAN Iupc Interface Signaling Transport
TS	25.453	UTRAN Iupc Interface Positioning Calculation Application Part (PCAP) Signaling
TR	25.854	Uplink Synchronous Transmission Scheme (USTS)
TR	25.856	High Speed Downlink Packet Access (HSDPA); Layer 2 and 3 Aspects
TR	25.857	UE Positioning Enhancements
TR	25.858	Physical Layer Aspects of UTRA High Speed Downlink Packet Access
TR	25.859	User Equipment (UE) Positioning Enhancements for 1.28 Mcps TDD
TR	25.860	Radio Access Bearer Support Enhancements
TR	25.867	Feasibility Study for Wideband Distribution Systems in 3rd Generation Networks
TR	25.868	Node B Synchronization for 1.28 Mcps TDD
TR	25.869	Transmitter Diversity Solutions for Multiple Antennas
TR	25.870	Enhancement on the DSCH Hard Split Mode
TR	25.875	NAS Node Selector Function
TR	25.876	Multiple-Input Multiple-Output Antenna Processing for HSDPA
TR	25.877	High Speed Downlink Packet Access (HSDPA)—Iub/Iur Protocol Aspects
TR	25.878	RL Timing Adjustment
TR	25.879	Separation of Resource Reservation and Radio Link Activation
TS	25.880	Traffic Termination Point Swapping
TR	25.881	Improvement of Radio Resource Management Across RNS and RNS/BSS
TR	25.882	1.28 Mcps TDD Option Base Station Classification
TR	25.883	Direct Transport Bearers Between SRNC and Node-B
TR	25.884	Iur Neighboring Cell Reporting Efficiency Optimization
TR	25.885	UMTS 1800/1900 MHz Work Items Report
TR	25.886	Small Technical Enhancements and Improvements Work Item
TR	25.888	Improvement of Inter Frequency and Inter System Measurement for 1.28 Mcps TDD
TR	25.890	High Speed Downlink Packet Access (HSDPA); User Equipment (UE) Radio Transmission and Reception (FDD)
TR	25.921	Guidelines and Principles for Protocol Description and Error Handling
TR	25.922	Radio Resource Management Strategies
TR	25.931	UTRAN Functions, Examples on Signaling Procedures
TR	25.933	IP Transport in UTRAN

TABLE B.1   (CONTINUED)

TYPE	NUMBER	TITLE
TR	25.942	RF System Scenarios
TR	25.943	Deployment Aspects
TR	25.945	RF Requirements for Low Chip Rate TDD Option
TR	25.951	Base Station Classification (FDD)
TR	25.952	Base Station Classification (TDD)
TR	25.956	UTRA Repeater: Planning Guidelines and System Analysis
TR	25.991	Feasibility Study on the Mitigation of the Effect of Common Pilot Channel (CPICH) Interference at the User Equipment
TS	26.071	AMR Speech Codec; General Description
TS	26.073	AMR Speech Codec; C-Source Code
TS	26.074	AMR Speech Codec; Test Sequences
TS	26.077	Minimum Performance Requirements for Noise Suppresser Application to the AMR Speech Encoder
TS	26.090	AMR Speech Codec; Transcoding Functions
TS	26.091	AMR Speech Codec; Error Concealment of Lost Frames
TS	26.092	AMR Speech Codec; Comfort Noise for AMR Speech Traffic Channels
TS	26.093	AMR Speech Codec; Source Controlled Rate Operation
TS	26.094	AMR Speech Codec; Voice Activity Detector for AMR Speech Traffic Channels
TS	26.101	Mandatory Speech Codec Speech Processing Functions; Adaptive Multi-Rate (AMR) Speech Codec Frame Structure
TS	26.102	AMR Speech Codec; Interface to Iu and Uu
TS	26.103	Speech Codec List for GSM and UMTS
TS	26.104	ANSI-C Code for the Floating-Point Adaptive Multi-Rate (AMR) Speech Codec
TS	26.110	Codec for Circuit Switched Multimedia Telephony Service; General Description
TS	26.111	Codec for Circuit Switched Multimedia Telephony Service; Modifications to H.324
TS	26.115	Echo Control for Speech and Multi-Media Services
TS	26.131	Terminal Acoustic Characteristics for Telephony; Requirements
TS	26.132	Narrow Band (3.1 KHz) Speech and Video Telephony Terminal Acoustic Test Specification
TS	26.140	Multimedia Messaging Service (MMS); Media Formats and Codes
TS	26.171	AMR Speech Codec, Wideband; General Description
TS	26.173	ANSI-C Code for the Adaptive Multi Rate (AMR) Wideband Speech Codec
TS	26.174	AMR Speech Codec, Wideband; Test Sequences
TS	26.190	Mandatory Speech Codec Speech Processing Functions AMR Wideband Speech Codec; Transcoding Functions

TABLE B.1   (CONTINUED)

Type	Number	Title
TS	26.191	AMR Speech Codec, Wideband; Error Concealment of Lost Frames
TS	26.192	Mandatory Speech Codec Speech Processing Functions AMR Wideband Speech Codec; Comfort Noise Aspects
TS	26.193	AMR Speech Codec, Wideband; Source Controlled Rate Operation
TS	26.194	Mandatory Speech Codec Speech Processing Functions AMR Wideband Speech Codec; Voice Activity Detector (VAD)
TS	26.201	AMR Speech Codec, Wideband; Frame Structure
TS	26.202	AMR Speech Codec, Wideband; Interface to Iu and Uu
TS	26.204	ANSI-C Code for the Floating-Point Adaptive Multi-Rate (AMR) Wideband Speech Codec
TS	26.226	Global Text Telephony; Transport of Text in the Voice Channel
TS	26.230	Global Text Telephony; Cellular Text Telephone Modem Transmitter C-Code Description
TS	26.231	Global Text Telephony; Cellular Text Telephone Modem Minimum Performance Requirements
TS	26.233	End-to-End Transparent Streaming Service; General Description
TS	26.234	End-to-End Transparent Streaming Service; Protocols and Codecs
TS	26.235	Packet Switched Conversational Multimedia Applications; Default Codecs
TS	26.236	Packet Switched Conversational Multimedia Applications; Transport Protocols
TR	26.911	Codec for Circuit Switched Multimedia Telephony Service; Terminal Implementer's Guide
TR	26.912	Codec for Circuit Switched Multimedia Telephony Service; Quantitative Performance Evaluation of H.324 Annex C Over 3G
TR	26.937	Transparent End-to-End Packet Switched Streaming Service (PSS); RTP Usage Model
TR	26.975	Performance Characterization of the Adaptive Multi-Rate (AMR) Speech Codec
TR	26.976	Results of the AMR Wideband (AMR-W) Selection Phase
TS	27.001	General on Terminal Adaptation Functions (TAF) for Mobile Stations (MS)
TS	27.002	Terminal Adaptation Functions (TAF) for Services Using Asynchronous Bearer Capabilities
TS	27.003	Terminal Adaptation Functions (TAF) for Services Using Synchronous Bearer Capabilities
TS	27.005	Use of Data Terminal Equipment—Data Circuit Terminating Equipment (DTE-DCE) Interface for Short Message Service (SMS) and Cell Broadcast Service (CBS)
TS	27.007	AT Command Set for 3G User Equipment (UE)
TS	27.010	Terminal Equipment to User Equipment (TE-UE) Multiplexer Protocol
TS	27.060	Packet Domain; Mobile Station (MS) Supporting Packet Switched Services
TS	27.103	Wide Area Network Synchronization

TABLE B.I   (CONTINUED)

TYPE	NUMBER	TITLE
TS	27.104	Objects and Other Constructs for Data Synchronization
TS	27.241	3GPP Generic User Profile Requirements; Stage 3; Access; Common Objects
TR	27.901	Report on Terminal Interfaces—An Overview
TS	28.062	In-Band Tandem Free Operation (TFO) of Speech Codecs; Service Description; Stage 3
TS	29.002	Mobile Application Part (MAP) Specification
TS	29.007	General Requirements on Interworking Between the Public Land Mobile Network (PLMN) and the Integrated Services Digital Network (ISDN) or Public Switched Telephone Network (PSTN)
TS	29.010	Information Element Mapping Between Mobile Station—Base Station System (MS—BSS) and Base Station System—Mobile-Services Switching Center (BSS—MCS) Signaling Procedures and the Mobile Application Part (MAP)
TS	29.011	Signaling Interworking for Supplementary Services
TS	29.013	Signaling Interworking Between ISDN Supplementary Services Application Service Element (ASE) and Mobile Application Part (MAP) Protocols
TS	29.016	Serving GPRS Support Node Sgsn—Visitors Location Register (VLR); Gs Interface Network Service Specification
TS	29.018	General Packet Radio Service (GPRS); Serving GPRS Support Node (SGSN)—Visitors Location Register (VLR); Gs Interface Layer 3 Specification
TS	29.060	General Packet Radio Service (GPRS); GPRS Tunneling Protocol (GTP) Across the Gn and Gp Interface
TS	29.061	Interworking Between the Public Land Mobile Network (PLMN) Supporting Packet Based Services and Packet Data Networks (PDN)
TS	29.078	Customized Applications for Mobile Network Enhanced Logic (CAMEL); CAMEL Application Part (CAP) Specification
TS	29.108	Application of the Radio Access Network Application Part (RANAP) on the E-Interface
TS	29.119	GPRS Tunneling Protocol (GTP) Specification for Gateway Location Register (GLR)
TS	29.120	Mobile Application Part (MAP) Specification for Gateway Location Register (GLR); Stage 3
TS	29.162	Interworking Between the IM CN Subsystem and IP Networks
TS	29.198-01	Open Service Access (OSA) Application Programming Interface (API); Part 1: Overview
TS	29.198-02	Open Service Access (OSA) Application Programming Interface (API); Part 2: Common Data
TS	29.198-03	Open Service Access (OSA) Application Programming Interface (API); Part 3: Framework
TS	29.198-04	Open Service Access (OSA) Application Programming Interface (API); Part 4: Call Control

TABLE B.1   (CONTINUED)

Type	Number	Title
TS	29.198-04-1	Open Service Access (OSA) Application Programming Interface (API); Part 4: Call Control; Subpart 1: Common Call Control Data Definitions
TS	29.198-04-2	Open Service Access (OSA) Application Programming Interface (API); Part 4: Call Control; Subpart 2: Generic Call Control Data SCF
TS	29.198-04-3	Open Service Access (OSA) Application Programming Interface (API); Part 4: Call Control; Subpart 3: Multi-Party Call Control Data SCF
TS	29.198-04-4	Open Service Access (OSA) Application Programming Interface (API); Part 4: Call Control; Subpart 4: Multimedia Call Control Data SCF
TS	29.198-05	Open Service Access (OSA) Application Programming Interface (API); Part 5: Generic User Interaction
TS	29.198-06	Open Service Access (OSA) Application Programming Interface (API); Part 6: Mobility
TS	29.198-07	Open Service Access (OSA) Application Programming Interface (API); Part 7: Terminal Capabilities
TS	29.198-08	Open Service Access (OSA) Application Programming Interface (API); Part 8: Data Session Control
TS	29.198-11	Open Service Access (OSA) Application Programming Interface (API); Part 11: Account Management
TS	29.198-12	Open Service Access (OSA) Application Programming Interface (API); Part 12: Charging
TS	29.198-13	Open Service Access (OSA) Application Programming Interface (API); Part 13: Policy Management SCF
TS	29.198-14	Open Service Access (OSA) Application Programming Interface (API); Part 13: Presence and Availability Management (PAM)
TS	29.202	Signaling System No. 7 (SS7) Signaling Transport in Core Network; Stage 3
TS	29.205	Application of Q.1900 Series to Bearer-Independent Circuit-Switched Core Network Architecture; Stage 3
TS	29.207	Policy Control Over Go Interface
TS	29.208	End-to-End Quality of Service (QoS) Signaling Flows
TS	29.228	IP Multimedia (IM) Subsystem Cx and Dx Interfaces; Signaling Flows and Message Contents
TS	29.229	Cx and Dx Interfaces Based on the Diameter Protocol; Protocol Details
TS	29.232	Media Gateway Controller (MGC)—Media Gateway (MGW) Interface; Stage 3
TS	29.240	3GPP Generic User Profile Requirements; Stage 3; Network
TS	29.278	Customized Applications for Mobile Network Enhanced Logic (CAMEL)—IP Multimedia System (IMS) Interworking; Stage 3
TS	29.328	IP Multimedia Subsystem (IMS) Sh Interface Signaling Flows and Message Contents
TS	29.329	Sh Interface Based on the Diameter Protocol
TS	29.414	Core Network Nb Data Transport and Transport Signaling

TABLE B.1 (CONTINUED)

TYPE	NUMBER	TITLE
TS	29.415	Core Network Nb Interface User Plane Protocols
TR	29.903	Feasibility Study on SS7 Signaling Transportation in the Core Network With SCCP-User Adaptation (SUA)
TR	29.993	Modifications To Be Incorporated in Equipment To Cater for Errors in the Standards
TR	29.994	Recommended Infrastructure Measures To Overcome Specific Mobile Station (MS) and User Equipment (UE) Faults
TR	29.998-01	Open Service Access (OSA) Application Programming Interface (API) Mapping for Open Service Access; Part 1: General Issues on API Mapping
TR	29.998-04-1	Open Service Access (OSA) Application Programming Interface (API) Mapping for Open Service Access; Part 4: Call Control Service Mapping; Subpart 1: API to CAP Mapping
TR	29.998-04-4	Open Service Access (OSA) Application Programming Interface (API) Mapping for Open Service Access; Part 4: Call Control Service Mapping; Subpart 4: Call Control Service Mapping; Subpart 4: Multiparty Call Control SIP
TR	29.998-05-1	Open Service Access (OSA) Application Programming Interface (API) Mapping for Open Service Access; Part 5: User Interaction Service Mapping; Subpart 1: API to CAP Mapping
TR	29.998-05-2	Open Service Access (OSA) Application Programming Interface (API) Mapping for Open Service Access; Part 5: User Interaction Service Mapping; Subpart 2: INAP Mapping
TR	29.998-05-3	Open Service Access (OSA) Application Programming Interface (API) Mapping for Open Service Access; Part 5: User Interaction Service Mapping; Subpart 3: MEGACO Mapping
TR	29.998-05-4	Open Service Access (OSA) Application Programming Interface (API) Mapping for Open Service Access; Part 5: User Interaction Service Mapping; Subpart 4: API to SMS Mapping
TR	29.998-06	Open Service Access (OSA) Application Programming Interface (API) Mapping for Open Service Access; Part 6: User Location and User Status Service Mapping to MAP
TR	29.998-08	Open Service Access (OSA) Application Programming Interface (API) Mapping for Open Service Access; Part 8: Data Session Control Service Mapping to CAP
TR	30.002	Guidelines for the Modification of the Mobile Application Part (MAP)
TS	31.048	Test Specification for Security Mechanisms for the (U)SIM Application Toolkit
TS	31.102	Characteristics of the USIM Application
TS	31.103	Characteristics of the ISIM Application
TS	31.111	USIM Application Toolkit (USAT)
TS	31.112	USAT Interpreter Architecture Description; Stage 2
TS	31.113	USAT Interpreter Byte Codes
TS	31.114	USAT Interpreter Protocol and Administration
TS	31.131	C-Language Binding for (U)SIM API

TABLE B.1   (CONTINUED)

TYPE	NUMBER	TITLE
TR	31.900	SIM/USIM Internal and External Interworking Aspects
TS	32.101	3G Telecom Management Principles and High Level Requirements
TS	32.102	3G Telecom Management Architecture
TS	32.111-1	Telecommunication Management; Fault Management; Part 1: 3G Fault Management Requirements
TS	32.111-2	Telecommunication Management; Fault Management; Part 2: Alarm Integration Reference Point: Information Service
TS	32.111-3	Telecommunication Management; Fault Management; Part 3: Alarm Integration Reference Point: CORBA Solution Set Version 1:1
TS	32.111-4	Telecommunication Management; Fault Management; Part 4: Alarm Integration Reference Point: CMIP Solution Set
TS	32.200	Telecommunication Management; Charging Management; Charging Principles
TS	32.205	Telecommunication Management; Charging Management; 3G Charging Data Description for the CS Domain
TS	32.215	Telecommunications Management; Charging Management; Charging Data Description for the Packet Switched (PS) Domain
TS	32.225	Telecommunication Management; Charging Management; Charging Data Description for the IP Multimedia Subsystem (IMS)
TS	32.235	Telecommunication Management; Charging Management; Charging Data Description for Application Services
TS	32.300	Telecommunication Management; 3G Configuration Management; Name Convention for Managed Objects
TS	32.301	Telecommunication Management; Configuration Management; Notification Integration Reference Point (IRP): Requirements
TS	32.302	Telecommunication Management; Configuration Management; Notification Integration Reference Point; Information Service Version 1
TS	32.303	Telecommunication Management; Configuration Management; Notification Integration Reference Point; CORBA Solution Set Version 1:1
TS	32.304	Telecommunication Management; Configuration Management; Notification Integration Reference Point: CMIP Solution Set Version 1:1
TS	32.311	Telecommunication Management; Generic Integration Reference Point (IRP) Management; Requirements
TS	32.312	Telecommunication Management; Generic Integration Reference Point (IRP) Management; Information Service
TS	32.321	Telecommunication Management; Test Management Integration Reference Point (IRP); Requirements
TS	32.322	Telecommunication Management; Test Management Integration Reference Point (IRP); Information Service
TS	32.324	Telecommunication Management; Test Management Integration Reference Point (IRP); CMIP Solution Set

TABLE B.1   (CONTINUED)

TYPE	NUMBER	TITLE
TS	32.401	Telecommunication Management; Performance Management (PM); Concept and Requirements
TS	32.403	Telecommunication Management; Performance Management (PM); Performance Measurements—UMTS and Combined UMTS/GSM
TS	32.600	Telecommunication Management; Configuration Management; 3G Configuration Management; Concept and Main Requirements
TS	32.601	Telecommunication Management; Configuration Management; Basic CM Integration Reference Point (IRP): Requirements
TS	32.602	Telecommunication Management; Configuration Management; Basic Configuration Management Integration Reference Point (IRP) Information Model
TS	32.603	Telecommunication Management; Configuration Management; Basic Configuration Management Integration Reference Point (IRP): CORBA Solution Set
TS	32.604	Telecommunication Management; Configuration Management; Basic Configuration Management Integration Reference Point (IRP) CMIP Solution Set
TS	32.611	Telecommunication Management; Configuration Management; 3G Configuration Management: Bulk CM Integration Reference Point (IRP) Requirements
TS	32.612	Telecommunication Management; Configuration Management; 3G Configuration Management: Bulk Configuration Management Integration Reference Point (IRP): Information Service
TS	32.613	Telecommunication Management; Configuration Management; 3G Configuration Management: Bulk Configuration Management Integration Reference Point (IRP): CORBA Solution Set
TS	32.614	Telecommunication Management; Configuration Management; 3G Configuration Management: Bulk Configuration Management Integration Reference Point (IRP): CMIP Solution Set
TS	32.615	Telecommunication Management; Configuration Management; 3G Configuration Management: Bulk Configuration Management Integration Reference Point (IRP): XML File Format Definition
TS	32.621	Telecommunication Management; Configuration Management; Generic Network Resources Integration Reference Point (IRP): Requirements
TS	32.622	Telecommunication Management; Configuration Management; Generic Network Resources Integration Reference Point (IRP): NRM
TS	32.623	Telecommunication Management; Configuration Management; Generic Network Resources Integration Reference Point (IRP): CORBA Solution Set
TS	32.624	Telecommunication Management; Configuration Management; Generic Network Resources: Integration Reference Point (IRP) CMIP Solution Set
TS	32.625	Telecommunication Management; 3G Configuration Management; Generic Network Resources Integration Reference Point (IRP): Bulk CM XML File Format Definition
TS	32.631	Telecommunication Management; Configuration Management; Core Network Resources Integration Reference Point (IRP): Requirements
TS	32.632	Telecommunication Management; Configuration Management; Core Network Resources Integration Reference Point (IRP): NRM

TABLE B.1   (CONTINUED)

TYPE	NUMBER	TITLE
TS	32.633	Telecommunication Management; Configuration Management; Core Network Resources Integration Reference Point (IRP): CORBA Solution Set
TS	32.634	Telecommunication Management; Configuration Management; Core Network Resources Integration Reference Point (IRP): CMIP Solution Set
TS	32.635	Telecommunication Management; 3G Configuration Management; Generic Network Resources Integration Reference Point (IRP): Bulk CM XML File Format Definition
TS	32.641	Telecommunication Management; Configuration Management; UTRAN Network Resources Integration Reference Point (IRP): Requirements
TS	32.642	Telecommunication Management; Configuration Management; UTRAN Network Resources Integration Reference Point (IRP): NRM
TS	32.643	Telecommunication Management; Configuration Management; UTRAN Network Resources Integration Reference Point (IRP): CORBA Solution Set
TS	32.644	Telecommunication Management; Configuration Management; UTRAN Network Resources Integration Reference Point (IRP): CMIP Solution Set
TS	32.645	Telecommunication Management; 3G Configuration Management; UTRAN Network Resources Integration Reference Point (IRP): Bulk CM XML File Format Definition
TS	32.655	Telecommunication Management; 3G Configuration Management; GERAN Network Resources Integration Reference Point (IRP): Bulk CM XML File Format Definition
TS	32.661	Telecommunication Management; 3G Configuration Management; Kernel CM Requirements
TS	32.662	Telecommunication Management; 3G Configuration Management; Kernel CM Information Service
TS	32.671	Telecommunication Management; 3G Configuration Management; State Management Integration Reference Point (IRP): Requirements
TS	32.672	Telecommunication Management; 3G Configuration Management; State Management Integration Reference Point (IRP): Information Service
TS	32.673	Telecommunication Management; 3G Configuration Management; State Management Integration Reference Point (IRP): CORBA Solution Set
TS	32.674	Telecommunication Management; 3G Configuration Management; State Management Integration Reference Point (IRP): CMIP Solution Set
TS	32.681	Telecommunication Management; Inventory Management; Inventory Management Integration Reference Point (IRP): Requirements
TS	32.682	Telecommunication Management; Inventory Management; Inventory Management Integration Reference Point (IRP): Information Service
TS	32.683	Telecommunication Management; Inventory Management; Inventory Management Integration Reference Point (IRP): CORBA Solution Set
TS	32.684	Telecommunication Management; Inventory Management; Inventory Management Integration Reference Point (IRP): CMIP Solution Set
TR	32.800	Management Level Procedures and Interaction With UTRAN
TR	32.802	Telecommunication Management; User Equipment (UE) Management Feasibility Study

TABLE B.1   (CONTINUED)

TYPE	NUMBER	TITLE
TS	33.102	3G Security; Security Architecture
TS	33.106	Lawful Interception Requirements
TS	33.107	3G Security; Lawful Interception Architecture and Functions
TS	33.108	3G Security; Handover Interface for Lawful Interception (LI)
TS	33.200	3G Security; Network Domain Security (NDS); Mobile Application Part (MAP) Application Layer Security
TS	33.201	Access Domain Security
TS	33.203	3G Security; Access Security for IP-Based Services
TS	33.210	3G Security; Network Domain Security (NDS); IP Network Layer Security
TR	33.800	Principles for Network Domain Security
TR	33.900	Guide to 3G Security
TR	33.903	Access Security for IP Based Services
TS	34.109	Terminal Logical Test Interface; Special Conformance Testing Functions
TS	34.123-1	User Equipment (UE) Conformance Specification; Part 1: Protocol Conformance Specification
TS	34.123-2	User Equipment (UE) Conformance Specification; Part 2: Implementation Conformance Statement (ICS) Specification
TS	34.124	Electromagnetic Compatibility (EMC) Requirements for Mobile Terminals and Ancillary Equipment
TR	34.926	Table of International EMC Requirements
TS	35.201	Specification of the 3GPP Confidentiality and Integrity Algorithms; Document 1: F8 and F9 Specifications
TS	35.202	Specification of the 3GPP Confidentiality and Integrity Algorithms; Document 2: Kasumi Algorithm Specification
TS	35.203	Specification of the 3GPP Confidentiality and Integrity Algorithms; Document 3: Implementers' Test Data
TS	35.204	Specification of the 3GPP Confidentiality and Integrity Algorithms; Document 4: Design Conformance Test Data
TR	35.205	3G Security; Specification of the MILENAGE Algorithm Set: An Example Algorithm Set for the 3GPP Authentication and Key Generation Functions F1, F1*, F2, F3, F4, F5 and F5*; Document 1: General
TS	35.206	3G Security; Specification of the MILENAGE Algorithm Set: An Example Algorithm Set for the 3GPP Authentication and Key Generation Functions F1, F1*, F2, F3, F4, F5 and F5*; Document 2: Algorithm Specification
TS	35.207	3G Security; Specification of the MILENAGE Algorithm Set: An Example Algorithm Set for the 3GPP Authentication and Key Generation Functions F1, F1*, F2, F3, F4, F5 and F5*; Document 3: Implementer's' Test Data

TABLE B.1   (CONTINUED)

TYPE	NUMBER	TITLE
TS	35.208	3G Security; Specification of the MILENAGE Algorithm Set: An Example Algorithm Set for the 3GPP Authentication and Key Generation Functions F1, F1★, F2, F3, F4, F5 and F5★; Document 4: Design Conformance Test Data
TR	35.909	3G Security; Specification of the MILENAGE Algorithm Set: An Example Algorithm Set for the 3GPP Authentication and Key Generation Functions F1, F1★, F2, F3, F4, F5 and F5★; Document 5: Summary and Results of Design and Evaluation

Useful Web Addresses

The Internet is a very useful information source, especially in the area of telecommunications. The following collection of Web pages, classified by subject, provided a great amount of information used in the development of this book.

Companies in 3G Business

Agilent Technologies	http://www.agilent.com
Alcatel	http://www.alcatel.com
Analog Devices	http://www.analog.com
ARM	http://www.arm.com
Ericsson	http://www.ericsson.com
Fujitsu	http://www.fujitsu.com
Golden Bridge Technology	http://www.gbtwireless.com
Hitachi	http://www.hitachisemiconductor.com
Interdigital Communications	http://www.interdigital.com
LG	http://www.lg.co.kr
Lucent	http://www.lucent.com
Matsushita	http://www.matsushita.co.jp
Mitsubishi	http://global.mitsubishielectric.com/bu/mobile/index.html
Motorola	http://www.motorola.com
NEC	http://www.nec.com
Nokia	http://www.nokia.com
Nortel	http://www.nortelnetworks.com
Philips	http://www.philips.com
Qualcomm	http://www.qualcomm.com
Samsung	http://www.samsung.com
Siemens	http://www.siemens.com
Sasken Communication Technologies	http://www.sasken.com
Sonera Smarttrust	http://www.smarttrust.com
Sony Ericsson	http://www.sonyericsson.com

STMicroelectronics	http://www.st.com/wireless
Symbian	http://www.symbian.com
Texas Instruments	http://www.ti.com
TTPCom	http://www.ttpcom.com

On-line Journals

BT Technology Journal	http://www.bt.com/bttj
Ericsson Review	http://www.ericsson.com/review

Operators

AT&T Wireless	http://www.attws.com
Bell Mobility	http://www.bellmobility.com
Cingular Wireless	http://www.cingular.com
J-Phone	http://www.j-phone.com
KDDI	http://www.au.kddi.com
Korea Telecom	http://www.kt.co.kr
KPN Mobile	http://www.kpnmobile.com
LG Telecom	http://www.Igtel.co.kr
mmO2 (ex. BT Cellnet)	http://www.o2.co.uk
NTT DoCoMo	http://www.nttdocomo.com
Orange	http://www.orange.com
Rogers Wireless	http://www.rogers.com
SK Telecom	http://www.sktelecom.co.kr
Sonera	http://www.sonera.com
Telecom Italia	http://www.tim.it
Telefonica Moviles	http://www.telefonicamoviles.com
Telenor	http://www.telenor.com
Telia	http://www.telia.com
T-Mobil	http://www.t-mobil.net
Verizon	http://www.verizonwireless.com
Vodafone	http://www.vodafone.com
VoiceStream	http://www.voicestream.com

Organizations

3G Americas	http://3gAmericas.org
3GPP (3G Partnership Project)	http://www.3gpp.org
3GPP2 (3G Partnership Project 2)	http://www.3gpp2.org
ARIB (Association of Radio Industries and Businesses), Japan	http://www.arib.or.jp
ATM Forum	http://www.atmforum.com
CEPT (Conference of European Post and Telecommunications)	http://www.cept.org
CWTS (China Wireless Telecommunication Standards group)	http://www.cwts.org
ETSI (European Telecommunications Standards Institute)	http://www.etsi.org
GSM Association	http://www.gsmworld.com
Hiperlan2 Global Forum	http://www.hiperlan2.com
IETF (Internet Engineering Task Force)	http://www.ietf.org
IPv6 Forum	http://www.ipv6forum.com
ITU (International Telecommunication Union)	http://www.itu.int
MeT (Mobile electronic Transactions)	http://www.mobiletransaction.org
Mobile Payment Forum	http://www.mobilepaymentforum.org
Mobey	http://www.mobey.org
Open Mobile Alliance	http://www.openmobilealliance.org
PayCircle	http://www.paymentgroup.org
PKI Forum	http://www.pkiforum.com
Radicchio	http://www.radicchio.org
T1 (Standards Committee T1 Telecommunications), USA	http://www.tl.org
TIA (Telecommunications Industry Association), USA	http://www.tiaonline.org
TTA (Telecommunication Technology Association), Korea	http://www.tta.or.kr
TTC (Telecommunication Technology Committee), Japan	http://www.ttc.or.jp
UMTS Forum	http://www.umts-forum.org
W3 Consortium (World Wide Web Consortium)	http://www.w3.org

Regulators

FCC (Federal Communications Commission)	http://www.fcc.gov
IGEB (Interagency GPS Executive Board)	http://www.igeb.gov
OFTEL	http://www.oftel.gov.uk

Satellite Systems

Globalstar	http://www.globalstar.com
GLONASS	http://www.rssi.ru/SFCSIC/english.html
ICO	http://www.ico.com
Iridium	http://www.iridium.com
Lloyd's Satellite Pages	http://www.ee.surrey.ac.uk/Personal/L.Wood/constellations
Orbcomm	http://www.orbcomm.com
Skybridge	http://www.skybridgesatellite.com
Teledesic	http://www.teledesic.com
Thuraya	http://www.thuraya.com

Universities, Research

Bell Laboratories, Inc.	http://www.bell-labs.com
Bristol University	http://www.bristol.ac.uk
BTexact (ex. BT Labs)	http://www.btexact.com
Cambridge University	http://www.eng.cam.ac.uk
Centre for Wireless Communications	http://www.cwc.oulu.fi
EU's IST programme	http://www.cordis.lu/ist
Helsinki University of Technology, Finland	http://www.hut.fi
Massachusetts Institute of Technology	http://www.mit.edu
Royal Institute of Technology/Radio Communications Systems (Sweden)	http://www.s3.kth.se/radio
Rutgers Winlab	http://www.winlab.rutgers.edu
Sheffield University	http://www.sheffield.ac.uk
Oulu University/Telecom Lab	http://www.telecomlab.oulu.fi
University of Surrey/Centre for Communication Systems Research	http://www.ee.surrey.ac.uk/CCSR

Nokia Communicator

The Nokia Communicator is an important device as it is a 3G-like terminal deployed in 2G networks. The model shown in Figure D.1 is the latest version of Nokia Communicator and includes, among other things, a color display. Nokia Communicator was the first successful non–voice-only terminal in 2G networks, and its concept has been studied considerably by other terminal manufacturers in preparing for the 3G and the data services it will bring to us.

Since Nokia Communicator cannot yet take advantage of unique 3G content (for which it has to wait a few years), it has a familiar look. When closed, it resembles a phone, although a quite large one. When open, it looks like a tiny PC. The PC part has all the features a PC has (for communications): a Web browser, a mail client, and so on.

FIGURE D.1
Nokia Communicator 9210. (Courtesy of Nokia Mobile Phones.)

APPENDIX E

Standardization Organizations and Industry Groups

There are many participants in the 3G development process. Various standardization organizations represent the technical, commercial, and political interests of their members. In addition to these organizations, there are special industry groups formed to promote the views of their members. The most important of these are briefly introduced in this appendix. This list is not exhaustive, and new industry groups are being set up all the time.

3G Americas (http://www.3g-americas.org). UWCC (Universal Wireless Communications Consortia) was a North American organization that promoted the usage of IS-136 TDMA systems. UWCC had its own 3G proposal (UWC-136), which was based on TDMA technology, and it was backward compatible with IS-136. It fits under IMT-2000, where it is named as IMT-SC, but it is incompatible with both 3GPP and 3GPP2 systems. However, in 2001 UWCC decided to adopt WCDMA as their 3G technology, and subsequently this organization was disbanded. Its work is now continued in a new organization, 3G Americas. The mission of 3G Americas is to support the migration of GSM and TDMA networks into WCDMA systems in the Americas.

3GPP (3G Partnership Project) (http://www.3gpp.org). This organization develops specifications for a 3G system based on the UTRA radio interface and on the enhanced GSM core network. In fact there are two IMT-2000 technologies being developed within 3GPP. IMT-DS (direct spread) technology is commonly known as the UTRA-FDD mode, and IMT-TC (time code) is the same as the UTRA-TDD mode. Furthermore, the UTRA-TDD mode includes two varieties: the wideband TDD mode and the narrowband TD-SCDMA. 3GPP's organizational partners include ETSI, ARIB, T1, TTA, TTC, and CWTS. It also has lots of other individual members, including market leaders in mobile telecommunications. Based on this strong backing, it seems to be quite certain that the standards to be developed by the 3GPP will be the most widely used 3G standards.

3GPP has also taken over the future GSM specification work from ETSI. This arrangement is rational as the core network in both the 3GPP and GSM systems is the same. Any changes to the GSM specifications would have an effect on 3GPP standardization and vice versa. In all, ETSI and 3GPP have a close relationship. For example, ETSI provides the support services for 3GPP.

3GPP2 (3G Partnership Project number 2) (http://www.3gpp2.org). This is another significant competing 3G standardization organization. It promotes the use of cdma2000 technology. Its membership includes ARIB, CWTS, TIA, TTA, and TTC. Cdma2000 is an advanced derivative of the IS-95B network currently deployed in some countries. Although both of these proposals (3GPP and 3GPP2) fit under the common IMT-2000 umbrella, they are technically incompatible. In IMT-2000 jargon this proposal is known as IMT-MC (multicarrier).

ARIB (Association of Radio Industries and Businesses) (http://www .arib.or.jp). ARIB is responsible for 3G radio interface standardization in Japan. Previously it was quite difficult for foreign international companies to participate in standardization work in Japan. Foreign participation was not especially encouraged; for example, the documentation language was Japanese. The situation changed when the Japanese PDC system could not succeed outside Japan, and today ARIB has a strong overseas membership. The 3GPP UTRA FDD mode air interface is based on a combination of proposals from both ETSI and ARIB. Note that ARIB specifies radio interface issues only. Core network issues in Japan are the responsibility of another organization, TTC.

CEPT (Conference of European Post and Telecommunications) (http:// www.cept.org). CEPT is a European body for telecommunication policy making and regulation. In 1988 CEPT decided to create ETSI, the European Telecommunications Standards Institute, into which all its telecommunication standardization activities were transferred. CEPT has established three committees:

1. CERP (Comité Européen des Régulateurs Postaux);

2. ERC (European Radiocommunication Committee);

3. ECTRA (European Committee for Regulatory Telecommunications Affairs).

The CERP handles issues related to postal matters; the ERC and the ECTRA discuss telecommunications issues. CEPT also has an important role in 3G as it allocates the radio frequencies used in Europe.

CWTS (China Wireless Telecommunication Standards group) (http:// www.cwts.org). CWTS is the standards development organization responsible for wireless standardization in China. It is a member of both the 3GPP and the 3GPP2. CWTS has been especially active in developing the TD-SCDMA specification, which is the narrowband version of the UTRA-TDD mode. In fact, the UTRA TDD mode specifications will also include the TD-SCDMA specifications as a TDD mode option. In 3GPP jargon TD-SCDMA is also called the UTRA TDD low-chip-rate option.

ETSI (European Telecommunications Standards Institute) (http://www.etsi.org). ETSI develops the telecommunications standards for Europe. It works in close cooperation with the European Union. Its members include governmental organizations, network operators, equipment manufacturers, and service providers. ETSI is headquartered in the Sophia Antipolis science park in France. ETSI has developed, in cooperation with ARIB, the UTRA-FDD (IMT-DS) proposal, which has become the most important "component" in the IMT-2000 standard family. This standard is currently under further development by the 3GPP organization.

FCC (Federal Communications Commission) (http://www.fcc.gov) The FCC is an independent U.S. government agency directly responsible to the U.S. Congress. The FCC is charged with regulating interstate and international communications by radio, television, wire, satellite, and cable. The FCC's jurisdiction covers the 50 states, the District of Columbia, and U.S. possessions. The tasks of the FCC include the regulation of spectrum usage in the United States and specifying the E911 rules (Enhanced Wireless 911 Services). The E911 rules state that the source of emergency calls must be located with a defined accuracy.

GSM Association (http://www.gsmworld.com). This group is responsible for the development, deployment, and evolution of the GSM standard for digital wireless communications and for the promotion of GSM open standards. Its members include GSM network operators, regulators, and administrative bodies. In addition, many manufacturers and suppliers have joined the GSM Association through associate membership. The GSM Association is based in Dublin, Ireland.

ITU (International Telecommunication Union) (http://www.itu.int). The ITU is the "parent" organization for all other telecommunications organizations. The ITU does not usually specify things itself, but merely coordinates the specification work of others. It may set the goals for some standardization work, but the actual work is often done by local organizations (ETSI, TIA, etc.). In many cases, the ITU works as a catalyst for future standardization work. The ITU is a U.N. organization headquartered in Geneva, Switzerland.

In the 3G area, the ITU's goal has been a global standard. The result may not be a single truly global standard, but a collection of standards called IMT-2000 by the ITU. An earlier name for IMT-2000 was the Future Public Land Mobile Telephone System (FPLMTS); nobody knew how to pronounce the acronym, hence the name change. IMT-2000 includes five "component" standards:

1. IMT-DS (direct spread) (a.k.a. UTRA FDD);
2. IMT-MC (multicarrier) (a.k.a. cdma2000);
3. IMT-TC (time code) (a.k.a. UTRA-TDD/TD-SCDMA);
4. IMT-SC (single carrier) (a.k.a. UWC-136);
5. IMT-FT (frequency time) (aka DECT).

MeT (Mobile electronic Transactions) (http://www.mobiletransaction .com). This is an initiative to develop an open and common industry framework for secure mobile electronic transactions. It aims to use and extend existing industry standards whenever possible and combine them into a single common framework. The component technologies include the following WAP security functions: wireless transport layer security (WTLS), wireless identity module (WIM), and wireless public key infrastructure (WPKI). The initial members of this initiative included Ericsson, Motorola, and Nokia. Siemens has also joined this initiative. MeT can be seen as a competing initiative to Radicchio.

The Mobey Forum (www.mobey.org). This is a consortium of several large banks (founding members are ABN AMRO Bank, Banco Santander Central Hispano, BNP Paribas, Deutsche Bank, HSBC Holdings, Nordea, and UBS; several other banks will have joined the consortium after this writing) and Nokia, Ericsson, and Siemens. The Mobey Forum aims to develop a mobile payment system. The membership list of the Mobey Forum suggests that the interests of banks will not be overlooked in their payment concept.

Mobile Payment Forum (www.mobilepaymentforum.org). The Mobile Payment Forum membership list includes credit-card companies and large operators amongst others. They intend to "complement the work that is already being done by other industry initiatives and organisations such as Mobey and MeT." Most probably they also want to make sure that the interests of credit-card companies and mobile operators are not forgotten.

MPT (Ministry of Posts and Telecommunications). MPT coordinates the work of ARIB and TTC in Japan, and also represents these organizations in the ITU. It also regulates the spectrum usage in Japan.

NTT (Nippon Telephone and Telegraph) (http://www.ntt.com). NTT is another important player in Japan. NTT's mobile-telecommunications branch is called NTT DoCoMo. Although NTT is not a standardization organization, but a telecommunications company, it is a global conglomerate and its decisions will have an impact on telecommunication standards. NTT is also a major contributor to ARIB's work. NTT is currently driving its own version of the WCDMA radio access network. It aims to launch this network in advance of the 3GPP-specified UTRAN. The short-time commercial advantages are obvious if it succeeds in this task, but the result would be an incompatible air interface technology. This NTT sole development is not part of the official IMT-2000 protocol family. Also note that the very popular i-mode system is an NTT DoCoMo proprietary technology.

Radicchio (http://www.radicchio.org). This is a global initiative to define a standard security platform for mobile e-commerce. This initiative competes with the MeT initiative. Radicchio is an older initiative, as Sonera SmartTrust, Gemplus, and EDS (Electronic Data Systems) launched it in September 1999. The MeT initiative was published only in April 2000. Radicchio aims to define and promote a standard based on a public key infrastructure. The number of Radicchio members had grown considerably since its launch to several dozen by the end of 2000. The members include mobile-telecommunications operators and e-commerce technology providers. Notably, MasterCard International and Visa International are members of this group. It must also be noted that Sonera SmartTrust owns IPR in some of the technology promoted by Radicchio.

T1 (Standards Committee T1 Telecommunications) (http://www.t1.org). T1 specifies standards related to interfaces for U.S. telecommunications networks. It is part of ATIS, which is accredited by ANSI. It has six subcommittees, and mobile/wireless standards are handled by subcommittee T1P1. T1P1 is closely related to TIA committee 46. T1P1 has made a standard proposal called WCDMA/NA, and it now contributes to 3GPP work.

TIA (Telecommunications Industry Association) (http://www.tiaonline .org) provides standardization services to its members in the United States. The TIA's work is quite market-driven, which means faster response, but also a larger number of standards. TIA specifications are accredited by ANSI. Within TIA there are two committees that set standards for modern mobile telecommunications, TR45 and TR46. Both committees have several subcommittees. Subcommittee 45.3 is responsible for IS-136 and its evolution into EGPRS-136. Similarly, subcommittee 45.5 is responsible for the IS-95B standard and the cdma2000 work. Finally, subcommittee 46.1 organizes the WIMS (Wireless ISDN and Multimedia Services) work. WIMS is compatible with ETSI W-CDMA.

TTA (Telecommunication Technology Association) (http://www.tta .or.kr). TTA specifies the 3G standards in Korea. TTA has been quite active in 3G work. It previously submitted two radio interface proposals to the ITU, called TTA I and TTA II (nowadays Global CDMA I and II). TTA II is quite similar to W-CDMA by ETSI/ARIB, and TTA I has similarities to cdma2000. TTA belongs to the 3GPP, so it is quite probable that Korea will also adopt the common standard developed by this organization.

TTC (Telecommunication Technology Committee) (http://www.ttc.or jp). TTC specifies the network standards in Japan. As the 3G concept is based on separate radio access and core networks, TTC could have in prin-ciple chosen to specify its own core network. However, it has chosen to base its 3G network on GSM, so it will also work in cooperation with ETSI to further develop this network. One reason to use the GSM MAP net-work, even if such does not exist yet in Japan, was the wide range of readily available services in GSM networks.

UMTS Forum (http://www.umts-forum.org). The UMTS Forum was established in 1996 to accelerate the process of defining the necessary stan-dards for UMTS. The UMTS Forum is mainly focused on the delivery of market-focused recommendations. It has over 200 member organizations drawn from the mobile operator, supplier, regulatory, consultant, IT, and media/content communities. The UMTS Forum works as a catalyst with other specialist organizations to examine issues such as technical standards, spectrum, market demand, business opportunities, terminal equipment cir-culation, and convergence between the mobile communications and com-puting industries.

UWCC (Universal Wireless Communications Consortia). See 3G Americas.

WAP Forum (http://www.wapforum.org). This is the industry associa-tion promoting the usage of WAR The WAP Forum has hundreds of members including network operators, infrastructure providers, software developers, and so on. The WAP Forum will have a challenging task ahead, once it tries to compete with the Japanese i-mode promoted by NTT DoCoMo.

WRC (World Radiocommunication Conference). This is a periodic meet-ing arranged by the ITU. This conference handles spectrum regulation, and allocates radio spectrum for specific purposes. The outputs of these meetings are worldwide recommendations. For example, in Europe the spectrum issues are further defined by CEPT, although CEPT usually follows ITU's recommendations quite closely. WRCs are rarely arranged; several years may elapse between meetings. WRC-2000 was held in Istanbul in May/June 2000.

Radio spectrum allocation is a slow process because typically all frequencies already have some users, and it is understandably difficult to get a common understanding about which services should be moved and to where. Moreover, it is necessary to allocate enough time for the old spectrum users to move their services to new spectrum allocations. These matters may take several years to accomplish. Furthermore, the military has been and will continue to be a significant user of radio spectrum in many countries. This may bring additional problems to the spectrum-allocation process, and civil and military requirements are often difficult to combine. At the regional level the spectrum issues are handled by CEPT in Europe, by the FCC in the United States, and by MPT in Japan.

About the Author

Juha Korhonen received an M.Sc. in data communications from the Lappeenranta University of Technology, Finland, in 1991.

He joined the Nokia Research Center in Espoo near Helsinki, first in 1988 as a research engineer. From 1993 to 1995, he worked as a design engineer at Nokia Mobile Phones in Camberley, England. In 1995, he returned to Nokia Research Center to work in advanced DECT, and thereafter in WCDMA projects.

In 1997 Mr. Korhonen moved to Cambridge, England, to join TTPCom, Ltd. to work with satellite handsets, and later in 3G projects.

In 2001 he received a Licentiate of Technology in telecommunications from the Lappeenranta University of Technology. Mr Korhonen is currently at the University of Cambridge, and his research interests include HSDPA and MIMO.

He lives in Cambridge, England, with his wife, Anna.

Index

Security mode control (continued)
    integrity protection, 174–76
    See also Radio resource control (RRC)
Series numbering, 484–85
Service capabilities, 399–401
    accessibility, 401
    defined, 401
    features, 401
Service providers, 298–99
Services. *See* 3G services
Service-Specific Connection-Oriented Protocol (SSCOP), 242
Service-Specific Coordination Function (SSCF), 242
Serving GPRS support node (SGSN), 212–13
Serving radio network subsystems (SRNS), 41
Session management (SM), 49, 195–96
    function, 295
    procedures, 196
Shannon, Claude, 122, 123
Shared channel control channel (SHCCH), 72
Shared channels, 78–80
    downlink, 79–80
    idea behind, 78
    resource use, 78
    uplink, 79
Shared control channel for HS-DSCH (HS-SCCH), 76, 77–78
Shared information channel for HS-DSCH (HS-SICH), 78
Short Message Peer-to-Peer Protocol (SMPP), 367
Short message service (SMS), 49, 195
    cell broadcast (SMS-CB), 429
    mobile-originated (SMS-MO), 195
    popularity, 362
    purpose, 195
Signaling ATM adaptation layer (SAAL), 242
Signaling Connection Control Part (SCCP), 243
Signaling connection establishment
    illustrated, 306
    procedures, 304
Signals
    despread, recovery, 30
    unrecoverable, 30
    wideband, 31
Signal-to-interference ratio (SIR), 35
Simple Mail Transfer Protocol (SMTP), 367
Site-selection diversity transmission (SSDT), 36, 56
    defined, 40, 329
    downlink, 87
    illustrated, 41
    principle, 40
    uplink, 86

See also Soft handovers (SHOs)
Skew
    cause, 421
    illustrated, 422
    removal, 421–22
    See also Multimedia applications
Skybridge, 453–54
Smart antennas, 386–92
    channel radius and, 389
    defined, 386
    MIMO, 391–92
    reasons for using, 386–87
    simplest, 387
    switched-beam, 387
    See also Antennas
Soft handovers (SHOs), 38–41
    branch, 38
    constraints, 39
    defined, 85
    employment, 39
    illustrated, 38
    management, 168
    parameters, 263
    procedures, 329–30
    successive, 41
    See also Handovers (HOs)
Software management, 288–89
    fault process, 288–89
    main process, 288
    See also Management
Spacetime-block-coding-based TX diversity (STTD), 88–89
Specifications, 479–86
    backwards compatibility, 486
    defined, 479
    list of, 491–508
    optional features, 479
    process, 480–82
    releases, 482–84
    series numbering, 484–85
    version numbering, 485–86
    See also 3GPP
Spectral efficiency, 255
Spectrum
    allocation process, 442
    new, 441–42
Spreading codes, 29, 82–83, 111–19
    defined, 29, 111
    illustrated, 83
    orthogonal codes, 58, 112–14
    PN codes, 58, 114–17
    scrambling codes, 59–60

Recent Titles in the Artech House
Mobile Communications Series

John Walker, Series Editor

Radio Resource Management for Wireless Networks, Jens Zander and
    Seong-Lyun Kim

RDS: The Radio Data System, Dietmar Kopitz and Bev Marks

Resource Allocation in Hierarchical Cellular Systems, Lauro Ortigoza-Guerrero and
    A. Hamid Aghvami

RF and Microwave Circuit Design for Wireless Communications,
    Lawrence E. Larson, editor

Sample Rate Conversion in Software Configurable Radios, Tim Hentschel

Signal Processing Applications in CDMA Communications, Hui Liu

Software Defined Radio for 3G, Paul Burns

Spread Spectrum CDMA Systems for Wireless Communications, Savo G. Glisic and
    Branka Vucetic

Third Generation Wireless Systems, Volume 1: Post-Shannon Signal Architectures,
    George M. Calhoun

Transmission Systems Design Handbook for Wireless Networks,
    Harvey Lehpamer

UMTS and Mobile Computing, Alexander Joseph Huber and Josef Franz Huber

Understanding Cellular Radio, William Webb

Understanding Digital PCS: The TDMA Standard, Cameron Kelly Coursey

Understanding GPS: Principles and Applications, Elliott D. Kaplan, editor

Understanding WAP: Wireless Applications, Devices, and Services,
    Marcel van der Heijden and Marcus Taylor, editors

Universal Wireless Personal Communications, Ramjee Prasad

WCDMA: Towards IP Mobility and Mobile Internet, Tero Ojanperä and
    Ramjee Prasad, editors

Wireless Communications in Developing Countries: Cellular and Satellite Systems,
    Rachael E. Schwartz

Wireless Intelligent Networking, Gerry Christensen, Paul G. Florack, and
    Robert Duncan

Wireless LAN Standards and Applications, Asunción Santamaría and
    Francisco J. López-Hernández, editors

Wireless Technician's Handbook, Andrew Miceli

For further information on these and other Artech House titles, including previously considered out-of-print books now available through our In-Print-Forever® (IPF®) program, contact:

Artech House
685 Canton Street
Norwood, MA 02062
Phone: 781-769-9750
Fax: 781-769-6334
e-mail: artech@artechhouse.com

Artech House
46 Gillingham Street
London SW1V 1AH UK
Phone: +44 (0)20 7596-8750
Fax: +44 (0)20 7630-0166
e-mail: artech-uk@artechhouse.com

Find us on the World Wide Web at:
www.artechhouse.com